Internet of Things, Threats, Landscape, and Countermeasures

Internet of Things, Threats, Landscape, and Countermeasures

Edited by Stavros Shiaeles and
Nicholas Kolokotronis

CRC Press
Taylor & Francis Group
Boca Raton London New York

CRC Press is an imprint of the
Taylor & Francis Group, an **informa** business

First edition published 2021
by CRC Press
6000 Broken Sound Parkway NW, Suite 300, Boca Raton, FL 33487-2742

and by CRC Press
2 Park Square, Milton Park, Abingdon, Oxon, OX14 4RN

CRC Press is an imprint of Taylor & Francis Group, LLC

© 2021 Taylor & Francis Group, LLC

The right of Stavros Shiaeles and Nicholas Kolokotronis to be identified as the authors of the editorial material, and of the authors for their individual chapters, has been asserted in accordance with sections 77 and 78 of the Copyright, Designs and Patents Act 1988.

Reasonable efforts have been made to publish reliable data and information, but the author and publisher cannot assume responsibility for the validity of all materials or the consequences of their use. The authors and publishers have attempted to trace the copyright holders of all material reproduced in this publication and apologize to copyright holders if permission to publish in this form has not been obtained. If any copyright material has not been acknowledged please write and let us know so we may rectify in any future reprint.

Except as permitted under U.S. Copyright Law, no part of this book may be reprinted, reproduced, transmitted, or utilized in any form by any electronic, mechanical, or other means, now known or hereafter invented, including photocopying, microfilming, and recording, or in any information storage or retrieval system, without written permission from the publishers.

For permission to photocopy or use material electronically from this work, access www.copyright.com or contact the Copyright Clearance Center, Inc. (CCC), 222 Rosewood Drive, Danvers, MA 01923, 978-750-8400. For works that are not available on CCC please contact mpkbookspermissions@tandf.co.uk

Trademark notice: Product or corporate names may be trademarks or registered trademarks and are used only for identification and explanation without intent to infringe.

Library of Congress Cataloging-in-Publication Data
Names: Shiaeles, Stavros, editor. | Kolokotronis, Nicholas, editor.
Title: Internet of things, threats, landscape, and countermeasures / edited by Stavros Shiaeles and Nicholas Kolokotronis.
Description: Boca Raton : CRC Press, 2021. | Includes bibliographical references and index.
Identifiers: LCCN 2020050793 (print) | LCCN 2020050794 (ebook) | ISBN 9780367433321 (hardback) | ISBN 9781003006152 (ebook)
Subjects: LCSH: Internet of things--Security measures.
Classification: LCC TK5105.8857 .I657 2021 (print) | LCC TK5105.8857 (ebook) | DDC 005.8--dc23
LC record available at https://lccn.loc.gov/2020050793LC ebook record available at https://lccn.loc.gov/2020050794

ISBN: 978-0-367-43332-1 (hbk)
ISBN: 978-0-367-76615-3 (pbk)
ISBN: 978-1-003-00615-2 (ebk)

Typeset in Times
by SPi Global, India

To my family for the magic they bring to my life, pushing me to go further. Also, to my father who left us too soon.

— Stavros Shiaeles

To my wife Mary and children, Athanasia and Manos, for their endless love and support.

— Nicholas Kolokotronis

Contents

Preface ix
    Who Should Read This Book xi
    What This Book Covers xi
Acknowledgements xiii
Editors xv
List of Contributors xvii

Chapter 1   Data Protection and Privacy Issues of the Internet of Things 1
                 O. Gkotsopoulou and P. Quinn

Chapter 2   IoT Reference Architectures 47
                 V. Kelli, E. G. Sfakianakis, B. Ghita, and P. Sarigiannidis

Chapter 3   Threats in Critical Infrastructures 97
                 A. Peratikou, S. Shiaeles, and S. Stavrou

Chapter 4   Threats in Industrial IoT 137
                 N. Scheidt and M. Adda

Chapter 5   Threats in IoT Supply Chain 167
                 S. A. Kumar, G. Mahesh, and Chikkade K. Marigowda

Chapter 6   Threats in IoT Smart Well-being 201
                 E. Darra, V. Mantzana, V. Giovana Bilali, and D. Kavallieros

Chapter 7   IoT Security Frameworks and Countermeasures 239
                 G. Bendiab, B. Saridou, L. Barlow, N. Savage, and
                 S. Shiaeles

Chapter 8   Cyber-resilience 291
                 E. Bellini, G. Sargsyan, and D. Kavallieros

Index 335

Preface

The emergent field of the Internet of Things (IoT) is accompanied by an unprecedented revolution in the physical and cyber world. Smart, always-connected IoT devices provide real-time contextual information with low overhead to optimize processes and improve how companies and individuals interact, work, and live.

As organizations worldwide, from governments to public and corporate enterprises, are trying to keep up with the advancements, new security challenges are coming into light. Evolving cyber-attacks following the revolution of IoT and the fact that there are literally billions of IoT devices globally, most of which are readily accessible and easily hacked, allows threat actors to use them as the cyber-weapon delivery system of choice in many today's cyber-attacks, e.g., from botnet-building for launching distributed denial of service attacks, to malware spreading and spamming. The first step towards developing an effective defense strategy towards these threats is to understand them and document them, including the existing vulnerabilities, in order to foster better detection and mitigation of threats.

This book establishes the tenets of the foundational concept of IoT security by presenting real-world examples in Critical Infrastructures (CIs), Industrial IoT, Supply Chain, and Smart Well-Being domains.

- Critical infrastructures span many sectors, ranging from energy, defense and the ICT sectors to information systems in space, civil protection, and health. They are important as they provide services that are essential for our social cohesion and economic growth; resilience as well as operational reliability and continuity are core requirements. Cyber-attacks against critical infrastructures have already shown the ability to cause harm and have adversarial effects on information systems' vital operations.
- Industrial IoT environments always involve several risks and dangers, and managers strive to find solutions for minimizing cyber-attacks' impact. IoT sensors can feed the industrial safety-related algorithms with real-time data and allow them to make instant decisions; e.g., upon detection of gas leakage, increased temperatures, etc., certain safety procedures should be initiated to manage the risks. In such systems, protection against cyber-attacks to ensure safety, security, and reliability is of paramount importance.
- IoT Supply Chain plays a major role in logistic management and it is used for locating, route management, inventory tracking, and warehousing. IoT for supply chain is all set to revolutionize the operational efficiencies and revenue opportunities of an organization. It does not only help an organization to keep track of its assets but also provides a way out to gain an edge over its competitors. Attacks on this area can disrupt the food supply and cause major issues to countries and people.
- Smart Well-being umbrella provides a number of applications from health, smart homes, smart cities, and various other functionalities for entertainment, assisted living, safety, remote control, etc. However, these smart appliances pose great risks to users' privacy as it is well-known that most of them lack

basic security features and can be easily compromised. The dependency (in most cases) on centralized cloud services, with a single access point for data storing, amplifies the security concerns

The application domains of IoT are vast and there is a great variance regarding the security-related characteristics under which different deployments operate. These differences may pertain to the technological measures deployed (e.g., existence or lack of firewalls), established procedures (e.g., taking backups), or human aspects (e.g., security awareness) and transcend all IoT application domains and Critical Information Infrastructures (CIIS). These differences are important to identify, since they are highly relevant to the degree that a deployment is vulnerable to certain threats or to the impact that a data breach may have.

The IoT can pose several risks for data protection and privacy rights and produce challenges for data controllers to keep their processing operations compliant. In 2016, the EU adopted a new data protection framework, which entered into force in 2018. After two years of implementation, the first studies coming from the European Commission, the national supervisory authorities and other relevant stakeholders show that implementation is slow. The implementation is further challenged by the need for research, innovation and business modelling and the urge for quick adoption of emerging and new technologies by the markets. To this end, this book analyzes the legislation and discusses the challenges raised from the point of view of European data protection law. Particular emphasis will be put on the principles of Data Protection by Design and by Default and the conduct of a Data Protection Impact Assessment as a theoretical exercise already at the conceptualization phase of each new device intended to be connected.

IoT architectural models will be also presented, and the corresponding application and network protocols will be further discussed to familiarize the reader with the subject. Hardware utilized for IoT technology, such as sensors, will be further analyzed, in order to cover a wide range of devices. After the introduction of IoT reference architecture, the book will focus on the four aforementioned domains where key security calls and relevant deployment architectures of IoT in these domains will be elaborated.

The analysis of the domains along with IoT security issues will foster the need for securing IoT devices and a question is raised to the reader on how could these domains be protected. Towards this direction, the book surveys the Cybersecurity threats landscape and the defense measures available. In this respect firewalls, honeypots, antivirus, and other existing solutions available from the industry are listed and explained along with best practices and policies that can be applied helping organization to tackle IoT security issues.

To complete readers' knowledge, cyber resilience is closing the book. The concept of "cyber resilience" has recently emerged in the Cybersecurity field as a result of the recognition that the traditional understanding of defense in cyberspace, built upon the notion that a system that must defend against all possible attacks, is unrealistic in the current rapidly evolving threat landscape with the increase of sophisticated levels of cyber-attacks. To maintain the balance between the need of reducing the malware outbreak while maintaining the network functionalities at an acceptable level is crucial and will be further explained in the book as well as the methods available to achieve resilience.

Preface

xi

WHO SHOULD READ THIS BOOK

This book targets IoT practitioners, including students, lectures, IoT researchers, security professionals, solution architects, developers, and business stakeholders. Anyone who needs to have a comprehensive understanding of the unique safety and security challenges of our connected world, and who needs to learn practical methodologies to secure devices and assets will find this book immensely helpful. This book is uniquely designed to benefit professionals for IT as well as individuals interested in advancing their knowledge in Cybersecurity.

WHAT THIS BOOK COVERS

This book provides the readers with a systematic overview of the European legislation on data protection and privacy issues, Cybersecurity issues in four large domains critical to our society, mitigation measures as well as the advancements made in cyber resilience, as there is not any 100% secure system. These topics are detailed in eight chapters:

Chapter 1 analyzes the applicable data protection and privacy framework in the European Union in relation to the Internet of Things. The Internet of Things and its enabling technologies both depend upon, and at the same time generate, a vast amount of information. When this information can lead to the identification of a unique individual, it is considered personal data and the EU data protection law will be applicable, also given the territoriality requirements. The Internet of Things can pose several risks for data protection and privacy rights and produce challenges for data controllers to keep their processing operations compliant. This Chapter, therefore, will outline the primary challenges and provide a set of recommendations and best practices, which can be useful for technologists and engineering students.

Chapter 2 provides an introductory overview of the most predominant IoT architectural models, analyzing the purpose and function of each layer, with the focus being the four-layer architectural model; thus, IoT application areas will be explored deeply. Additionally, an extensive overview of IoT application layer and network layer protocols will be presented, and the way IoT devices talk and connect will thoroughly be explored. The physical aspect of intelligent devices will be explained as well, presenting a plethora of sensors utilized in IoT environments. Finally, a brief introduction to ubiquitous computing, followed by security issues IoT devices may encounter is closing this chapter.

Chapter 3 focuses on defining the term critical infrastructure and specifying the sectors and domain of CIIs. Cybersecurity is of vital importance, especially in the context of Critical infrastructures as cyber-attacks could potentially disrupt communications, access to control systems, military and have catastrophic effects on the governments and the well-being of the citizens. This chapter addresses these issues first by analyzing the attack incidents to Critical infrastructures in recent years, then by providing an overview of the type of attacks and threats along with the mitigation techniques.

Chapter 4 elaborates the key security goals as well as relevant deployment architectures of industrial IoT in areas such as agriculture and farming, industry safety, and preventive maintenance. Specific IIoT domains as well as technologies are elaborated. Possible threats endangering these domains as well as incidents of such in the past have been outlined to establish the threats within Industrial IoT and achieve a better understanding of such.

Chapter 5 focuses on the IoT domain for supply chain. A detailed description about each of the subdomains under supply chain, logistics management, fleet management, asset tracking, and manufacturing process management is provided. Use of IoT in these subdomains makes them vulnerable to security attacks and hence building a secure architecture would be one of the important objectives in the design of applications for all of these subdomains. The security goals in each of these subdomains of supply chain along with the relevant deployment architectures for various sectors in the domain are discussed.

Chapter 6 describes the integration of IoT devices in three ecosystems falling under the greater umbrella of Smart Well-being. Firstly, the concept of Smart Cities is presented along with the respective components, threats, vulnerabilities, opportunities, and benefits for the society. Then we focus on two domains under the Smart Cities concept, Smart Homes and Smart Healthcare. In the Smart Home domain, emphasis in given to the automations, applications, and possible architecture of deployment while an extended analysis regarding security challenges, threats as well as security measures are discussed. The last part of the chapter focuses on the Smart Healthcare domain. This domain has appeared to be a business transformation in the next years and it will lead to enhanced services provision that will provide benefits to life, society, and environment. Despite the benefits they provide, multiple barriers have been reported, with security having been characterized as a top challenge for healthcare organizations planning, implementing, and adopting IoT solutions.

Chapter 7 provides an overview of the current threat landscape in IoT ecosystem, along with the most relevant security measures, either technical or non-technical, that can be applied to protect from potential cyber-threats. For each security measure, an explanation on how it may alter the threat exploitability level and/or the level of the technical impact is given. This will help the reader have a comprehensive view of the essential security countermeasures that should be applied for protecting IoT infrastructures from potential cyber threats.

Chapter 8 aims at formalizing the concept and exploring current promising practices for resilience implementation in the context of the Internet of Things. The Cyber Resilience paradigm provides a comprehensive framework to define evidence-driven strategies to balance between the need of reducing malware spread while maintaining the network functionalities at an acceptable level. In this game, a number of assets and actors come into play with different rules and contributions to the overall resilience. In this respect, a holistic view on resilience is able to reconciliate different perspective and definitions in resilience domain is provided.

Stavros Shiaeles
Portsmouth, UK
Nicholas Kolokotronis
Tripolis, Greece
October 2020

Acknowledgements

Many people had been involved with great enthusiasm and supported us during the preparation of this book. First of all, we would like to express our sincere gratitude to all the contributors; without their valuable help, this book would not have reached this state and quality. We would like to warmly thank CRC press and the editorial team for entrusting the preparation of this book and for doing an excellent job in guiding us at each step of the process. The authors express their heartfelt gratitude to Dr Gueltoum Bendiab and Dr Maria Shiaele for their invaluable help during the preparation of this book.

Editors

Stavros Shiaeles, Ph.D., is an assistant professor in Cybersecurity at the University of Portsmouth, UK. He worked as an expert in Cybersecurity and digital forensics in the UK and EU, serving companies and research councils. His research interest span in the broad area of Cybersecurity and more specifically in OSINT, social engineering, distributed denial-of-service attacks, cloud security, digital forensics, network anomaly detection, and malware mitigation. Dr. Shiaeles has authored more than 60 publications in academic journals and conferences, co-chaired many workshops and conferences and actively involved in research projects as Principal Investigator leading his Cybersecurity research team.

He is currently a guest editor in the topical collection "Cybersecurity, digital forensics and resilience" at Springer's *Applied Sciences* Journal (since 2019), topic editor at MDPI's *Forensic Sciences Journal* (since 2020), guest editor in the Special Issue "Advancements in Networking and Cybersecurity" at MDPI's *Electronics Journal* (2020), guest editor in the "Special Issue on Novel Cybersecurity Paradigms for Software-defined and Virtualized Systems" at Elsevier's *Computer Networks Journal* (2020), active member at IEEE Technical Committee on Information Infrastructure and Networking (TCIIN) and a regular reviewer for several prestigious journals.

Further to his academic qualifications, he holds a series of professional certifications, namely EC-Council Certified Ethical Hacker (CEH), EC-Council Advanced Penetration Testing (CAST611), ISACA Cobit 5 Foundation and a Cyberoam Certified Network and Security Professional (CCNSP), and he is an EC-Council accredited instructor providing professional certifications training on Cybersecurity and penetration testing. He is also a fellow of the BCS and a fellow of the Higher Education Academy in the UK.

Before entering academia, Dr. Shiaeles was in the industry, where he has more than ten years of experience, and he has worked on various aspects of IT and Cybersecurity, gaining invaluable hands-on knowledge on various systems and software developing.

Nicholas Kolokotronis, Ph.D., is an associate professor and head of the Cryptography and Security Group at the Department of Informatics and Telecommunications, University of the Peloponnese. He received his B.Sc. in mathematics from the Aristotle University of Thessaloniki, Greece, in 1995, an M.Sc. in highly efficient algorithms (highest honors) in 1998 and a Ph.D. in cryptography in 2003, both from the National and Kapodistrian University of Athens.

Since 2004, he has held visiting positions at the University of Piraeus, University of the Peloponnese, the National and Kapodistrian University of Athens, and the Open University of Cyprus. During 2002–04, he was with the European Dynamics S.A., Greece, as a security consultant. He has been a member of working groups for the provisioning of professional Cybersecurity training to large organizations, including the Hellenic Telecommunications and Posts Commission (EETT). He has published more than 85 papers in international scientific journals, conferences, and books and has participated in more than 20 EU-funded and national research and innovation projects. He has been a co-chair of conferences (IEEE CSR 2021), workshops (IEEE SecSoft 2019, IEEE CSRIoT 2019, 2020, and ACM EPESec 2020), and special sessions focusing on IoT security. Moreover, he has been a TPC member in many international conferences, incl. IEEE ISIT, IEEE GLOBECOM, IEEE ICC, ARES, and ISC.

He is currently a guest editor in "Engineering – Cybersecurity, digital forensics and resilience" area of Springer's *Applied Sciences Journal* (since 2019) and in the Reviewer Board of MDPI's *Cryptography Journal* (since 2020), whereas he has been an associate editor of the EURASIP *Journal on Wireless Communications and Networking* (2009–2017) and a regular reviewer for a number of prestigious journals, incl. IEEE TIFS, IEEE TIT, Springer's DCC, etc. His research interests span the broad areas of cryptography, security, and coding theory.

Contributors

Adda, Mo
University of Portsmouth
Portsmouth, UK

Barlow, Luke
University of Plymouth
Plymouth, UK

Bellini, Emanuele
University of Campania "Vanvitelli"
Caserta, Italy

Bendiab, Gueltoum
University of Portsmouth
Portsmouth, UK

Bilali, V. Giovana
Center for Security Studies
Athens, Greece

Darra, Eleni
Center for Security Studies
Athens, Greece

Ghita, Bogdan
University of Plymouth
Plymouth, UK

Gkotsopoulou, Olga
Vrije Universiteit Brussel
Brussels, Belgium

Kavallieros, Dimitris
University of the Peloponnese
Tripolis, Greece
Center for Security Studies
Athens, Greece

Kelli, Vasiliki
University of Western Macedonia
Kozani, Greece

Mahesh, Gangathimmappa
BMSIT-Bangalore
Bengaluru, India

Mantzana, Vasiliki
Center for Security Studies
Athens, Greece

Marigowda, Chikkade K.
AcIT- Bangalore
Bengaluru, India

Peratikou, Adamantini
Open University of Cyprus
Nicosia, Cyprus

Quinn, Paul
Vrije Universiteit Brussel,
Brussel, Belgium

Scheidt, Nancy
University of Portsmouth
Portsmouth, UK

Sfakianakis, Evangelos
Directorate of OTE group Technology &
 Operations OTE S.A
Athens, Greece

Sargsyan, Gohar
CGI
Amsterdam, Netherlands

Saridou, Betty
Democritus University of ThraceXanthi,
 Greece

Sarigiannidis, Panagio
University of Western Macedonia
Kozani, Greece

Savage, Nick
University of Portsmouth
Portsmouth, UK

Shiaeles, Stavros
University of Portsmouth
Portsmouth, UK

Stavrou, Stavros
Open University of Cyprus
Nicosia, Cyprus

Subramanian, Ananda Kumar
VIT University
Vellore, India

1 Data Protection and Privacy Issues of the Internet of Things

O. Gkotsopoulou and P. Quinn

CONTENTS

1.1 Introduction 2
1.2 Internet of Things and Specific Focus 4
1.3 The Legal Framework: Privacy and Data Protection 6
    1.3.1 Background 7
    1.3.2 The United Nations Approach 8
    1.3.3 Privacy and Data Protection in Europe 8
        1.3.3.1 The Council of Europe Framework 8
        1.3.3.2 The EU Framework 9
1.4 Data Protection and Privacy Challenges raised by the IoT
Technologies, Applications, and Ecosystem 22
    1.4.1 Insufficient Security Due to Heterogeneity 23
    1.4.2 Knowledge Asymmetries, Human Errors and Unregulated Access 23
    1.4.3 Insufficient Transparency 23
    1.4.4 Forced Consent Mechanisms, "Consent Fatigue" and
Other Legal Grounds...24
    1.4.5 Exercising Data Subjects' Rights 25
    1.4.6 Incidental Collection of Personal Data, Including
Processing of Nonusers' Data 25
    1.4.7 Mixed Datasets 25
    1.4.8 Profiling and Discrimination 26
1.5 Salient Use Cases 27
    1.5.1 Domotics: Smart Home Appliances 27
    1.5.2 Smart Glasses 28
    1.5.3 Connected Vehicles and Mobility Related Applications 28
    1.5.4 Pandemics and Other Emergencies: Lessons Learnt by
COVID-19 29
1.6 Best Practices and Recommendations 32
    1.6.1 Human-Centric Approach Through Awareness and Education 32

         1.6.2   Incentives and Audits 32
         1.6.3   Information by Design for the Sake of Transparency 33
         1.6.4   Documenting the Operations and Mapping the Partnerships 34
         1.6.5   Data Protection Impact Assessment 34
         1.6.6   Appointment of a Data Protection Officer 35
         1.6.7   Privacy and Data Protection by Design and by Default 35
         1.6.8   Standardization and Certification 36
         1.6.9   Investing in Security by Design 36
         1.6.10   New Legislative Initiatives and Review of Existing Legislation 38
    1.7 Conclusions 38
Notes 39
References...40

1.1 INTRODUCTION

The first toaster was connected to the Internet in 1990 [1]. Since then, more and more devices are being configured to connect to the Internet, giving existence to the Internet of Things (IoT)—a term often attributed to Ashton [2]—, or the so-called Internet of Everything—a term suggested to have been coined together by two corporations, Cisco and Qualcomm—popularizing the "everything (or anything) as a service" concept. By 2025, it is estimated that 80% of the total data processing will take place by smart connected objects [3]. Even though the term IoT has gained large popularity, in particular with respect to the Consumer IoT, a term like "Internet of Things which are connected to people" would seem more appropriate, given that many of them are equipped with tools aimed to record the sound, movement, image and other physical parameters of those owning them and/or those using them [4, p. 9].

Those devices, which did not traditionally connect to the Internet, are now capable to generate and share a large amount of data and therefore, be engaged in large-scale data processing. Thus, IoT in combination with another new technology, the fifth generation of connectivity technologies (5G), increases the possibility for deployment, innovation, and business. Several speak about a revolution that could finally lead to wholesome smart cities. Furthermore, according to the United Nations 2030 Sustainable Development Goals (SDGs), Smart Networks will lead the path toward the digitalization of the society and the economy both in developing and developed countries. The European Commission recognizes it as the next major breakthrough for the Information and Communication Technologies [5, p. 4]. This interconnection of everything supported by Smart Networks appears to be unavoidable and will be characteristic of competitiveness and advancement [6, p. 4]. Of course, 5G is not necessary for all IoT applications and other technologies can be used too. Radio Frequency Identification Technology (RFID) tags for the tracking of physical objects which constituted actually some of the first applications of IoT to track locations, date back a long time. RFID applications have already been widely used in healthcare, transportation, security control and retail [7, pp. 4–5].

Despite the benefits stated by those researching for, and investing in, this kind of technologies that would enable IoT, this unprecedented connection to the Internet, in particular of devices which were not originally intended for such use, can create several implications in relation to fundamental rights, and in particular the rights to privacy and data protection, not only for users of IoT services but also for nonusers. Let us take for example an area where a sensor network is installed; persons entering this area will not have control over the information being collected over their person [8, p. 2802]. The A29 WP calls those data subjects "people interacting" with the device as opposed to actual users [9].

This is more so apparent, when considering that the IoT does not refer only to technologies, but also to applications and ecosystems, having as a result that more and more people from every corner of the world, obtain access to connected devices, often without realizing it. Data subjects can be of different ages, can come from different socioeconomic, cultural, and educational backgrounds, and may live in democratic societies or under authoritarian regimes. What they have in common, though, is that consciously or unconsciously, they generate every moment hundreds of pieces of information about themselves and their contacts landscape. In a human-centric Internet, information can have several receivers, known or even unknown to the data subject.

In 2016, the European Union (EU) adopted a new data protection framework, which entered into force in 2018. After 2 years of implementation, the first studies coming from the European Commission, the national supervisory authorities, and other relevant stakeholders show that implementation is slow [10]. The implementation is further challenged by the need for research, innovation and business modeling, and the urge for quick adoption of emerging and new technologies by the markets. For innovation to work, however, trust is key [8, p. 2801]. In the Impact Assessment of the draft General Data Protection Regulation [11], one reads:

> Building trust in the online environment is key to economic development. Lack of trust makes consumers hesitate to buy online and adopt new services [...] This is why data protection plays a central role in the Digital Agenda for Europe [...].

[12]

In his Keynote speech to the Privacy and Security Conference 2016, the late European Data Protection Supervisor Giovanni Buttarelli stated: "Processing personal information is not prohibited, but it comes at a cost. It affects the rights and interests of the individual concerned by the data. So it is right for anyone who profits from the data to give account for what they have done and why" [13].

Trust goes hand in hand with trustworthiness, which in the IoT context is being understood as encompassing several principles namely, security, privacy, reliability, safety, availability, resilience, ability to connect, inclusivity, transparency, and accountability [14, p. 16]. Trust and trustworthiness are seen as indispensable for growth in complex value chains related to innovation clusters, urging for cross-disciplinary contemplations: technological (human–machine interaction), cultural, societal, behavioral, psychological, and so forth [14, p. 17].

Apart from trust and trustworthiness, the IoT is dependent upon the free flow of data and open access to allow mutual cooperation among stakeholders in the so-called open digital environments [14]. The free flow of data is the primary aspiration of the EU Digital Single Market. IoT services and applications are in need of both personal and nonpersonal information.[1] Nevertheless, even non personal data could be used to disclose personal information, for instance, a person's or a family's current location, travel patterns, or even behavioral characteristics.

The data protection law is applicable only on personal data. It is important to remember that the General Data Protection Regulation regulates both the protection of natural persons with regard to the processing of personal data and the free movement of such data. Manufacturers and developers as well as data controllers must ensure compliance with the envisaged legal obligations in order to safeguard a data subject against abuse and harm. A Next Generation Internet requires what Cunningham called already in 2014 a "next generation privacy" [15, p. 132].

This chapter will address the following topics. First, it will introduce the common definition of IoT to be used in this work and a basic taxonomy. It will focus explicitly on consumers' devices, such as wearables and Smart Home equipment. It will then discuss the challenges raised from the point of view of European data protection law. Particular emphasis will be put on the principles of Data Protection by Design and by Default and the conduct of a data protection impact assessment as a theoretical exercise already at the conceptualization phase of each new device intended to be connected.

1.2 INTERNET OF THINGS AND SPECIFIC FOCUS

Given its complexity and some purposeful vagueness, several definitions of IoT have been proposed over the years [1]. Already in its report from 2005, the International Telecommunications Network states that to the world of information and communication technologies has been added a new dimension: "from anytime, any-place connectivity for anyone, we will now have connectivity for anything." It also recognizes that an increasing "availability" of processing power will be accompanied by decreasing "visibility" [16]. In most instances, IoT appears to be: (a) things-oriented, (b) Internet-oriented and (c) semantic-oriented [8, p. 2789]. Thus, in plain words, the IoT paradigm appears to result from the convergence of those "three different visions" [8, p. 2789]. The Internet of Everything, on the other hand, focuses on *four* elements: people, process, things, and data [1].

To comprehend the various IoT facets, apart from its inherent elements, the reader has to also understand its surroundings. In the Report produced by the IoT Security and Privacy Workshop which was held by the European Commission on 13 January 2017, the stakeholders recognized that when assessing security and privacy requirements in the IoT landscape, it is necessary to identify four life cycles: [17] (a) the IoT device/product life cycle, (b) the stakeholders life cycle, (c) the data life cycle, and (d) the contextual life cycle. Each life cycle corresponds to different characteristics. The first one refers to the duration of the connection, the user expectations, and the possibility to keep a device up-to-date [17]. The second one refers to the role of the stakeholders involved in the IoT ecosystem and their in-between dynamics, including accountability and incident management [17]. The data life cycle refers to all the matters concerned by data protection law, whereas the contextual life cycle refers to the context (dynamic or static) of use of a device and how changes on the context could affect the obligations of stakeholders and the rights of the persons involved [17]. Those life cycles further correspond to five layers, namely: (a) the tangible element (e.g., the hardware), (b) the embedded software, (c) the software maintenance, (d) the supply of digital infrastructures or services (with long-term contract), and (e) the processing and exploitation of user data [17]. Moreover, IoT depends on a spectrum of enabling technologies, which include authentication and security, artificial intelligence (AI) and machine learning, wearables, sensors, IoT Platforms and Middlewares, broadband, and communications [5, p. 36].

Based upon the contextual life cycle and the five additional layers, IoT devices can take different forms and shapes, are used for numerous purposes, and have several application domains [5, p. 34]. As of now, IoT is used for industrial purposes, including logistics, tracking, fleet management (e.g., transmitters installed on pallets and trolleys to report data about temperature, humidity, transport conditions to improve the safety of goods; devices used to monitor a warehouse status); assets tracking/preventative maintenance; for manufacturing; for agriculture and farming; for shopping and retail; for industrial safety and physical security. IoT devices, however, can also be used for commercial purposes in smart homes and smart devices (home goods and services), smart meters, and connected vehicles (through capillary networks). IoT lay in the heart of smart cities and they can have several applications in healthcare and in education, as well.

Last but not least, IoT is driven by complex multidimensional partnerships [18, p. 259]. Those partnerships may include manufacturers, enterprises, resellers, system integrators, and vendors [18, p. 259]. Roles are not always distinct. A vendor, for instance, may sell and/or lease IoT devices but also manufacture them. An enterprise may purchase vendors' services or lease devices. The vendors may have as well several third-party partners offering and/or receiving services. Those services may include access to device embedded intelligence, diagnostics, device locations, and health check-ups which can turn into business intelligence, proactive support, marketing and product development, and new customer services [18, p. 259]. To this image, someone should add the subscriber or end-user of a network operator, who owns a device as well as a third person who is neither a subscriber nor an end-user but their information may be still gathered and processed.

Provided the complexity of the IoT landscape and the importance of the field for economy and innovation, in 2015 the European Commission established the Alliance for Internet of Things Innovation (AIOTI), consisting of European IoT stakeholders ranging from large companies to SMEs, universities, and other research institutions. The aim was "to strengthen the dialogue and interaction among IoT players in Europe."[19] The work of the Alliance is divided into several horizontal Working Groups regarding IoT research, innovation ecosystems, standardization, and policy as well as vertical Working Groups focusing on specific sectors, including Smart Living Environment for Ageing Well, Smart Farming and Food Security, Smart Cities, Smart Mobility, Smart Water Management, Smart Manufacturing, Smart Energy, Smart Buildings, and Architecture. The Alliance has also prioritized work on SME interests and Distributed Ledger Technologies (DLTs). In 2019, AIOTI suggested that the IoT represents one of the largest innovation areas of the forthcoming decade [6, p. 4].

All in all, an IoT device can be in principle seen as any device with an IP address, which can connect to the Internet and transfer and share data. Wachter refers to IoT as "a network of sensing objects that monitor and record aspects of their environment and the behaviours of users within it. Alongside well-established RFID tags, wireless sensor networks and Bluetooth-enabled devices have emerged as IoT sensors" [20, p. 2 (Footnote 1)]. IoT shares several characteristics with other innovative technologies, for instance, robotics and AI. This is because IoT may depend on those technologies as seen earlier. Common characteristics include connectivity, autonomy, data dependency, a degree of complexity with regard to their supply chain and their numerous components, openness to continuous updates and upgrades, reliance on algorithms which may sometimes be rather opaque, and vulnerability to new cyber threats [21, p. 2].

Provided the diversity of the application domains and that of the enabling technologies including those emerging, this chapter will focus on the same three categories as did the A29WP in its Opinion 8/2014, which fall under the so-called Consumer Internet of Things, as opposed to the Industrial IoT, which refers to applications such as manufacturing, logistics, agriculture, automotive, and industrial markets [22], namely: Wearable Computing, Quantified Self and domotics. As the A29WP suggests, those three categories are directly interfaced to the end-user and correspond to devices already widely in use. Yet while reading this chapter, the reader will also find useful insights for more global IoT applications, such as smart cities and smart transportation, for a better understanding of the data protection and privacy concerns. Exceptionally, due to the ongoing corona pandemic, a section will be dedicated to corona tracing and containment initiatives based on IoT.

As for the definitions, *wearables* are devices that often look similar to their non-connected predecessors and can be used in everyday life, such as fitness trackers, smart watches, clothes with sensors, smart glasses, and so forth [9, p. 5]. The *Quantified self* refers to applications and devices "designed to be regularly carried by individuals who want to record information about their own habits and lifestyles," for instance, sleep trackers or personal assets trackers [9, p. 5]. *Domotics* refer to the so-called smart home automation appliances, such as smart lamps, smart plugs, home security, digital assistants, smoke alarms and smart locks [9, p. 5]. As seen, the IoT devices can be equipped with sensors, which can be understood similarly to the human senses: tools that can detect changes in real time and deliver digital or analog output about the physical environment, for example, changes in weather or light conditions [18, p. 69].

1.3 THE LEGAL FRAMEWORK: PRIVACY AND DATA PROTECTION

IoT, like other technological innovations and smart information systems, is governed by a multitude of regulatory frameworks, including intellectual property rights, product liability and accountability, contract and business law, consumer law, and so forth. In this particular chapter, the authors will exclusively examine the implications with respect to privacy and data protection in relation to Consumer IoT.

1.3.1 BACKGROUND

At the United Nations and Council of Europe (CoE) level, only the right to private life is expressly protected in the body of the conventions. At EU level, the rights to personal data protection and respect for private life are closely related, but they constitute two different rights enshrined in the EU treaties and in the EU Charter of Fundamental Rights (EUCFR). Neither right is absolute, which means that they can be limited under specific conditions in line with the EUCFR [23]. The rights may have to be balanced toward other EU values, or other fundamental rights such as the freedom of expression and the access to information, and private or public interests, such as national security, which constitutes sole responsibility of each Member State [23].

Both rights safeguard personal autonomy and are considered prerequisite for the exercise of other fundamental freedoms. Often in practice, the right to data protection is considered a subset of the right to privacy and the terms tend to be used interchangeably. However, they diverge in formulation and scope. The right to data protection refers to the protection of information relating to an identified or identifiable natural living person. It equips individuals with further rights and control over their data, particularly given new risks emerging in the digital era [23].

A long scholarly discussion has taken place around privacy [24, p. 22]. The right to privacy or the right to a private life refers to the right to be autonomous, to be let alone or to be able to avoid state intrusion. It further refers to the inviolability of the home and the confidentiality of the communication [24, p. 23]. In some jurisdictions, it has been associated with freedom, the notion of human dignity, and that of personhood or being human [24, p. 23]. Other theories have also proposed the so-called informational privacy or privacy of information, a discussion which was re-enforced with the introduction of the computers and seemed to have led in the adoption of the first data protection laws around the world in the 1970s.

Provided the massive amount of data being generated, collected, and processed, IoT developers, providers as well as consumers will have several questions to consider. What types of data are collected and by whom? Why are these data collected? Could the device function properly without those data? Where are data sent to, where are they stored, and for how long are they retained? How are the data processed and who has access to them? Does the device collect information about others, third users, or nonusers or does it collect more information than what it is needed to provide the service? How are the data kept secure and how do users get notified in case of a data breach or who they should address when they exercise their rights?

Intelligent "things" can revolutionize one's daily life. Yet, they are objects found in the most intimate personal sphere of an individual. A striking characteristic of some of those smart objects is that many of them do not have an own interface. This lack poses challenges in providing sufficient information to the user about the respective data processing operations and often causes misconceptions about the actual collection of personal data, which may occur any time at any space even when someone does not actively interact with a device. This is why, in the IoT landscape, some authors speak about the need for a unicum between data protection and privacy, a so-called data protecy [25], simultaneously ensuring both rights, through tools which—in addition to a minimum protection threshold—allow the user to self-activate different forms of protection.

In the next sections, the authors will present the different frameworks in relation to data protection and privacy at the international and European levels, paying particular attention to the EU framework.

1.3.2 THE UNITED NATIONS APPROACH

In the international human rights law, privacy is a fundamental right and can be found in all the international human rights instruments, including primarily the United Nations Declaration of Human Rights (UDHR) 1948, Article 12: *"No one shall be subjected to arbitrary interference with his privacy, family, home or correspondence, nor to attacks upon his honour and reputation. Everyone has the right to the protection of the law against such interference or attacks."* UDHR is a nonbinding instrument, but it is of considerable symbolic importance, provided its historic significance.

The right to privacy is protected as well in the International Covenant on Civil and Political Rights (ICCPR) 1966. Article 17 ICCPR states:

1. No one shall be subjected to arbitrary or unlawful interference with his privacy, family, home or correspondence, nor to unlawful attacks on his honour or reputation.
2. Everyone has the right to the protection of the law against such interference or attacks.

The ICCPR Signatory States are themselves bound to uphold the standards enhanced in the Covenant and despite the lack of a judicial body, a monitoring and complaints-handling committee—the Human Rights Committee—has been assigned for the implementation oversight. Individuals can launch complaints upon exhaustion of domestic remedies, given that the concerned state party has signed and ratified the First Optional Protocol to the Covenant. Even though the views, the reports and the interpreting comments of the Committee are not binding either, they carry a special weight.

The right to data protection is not explicitly protected at the United Nations level. Albeit, the right to data protection is considered to be protected under the right to privacy and this has also become obvious in the Report of the High Commissioner for Human Rights on the right to privacy in the digital age [26].

1.3.3 PRIVACY AND DATA PROTECTION IN EUROPE

1.3.3.1 The Council of Europe Framework

The CoE is an international organization based in Strasbourg (France) and comprises 47 countries as of today. It was founded in 1949 in order to promote democracy and protect human rights and the rule of law in Europe. The CoE adopted the European Convention on Human Rights (ECHR) in 1950, and the latter entered into force in

1953. The European Court of Human Rights (ECtHR), which was founded in 1959, ensures that the Parties to the Convention observe their conventional obligations.

The right to respect one's private and family life, home, and correspondence is enshrined in Article 8 ECHR. The ECtHR has interpreted this right as encompassing also a right to personal data protection in its case law [27, p. 38]. Specifically, the processing of information relating to an individual's private life may amount to an interference within the meaning of Article 8. The rulings of the ECtHR are binding for the legal systems of the Member States and set precedents for the interpretation of their national law.

Article 8 ECHR reads:

1. *Everyone has the right to respect for his private and family life, his home and his correspondence.*
2. *There shall be no interference by a public authority with the exercise of this right except such as is in accordance with the law and is necessary in a democratic society in the interests of national security, public safety or the economic well-being of the country, for the prevention of disorder or crime, for the protection of health or morals, or for the protection of the rights and freedoms of others.*

In 1981, CoE opened for signature a legal instrument specifically concerning personal data protection, the Convention for the protection of individuals with regard to the automatic processing of personal data (Convention 108).[2] The Convention 108+ which underwent modernization in parallel with the EU reform of the data protection framework, in order to ensure consistency, applies both to the private and the public sector, as well as the judiciary and law enforcement matters, unless a Member State explicitly opts outs. The significance of the Convention 108+ lies on the fact that up to today, it remains the only international legally binding document for the protection of personal data and its principles have been taken into consideration by the ECtHR, even though the Convention is not under its judicial supervision. The CoE has also issued and adopted several nonbinding recommendations.

This is the case of the CoE framework, where Article 8 (right to private life) encompasses the right to data protection as well. At the EU level, however, the two rights are distinct, as we will see below.

1.3.3.2 The EU Framework

First, some considerations to guide the nonlegal reader through the discussion to follow. The EU is a separate organization from the CoE, even though all EU Member States participate at the CoE as well. EU is based on the rule of law, meaning that every action taken is founded on the adopted Treaties. Treaties, i.e., binding agreements among EU Member States, set out the EU objectives and principles as well as rules for the EU institutions, and define the relationship between the EU and its Member States. Those treaties constitute primary EU law and the basis for the secondary law. Regulations, directives, decisions, recommendations, and opinions belong to secondary law.

Further, according to Article 288 of the Treaty on the Functioning of the European Union (TFEU), the secondary law instruments are divided into soft law and hard law, depending on their binding nature. Specifically:

> A regulation shall have general application. It shall be binding in its entirety and directly applicable in all Member States. A directive shall be binding, as to the result to be achieved, upon each Member State to which it is addressed, but shall leave to the national authorities the choice of form and methods. A decision shall be binding in its entirety. A decision which specifies those to whom it is addressed shall be binding only on them. Recommendations and opinions shall have no binding force.

1.3.3.2.1 Primary Law

The entry into force of the Lisbon Treaty in 2009, amending the Treaty of the European Union and the TFEU, is a significant moment in the history of the EU data protection law. First, the Lisbon Treaty introduced Article 16 TFEU which established the principle that everyone has the right to the protection of personal data concerning him or her and a specific legal basis for the adoption of rules on the protection of personal data. Second, the Lisbon Treaty elevated the 2000 Charter of the Fundamental Rights of the European Union (EUCFR) at the level of primary law, making its provisions binding. The Charter provides for two separate fundamental rights, the right to private and family life (Article 7 EUCFR) and the right to the protection of personal data (Article 8 EUCFR), as they read below:

Article 7 EUCFR—Respect for private and family life

Everyone has the right to respect for his or her private and family life, home, and communications.

Article 8 EUCFR—Protection of personal data

1. *Everyone has the right to the protection of personal data concerning him or her.*
2. *Such data must be processed fairly for specified purposes and on the basis of the consent of the person concerned or some other legitimate basis laid down by law. Everyone has the right of access to data which has been collected concerning him or her, and the right to have it rectified.*
3. *Compliance with these rules shall be subject to control by an independent authority. On the other hand, the Lisbon Treaty provided for the right to the protection of personal data. Article 16 of the TFEU introduces the right explicitly and creates a new independent legal basis, for the adoption of comprehensive EU data protection legislation.*

Concerning the relation between the EU and the CoE framework, it is important to note that Article 52(3) EUCFR states that the meaning of the rights guaranteed in the Charter is the same as in the European Convention on Human Rights (ECHR). Moreover, for any limitation on the exercise of the fundamental rights protected by the Charter to be lawful, it must comply with the following criteria, laid down in the first paragraph of Article 52 EUCFR: (a) it must be provided for by law; (b) it must respect the essence of the rights, (c) it must genuinely meet objectives of general interest recognized by the Union or the need to protect the rights and freedoms of others, (d) it must be necessary, and (e) it must be proportional.

The first condition is very straightforward and refers to the existence of an accessible and foreseeable law. Respect for the essence of the right means that the limitation should not go so far as to void the exercise of the right. If the essence of the right is nullified, the measure is unlawful, and no further examination is necessary. If not, then the objectives of the intended measure will be assessed and will have to be explained in detail, since they will serve as the basis for the assessment of necessity.

The test of necessity assesses whether a measure is indeed necessary, based on a strict necessity criterion: *"derogations and limitations in relation to the protection of personal data must apply only in so far as is strictly necessary"*; moreover, *"the measure should be effective and the least intrusive"* [28, pp. 16–17]. If the test of necessity is not satisfied, the measure will be considered unlawful and the assessment ends; if on the other hand, it is satisfied, the test of proportionality will take place next. Both the necessity and the appropriateness are engulfed in proportionality in broad sense. The advantages of this measure should not be outweighed by the disadvantages for the exercise of the right and a balance should be achieved between the means and the cause [28, p. 9].

1.3.3.2.2 Secondary Law

Under the motto "one continent, one law"[29], the goals of the European Data Strategy, as part of the Digital Single Market [30] is to facilitate data flows within the EU and across sectors, for the benefit of all; ensure full implementation of the EU rules, in particular privacy and data protection, as well as competition law; and allow for fair, practical and clear rules for data access and use [3].

1.3.3.2.2.1 The General Data Protection Regulation From 1995 until May 2018, the principal EU legal instrument was the Directive 95/46/EC of the European Parliament and the Council of 24 October 1995 on the protection of individuals with regard to the processing of personal data and on the free movement of such data (Data Protection Directive or DPD) [31]. The Directive was repealed by the Regulation 2016/679 on the protection of natural persons with regard to the processing of personal data and on the free movement of such data and repealing Directive 95/46/EC [11]. It is more popularly known as the General Data Protection Regulation and as a Regulation, it took force in the Member States directly, even though the latter could opt for derogations with regard to specific provisions. The GDPR primarily addresses Article 8 EUCFR.

The GDPR was part of the data protection reform package, along with the Data Protection Directive for the police and criminal justice sector, areas which are outside of the GDPR's scope as is the case for processing taking place for national security purposes. Purely household activities are also exempted from the scope. The reform was considered necessary in order to strengthen fundamental rights in the digital age and facilitate innovation and business in the internal market. Namely, the main objectives of GDPR are to protect the fundamental rights and freedoms of natural persons and in particular, their right to the protection of personal data, and to ensure the free movement of personal data within the Union. GDPR lays down rules for consistent and effective protection, aiming to a balance between the benefits and drawbacks emerging from the data processing operations [32, p.8].

1.3.3.2.2.2 ePrivacy Directive and the Future Regulation The Directive 2002/58/EC of the European Parliament and the Council of July 12, 2002 concerning the processing of personal data and the protection of privacy in the electronic communications sector (hereinafter, ePrivacy Directive) was adopted in 2002 and amended in 2006 and 2009. The ePrivacy Directive is part of the Regulatory Framework for Electronic Communications and as a Directive, Member States had to transpose it into national law (as opposed to a Regulation). In January 2017, the Commission adopted a new proposal for an ePrivacy Regulation, to replace the old Directive and enforce a harmonized approach across EU and in accordance with the GDPR. The new proposal would include explicitly electronic communications content and metadata. However, in November 2019 no compromise was achieved between the co-legislators, the discussions reached a stall, and the procedure is still pending (as of September 2020).

The main objective of the Directive is to harmonize the national provisions, to ensure an equivalent level of protection of fundamental rights and freedoms, and in particular the right to privacy and confidentiality, with respect to the processing of personal data in the electronic communication sector and to safeguard the free movement of such data and of electronic communication equipment and services in the internal market [32, p. 8]. The ePrivacy Directive guarantees the right provided in articles 7 EUCFR and 8 ECHR.

The ePrivacy Directive applies to *"the processing of personal data in connection with the provision of publicly available electronic communications services in public communications networks in the Community, including public communications networks supporting data collection and identification devices."* More specifically, the ePrivacy Directive applies when there is an electronic communications service, which is offered over an electronic communications network, both of them are publicly available and are offered in the EU. Article 2(d) ePrivacy Directive specifies that *"communication"* means *"any information exchanged or conveyed between a finite number of parties by means of a publicly available electronic communications service"* and excludes broadcasting services. The term *"electronic communications service"* is currently defined by article 2(d) Framework Directive, though with effect from 21 December 2020 it shall be defined by article 2(4) of the Electronic Communications Code.

According to Art. 3, the Directive is applicable *"to the processing of personal data in connection with the provision of publicly available electronic communications services in public communications networks in the Community."* This entails that *"only services consisting wholly or mainly in the conveyance of signals—as opposed to, e.g., the provision of content or other value-added services"* fall within the scope of the Directive.

As for the new ePrivacy regulation debate, the European Data Protection Supervisor states that it is particularly important for the new regulation to cover all current and future tracking techniques via IoT devices and applications, specifically device fingerprinting and all *"passive tracking,"* in other words, the use of identifiers and other data transmitted from the devices [4, p. 28]. The Supervisor also recommends that the new regulation covers machine-to-machine communication in the context of IoT, which includes data transfers and communications not triggered by the device owner/user [4]. Last but not least, emphasis is put on consent with the exceptions for transmission and provision of a service, and for a service requested by the user and performed exclusively by the service provider [4].

1.3.3.2.2.3 The Relation Between GDPR and ePrivacy It is apparent that there is an overlap between the ePrivacy Directive and the GDPR, with respect to their material scope. Article 1(2) of the ePrivacy Directive provides that *"the provisions of this Directive particularize and complement Directive 95/46/EC(...)."* The Directive 95/46/EC as mentioned earlier is the predecessor of GDPR.

"To particularize" refers to the fact that the ePrivacy Directive includes a number of provisions that further stipulate the provisions of the GDPR with respect to the processing of personal data in the electronic communication sector [32, p.13]. To that end, the specific provisions of the ePrivacy Directive function as *lex specialis* and take precedence over the more general GDPR provisions, which function as *lex generalis*. For instance, the ePrivacy Directive *"particularizes"* the provisions of the GDPR in relation to traffic data or the prior consent required for the storing of information, or the gaining of access to information already stored, in the terminal equipment of a subscriber or user [32, p.13]. For whatever is not explicitly covered by the Directive, the Regulation still applies, in particular with relation to the data controller's obligations and data subjects' rights. *"To complement"* refers to the provisions of the ePrivacy Directive which offer a complementary, additional level of protection, for instance by specifically protecting the subscribers and users of a publicly available electronic communications service, including both natural and legal persons (as opposed to GDPR that protects only natural persons) [32, p.14].

Article 95 GDPR states the GDPR *"should not impose additional obligations on natural or legal [...] in relation to matters for which they are subject to specific obligations with the same objective set out in Directive 2002/58/EC."* The aim of this article is to avoid the imposition of unnecessary extra administrative burdens upon controllers, for instance, a double obligation for notification in case of a data breach [32, p.15].

Another important element of this co-existence between the ePrivacy Directive and the GDPR, with relevance for the IoT ecosystem, is that when the provider of an electronic communications service or of a value-added service subcontracts the processing of personal data necessary for the provision of these services to another entity, such subcontracting and subsequent data processing should be in full compliance with the GDPR requirements [32, p.15].

1.3.3.2.2.4 Bodies and Authorities For the better understanding of the nonlegal reader, it is important to clarify the difference between the different bodies and authorities which are mentioned in the text. The *Article 29 Working Party (A29 WP)* was the independent working party at EU level that handled issues relating to the protection of privacy and personal data until GDPR entered into force [33, p. 29]. The *European Data Protection Board (EDPB)* was then established, having its basis in Brussels [34]. The EDPB, composed by representatives of the national supervision authorities and the European Data Protection Supervisor, is an independent European body, which contributes to the consistent application of data protection rules throughout the EU and promotes the cooperation among the national authorities.

National Supervisory Authorities (also known as Data Protection Authorities) are independent public authorities that are given investigative and corrective powers in order to supervise the implementation of the data protection law in a Member State [35]. The supervisory authorities are the main points of reference in relation to data protection issues; they provide advice and handle complaints. There is one supervisory authority per EU Member State. The *European Data Protection Supervisor (EDPS)* is the EU's independent data protection authority [36]. The statements, opinions, and guidelines issued by all those bodies and authorities are valuable for the interpretation and implementation of the data protection law provisions at EU and national level.

1.3.3.2.3 Definitions

The following definitions are necessary in order to understand the material scope. The definitions are mainly as depicted in Article 4 GDPR, followed by a short commentary.

1.3.3.2.3.1 Personal Data   In line with Article 4(1) GDPR:

> [p]ersonal data means any information relating to an identified or identifiable natural person ('data subject'); an identifiable natural person is one who can be identified, directly or indirectly, in particular by reference to an identifier such as a name, an identification number, location data, an online identifier or to one or more factors specific to the physical, physiological, genetic, mental, economic, cultural or social identity of that natural person.

An online identifier refers, for example, to an IP address or aggregate network indicators and network flow data. A person may be identified by a single or a number of identifiers. For data to relate to an individual, it is sufficient if there is a link in relation to the content of the data or the purpose or the result of the data processing [37, p. 11].

An entity may hold information which only indirectly identifies an individual. Even if additional information is needed from other sources to identify an individual, this information may still be considered personal data. In the context of IoT, data generated from "things" can lead to the identification of individuals or families living together [9, pp. 10–11]. Pseudonymization is recognized as a safeguard that the data controllers can adopt. However, pseudonymized data remain personal data and fall under the EU data protection law scope. Truly anonymized or anonymous data, on the other hand, are exempted. It is, however, to keep in mind that the large-scale processing occurring in the IoT context in an automated fashion, entails high risks of re-identification, even when it concerns anonymized data and thus data protection law may still be applicable in some cases [9, p. 11].

1.3.3.2.3.2 Special Catgories of Data According to Article 9 paragraph 1 GDPR, special categories of data are:

> personal data revealing racial or ethnic origin, political opinions, religious or philosophical beliefs, or trade union membership, and the processing of genetic data, biometric data for the purpose of uniquely identifying a natural person[3], data concerning health[4] or data concerning a natural person's sex life or sexual orientation.

In the IoT context, special categories of data may be relevant for two reasons. Special categories of data may be collected directly (for instance, a fingerprint for locking and unlocking a device, information provided for the use of a menstrual cycle application or a dating application) or may be derived or inferred by other raw (not initially sensitive) data. This would be often the case of Quantified Self devices [9, p. 5].

The processing of such data is in principle prohibited unless an exemption—one of those laid out in Article 9 (2) GDPR—applies. For the processing of special categories of personal data, stricter safeguards are imposed by the EU data protection law framework due to the sensitive character of the data which would lead to adverse effects and discrimination for the data subjects. This would mean, for example, that the data subject would have to provide their explicit consent in addition to the requirements for the legal ground in relation to the general personal data processing, or to have made their personal data manifestly public.

1.3.3.2.4 Processing of Personal Data

According to Article 4(2) GDPR, processing refers to:

> any operation or set of operations which is performed on personal data or on sets of personal data, whether or not by automated means, such as collection, recording, organisation, structuring, storage, adaptation or alteration, retrieval, consultation, use, disclosure by transmission, dissemination or otherwise making available, alignment or combination, restriction, erasure or destruction.

The definition of processing is rather broad and nonexhaustive and includes any kind of processing in relation to personal data.

1.3.3.2.4.1 Controllers, Processors, Recipients and Third Parties Data controller, according to Article 4(7) GDPR is:

> the natural or legal person, public authority, agency or other body which, alone or jointly with others, determines the purposes and means of the processing of personal data; where the purposes and means of such processing are determined

by Union or Member State law, the controller or the specific criteria for its nomination may be provided for by Union or Member State law.

Data processor is *"a natural or legal person, public authority, agency or other body which processes personal data on behalf of the controller"* (Article 4(8) GDPR).

The idea of a principal data controller and multiple data processors as secondary actors reflects a centralized perception of data processing which was the case during the adoption of the Directive preceding the GDPR [38, p. 41]. Based on the current practice though in a networked environment introduced by IoT and recent decisions of the Court of Justice of the EU, it appears that *"any actor who has a purpose for a data processing operation, and can directly influence that processing, can be considered a data controller."* [38, p. 40] Extending the role of the controller is seen as a way to ensure *"effective and complete protection"* of the personal data protection right [38, p. 48].

Apart from the data controllers and the data processors, in an interconnected context, recipients and third parties also play a role. Recipient is *"a natural or legal person, public authority, agency or another body, to which the personal data are disclosed, whether a third party or not [...]."* Third party is *"a natural or legal person, public authority, agency or body other than the data subject, controller, processor and persons who, under the direct authority of the controller or processor, are authorised to process personal data."* Third parties may be device manufacturers, third-party application providers as well any other recipients, who even though they have no control over the type of data collected, yet they can be considered as data controllers for receiving, storing, and processing data generated by IoT devices for purposes they determine [9, p. 12].

1.3.3.2.5 Data Processing Principles and Respective Data Controllers' Obligations

The fundamental data protection principles are found in Article 5 of the GDPR (and in a number of Articles in the Convention 108+). These key principles must be applied in all instances of processing of personal data. Restrictions are allowed only if they are provided for by law, pursue a legitimate aim, and are necessary and proportionate in a democratic society. The GDPR sets out seven principles.

First, the principle of lawfulness requires that the processing is based on one of the six grounds provided in the data protection legislation (Article 6 GDPR). This entails that any processing of personal data shall be founded upon:

- consent
- contract
- legal obligation
- legitimate interest
- public interest or public task
- protection of vital interests

It is of paramount important to keep in mind that only one legal basis must be identified for a given purpose of processing and that different legal bases may have to be defined for the same data processing operation if the latter pursues more purposes. Depending on the kind of the processing operation, the data controller, and the nature of the data, a different legal basis may be or may not be appropriate. For example, in the case of a public authority as data controller, the most appropriate legal bases will be the legal obligation and the public interest/public task, whereas when the processing is necessary for the performance or preparation of a contract, the contract will most likely be the most appropriate legal basis. The data controller is encouraged, as part of their recording activities, to document their choice of legal basis in a processing map or in their technical documentation [39]. The choice of legal basis will further impact the rights of the data subject, for example, in the case of legal obligation the data subject will not have the right to erasure, the right to data portability, and the right to object or in the case of contract the data subject will not have the right to object.

Futher, the principle of fairness is perceived first as the reasonable expectation of the data subject to become aware of the data processing [40, p. 7] and second as the ability of the controller to demonstrate full compliance with the existing legislation. Transparency means that the controller shall keep the data subjects informed about the processing risks, their rights, and the policies applicable to the processing.

The principle of purpose limitation requires that any processing must be carried out for a specific, well-defined purpose and only for additional purposes that are compatible with the initial purpose. Any new purpose which is not compatible with the original one or is different, requires its own legal basis. Further processing for archiving purposes in the public interest, for scientific or historical research purposes or statistical purposes, is always considered compatible with the original purpose. In that case, appropriate safeguards must be put in place, such as anonymization, encryption or pseudonymization and the data subject should be informed.

According to the principle of data minimization, data should be adequate, relevant, and limited to what is necessary in relation to the purposes for which they are processed. As for accuracy, data controllers must take measures to make sure that the data they keep are accurate and up to date, with respect to the purpose of the data processing. Inaccurate data must be corrected or erased without delay. Concerning storage limitation, personal data must be erased or anonymized once they are no longer needed for the purposes which they were collected for. Article (5)(1)(e) GDPR provides that archiving data for public interest, scientific or historical purposes, or for statistical use, may be stored for more extended periods.

With respect to security (integrity and confidentiality), appropriate technical and organizational measures should be taken so as to protect personal data against accidental, unauthorized or unlawful destruction, loss, alteration, disclosure, damage, or access. Article 25 GDPR explicitly mentions pseudonymization, as an example of appropriate technical and organizational measures. Other solutions include the storage of the data in a secure physical environment and strong cryptography. Measures must be regularly reviewed, and their efficiency tested, and personal data breaches must be notified to the national supervisory authority and in some situations, the data subject itself. Lastly, concerning accountability, the controller is responsible for and must be able to demonstrate compliance with all the aforementioned data protection principles. There are many ways that the controllers can ensure their compliance, for instance by implementing data protection by default and by design or by designating a Data Protection Officer.

1.3.3.2.6 Data Subjects' Rights

User empowerment is considered a success factor and a strong competitive advantage: *"[...] users must remain in complete control of their personal data throughout the product lifecycle [...]"* [9, p. 3]. This conception is materialized in practice through the data subjects' rights, which in GDPR can be found in Chapter 3. Laconically, those rights are as follows:

- The right to be informed (Articles 12, 13, and 14 GDPR): The data subject has the right to receive information from the data controller about the data processing operation, such as the contact details of the data controller, the legal basis, and the type of data being processed. Depending on whether the data are collected directly from the data subject or other sources, small differences may exist on the kind of information the data controllers are obliged to provide.
- The right of access (Article 15 GDPR): The right to be informed should not be mixed with the right to access his or her personal data. The latter means the right to obtain from the controller confirmation as to whether or not personal data concerning him or her are being processed, and, where that is the case, access to the personal data and a number of other information.
- The right to rectification (Article 16 GDPR): The data subject shall have the right to request from the controller the rectification of inaccurate or incomplete personal data concerning him or her, for example by submitting a supplementary statement.
- The right to erasure (*"the right to be forgotten"*) (Article 17 GDPR): The data subject shall have the right to request the controller to delete personal data concerning him or her, when the data are no longer necessary for the purposes for which they were initially collected, the processing is unlawful, the data subject withdraws her consent or objects, or erasure constitutes a legal obligation of the data controller.
- The right to restrict the processing: The data subject can request the restriction of the processing when the accuracy of the data is challenged, the processing is unlawful, the data are no longer necessary for the purposes of the processing but the data subject needs them for exercising a legal claim and when the data subject objects.
- The right to data portability (Article 20 GDPR): The data subject has the right to receive the personal data concerning him or her, which he or she has provided to a controller, in a structured, commonly used and machine-readable format—or to have them directly transferred to another data controller—and has the right to transmit those data to another controller, if the processing is based on consent or contract or is carried out by automated means.
- The right to object (Article 21 GDPR): The data subject has the right to object at any time the processing of personal data concerning him or her if the processing is based on legitimate interest or public interest/task. The controller shall no longer process the personal data unless the controller demonstrates

Data Protection and Privacy Issues

compelling legitimate grounds which override the interests, rights, and freedoms of the data subject or for the establishment, exercise or defense of legal claims.

- Rights in relation to automated decision making and profiling: The data subject further has the right not to be subject to a decision based solely on automated processing, including profiling, which produces legal effects concerning him or her or similarly significantly affects him or her. This prohibition can only be lifted under specific circumstances, for instance, if the data subject provides their explicit consent or this is part of entering into a contract.
- The rights to remedy and to compensation: The data subject has the right to lodge a complaint with a supervisory authority or/and bring their case before a court. For the right to remedy to be effective, the Regulation gives individuals the right to receive compensation from the controller for material and nonmaterial damages. In addition to that, GDPR has equipped the national supervisory authorities with the power to impose administrative fines (Article 83 GDPR) and other corrective powers (Article 58 GDPR).

1.3.3.2.7 Data Transfers

The production of IoT devices as well as their maintenance often takes place outside the EU, whereas the data and metadata generated by the device may be processed by a cloud-based middleware or be stored in remote servers, which would require cross-border transfer given all the different third party services [25, p. 182].

EU law aims to facilitate the free flow of personal data inside the internal market, by harmonizing the level of protection and allowing organizations to process and transmit data freely as long as they comply with the defined rules and mitigate any risk such processing may entail [41]. In addition to the EU Member States, the area of free flow is also expanded to Iceland, Liechtenstein, and Norway.

However, when transfers take place to third countries, i.e., outside the EU or to international organizations, the legislator has imposed stricter rules, based on the assumption that those third countries do not guarantee the same level of protection. For transfers to take place, two ways have been foreseen: either on the basis of an adequacy decision in line with Article 45 GDPR or in the absence of such a decision, where the controller or processor provides appropriate safeguards in line with Article 46 GDPR. The European Commission has concluded so far adequacy decisions with Andorra, Argentina, Canada (commercial organizations), Faroe Islands, Guernsey, Israel, Isle of Man, Japan, Jersey, New Zealand, Switzerland, and Uruguay, whereas negotiations are ongoing with a number of other countries [42].[5] Provided that none of the above can be achieved, personal data can be transferred to third countries only if additional conditions are met, as provided in Article 49 GDPR.

Specifically, about the United Kingdom, after the Brexit on January 31, 2020, a transition period until December 2020 has agreed, which includes as well the implementation of EU data protection laws as it was until now. What will happen after December 2020, in other words at the end of the transition period, will depend on the negotiations between EU and the United Kingdom. The latter will then have the independence to review their data protection framework. The Information

Commissioner's Office follows closely the negotiation process and provides guidelines to that matter [43].

1.3.3.2.8 Children and Other Vulnerable Data Subjects

There is a market of IoT devices, which clearly targets children, parents, or child-carers: from smart cameras and smart toys to wearables, tablets, and smartphones. From a very young age, children are exposed to a vast range of Internet-connected devices, which can, for example, keep them entertained or help them with learning [44]. Authors speak about the *"datafication"* and *"quantification"* of a child's life, which may start even before they are born [44, p. 3] and can lead to their continuous *"dataveillance"* [45, p. 285]. Stories about malicious actors gaining access to baby monitors and Internet-connected dolls have made headlines the last years [45, p. 286] and have led to the ban of such products in some countries over concerns for children's safety [21, p. 5]. A smart watch for kids, for instance, offering localization features may not directly pose risks for children's safety but it could be hijacked and used for tracking or contacting the kids by malevolent or other third party actors [21, p. 5].

At the international human rights framework, children[6] are entitled to the protection of their private life, on the top of Article 12 UDHR and Article 12 ICCPR, also explicitly on Article 16 of the United Nations Convention on the Rights of the Child (UNCRC). Those rights apply equally both on offline and on an online, digital environment [45, p. 287]. At CoE level, the children's rights to privacy and data protection are considered part of the instruments we discussed earlier (Article 8 ECHR and Modernized Convention 108+). The same goes for the EU level (Articles 7 and 8 EUCFR). It is important to highlight that children's rights are further protected in Article 24 EUCFR, where children are entitled to protection and care which ensures their well-being. Moreover, all actions that concern them should take the child's best interests into primary account [45, pp. 287–288].

Even though the Data Protection Directive did not include specific provisions for the protection of children, already in 2009 the A29WP assessed the issues in relation to children's data protection in its Opinion [46]. The Party reinstated its positions later on in relation to apps on smart devices and the IoT. Under the GDPR, special protection provisions were included to reduce online risk for children, in particular for those under 16 years old. In GDPR, children are considered vulnerable data subjects [47] and their protection is assessed through requirements for consent in Article 8, specific transparency obligations for the data controllers in Article 12 and risk management with particular focus on children in Recital 75. Recital 38 GDPR, read under the light of Recitals 58 and 65, provides that *"children merit specific protection with regard to their personal data, as they may be less aware of the risks, consequences and safeguards concerned and their rights in relation to the processing of personal data"* [47].

Recital 38 further states that:

> [s]uch specific protection should, in particular, apply to the use of personal data of children for the purposes of marketing or creating personality or user profiles and the collection of personal data with regard to children when using services offered directly to a child.

Data Protection and Privacy Issues   21

However, this is to be understood as the specific protection is not only related to marketing or profiling but refers to *"a wider 'collection of personal data with regard to children'"* [48, p. 24]. According to Article 8(1) GDPR, where information society service providers[7] rely on consent for the provision of such services directly to a child, the processing of the child's personal data is only lawful if the child is at least 16 years old. Specifically, for minors under 16 years old, the GDPR requires the legal guardian's consent when information society service providers process children's personal data, aiming to address the unique needs of children as data subjects [49], even though the effectiveness of such scheme has been disputed [50]. The consent must be given or authorized by the holder of parental responsibility, which can be more than one persons [48, p. 24]. The age of digital consent can be lowered individually in Member States, but it should not be less than 13 years old.

In the US, the Children's Online Privacy Protection Act (COPPA) was put in place almost two decades before GDPR and also offers specific provisions for the processing of children's data, in particular for minors under the age of 13 [45, p. 288]. Studies provide extensive comparative analyses of both legal instruments; in principle, some differences can be found in definitions and the application scope [49].

Complying with the principle of transparency proves to be rather challenging not only toward adult users, but foremost minor users [51, p. 17]. Providing meaningful information about the data processing in plain language and in a way that can be easily understood by children requires effort, which could include both textual and visual information, less links to other webpages, and more concise texts [51, p. 17]. Empirical studies have demonstrated that data protection policies are often opaque, inadequate, and nonchild friendly [49]. Excessive collection of personal data from children and subsequent disclosure to third parties has also been observed [49]. This might entail that other nonconventional ways for providing information shall be adopted, including cartoons, pictograms, and animations, depending on the age and maturity of the child [52, p. 12].

Thus, all actors involved in the IoT ecosystem should "go the extra mile" to make sure that children's rights are sufficiently protected. For IoT products directly targeting children, suggestions have been made for the inclusion of data protection policies on the packages both for children and parents, including audio and visual notifications that inform children and parents of the data processing activities [45, p. 299]. The implementation of the highest security standards and reconsideration of the need to collect data from children and further, of who should have access to the collected data, as well as reshaping data storage solutions, should be of primary concern of actors involved in IoT targeting children, parents and educators [45, p. 299]. Industry code of conducts, tackling the particular issues in a self-regulatory manner, are also encouraged [45, p. 300].

Specifically, in relation to automated decision making and profiling, such provision cannot concern children [44, p. 5]. Nevertheless, A29WP suggests that even though no absolute prohibition exists, yet entities involved in children's personal data processing should refrain from children's profiling for marketing purposes.

Apart from children, vulnerability can be attributed to other data subjects as well, including adults. The key element of vulnerability appears to be a power imbalance between the data subject and the data controller. Thus, vulnerable data subjects can include children, as seen earlier, employees toward their employer, and other individuals belonging to population groups which require special protection, for instance, *"mentally ill persons, asylum seekers, or the elderly, patients, etc."* [47] Malgieri and Niklas argue that based on a layered analysis of vulnerability in GDPR, *"everyone is potentially vulnerable, but at different levels and in different contexts."* In the IoT context, this would entail that IoT actors have to go two or more extra miles, to take into consideration the particularities of each application. A way to identify and implement specific safeguards for vulnerable data subjects could be the conduct of data protection impact assessments and the full implementation of the principle of data protection by design [47]. If the system is not designed to protect the most vulnerable or there are remaining risks after the risk management process, those risks should be made known in a transparent manner.

Lastly, during the corona outbreak, EU Member States considered standalone devices and wearables specifically for children and other vulnerable groups which would constitute an alternative to smartphones, deploying directly the nationally adopted contact tracing apps [53, p. 20]. The possibility of using domotics and other home-based solutions was also explored [53, p. 20].

To sum up, IoT actors will have to observe the relevant provisions of the forthcoming ePrivacy Regulation, as the latter will have a significant impact also in relation to children's and other vulnerable individuals' protection [45, p. 290].

1.3.3.2.9 Inferences

Another issue, which is very complex in the IoT ecosystem, is that of inferred data. Personal data initially collected may not as such reveal sensitive aspects of someone's private life. Nonetheless, those data tracked over a period of time, combined with other data and examined through advanced data analysis tools may lead to inferences, which could constitute personal data, even belonging to a special category, despite the fact that the initial collected data did not [54].

Article 29 WP in its Opinion from 2014 concerning IoT brings the example of applications which track data subjects' movements in order to extract the number of daily steps and display information about an individual's physical and mental condition [9, pp. 7–8]. Even though the user of the smart device may have been *"comfortable"* sharing the original information, they may not be comfortable with sharing the secondary (derived or inferred) information [9, pp. 7–8]. The concerned stakeholder should take into account the sensitivity of inferred data and ensure that all the purposes of the processing concerning the raw, the extracted and the displayed data are known to the data subject, in line with the principles of transparency and purpose limitation [55].

1.4 DATA PROTECTION AND PRIVACY CHALLENGES RAISED BY THE IOT TECHNOLOGIES, APPLICATIONS, AND ECOSYSTEM

The inherent characteristics of IoT devices and services create a number of opportunities. Those same characteristics though lead to several challenges and risks [17]. The most important of those challenges will be discussed below.

1.4.1 Insufficient Security Due to Heterogeneity

First of all, IoT devices are vulnerable to attacks, which could have as a result not only disruption of the continuity and availability of a service but could potentially also cause material damage and physical harm. IoT devices may further enable unauthorized access to personal information, facilitate attacks on other systems, and even bring about general safety risks [56, p. 11]. One of the reasons for this vulnerability is that information security features are not embedded in the products by design, but they are configured later once the desired functionality has been achieved. Moreover, a number of sensors do not allow for the establishment of encrypted links and there is no automated updates service available. Another reason is the lack of harmonization, due to the development of IoT devices and services by disparate communities, and lack of standardization, due to the use of proprietary standards and standards applicable in different sectors. Therefore, the security risk analysis, assessment, and mitigation are more difficult than in the case of coordinated ecosystems [1].

Security vulnerabilities and flaws have been reported by a number of consumer rights organizations in EU Member States concerning several interconnected products available in the market, which raise questions about the compliance of those products with data protection requirements in relation to security [10, p. 30].

1.4.2 Knowledge Asymmetries, Human Errors and Unregulated Access

App developers and device manufacturers are often unaware of the data protection requirements[9], even though this seems to slowly change given that companies may be obliged to offer trainings to their employees and more and more public debates take place on the matter. Given the large processing which can lead to extensive monitoring [57], the intrusive use of IoT devices cannot be excluded as well as any unlawful surveillance.

Without proper training, additional risks may occur from human errors, inside threats and internally or externally caused personal data breaches [9, p. 29].

1.4.3 Insufficient Transparency

The IoT landscape is characterised by the complexity of the relations among the several involved entities which are a lot more in number than in a traditional context. The vast range of actors can include hardware manufacturers, device manufacturers, device vendors, operating system and other software vendors, telecommunications and network providers, third-party app developers and vendors, Cybersecurity experts, end-users (individuals), including subscribers and owners, other third users, and people who are incidentally captured by the device or the service. Such a constellation of actors creates questions as to when a stakeholder acts as a data controller or as a data processor.

The distinction is of high importance, since GDPR, even though it has introduced obligations for the data processors too, considers the data controller the principal entity for taking care of the main responsibilities in relation to data processing operations, including partnering with data processors who are GDPR-compliant. For instance, it is the data controller who has to provide the data subject with information about the processing of the personal data and it is the data controller that the data subject will contact to revoke his/her consent, even if data processors are more or substantially involved in the processing of the data.

In addition, tracing data flows and functions may be proven rather difficult, causing uncertainty as to who the owner of the generated data is and who the recipient of those data is. The individual appears to have little control over the dissemination and flows of data, which could lead to excessive self-exposure. Moreover, in practice, there is limited possibility to use services anonymously [58].

1.4.4 FORCED CONSENT MECHANISMS, "CONSENT FATIGUE" AND OTHER LEGAL GROUNDS

Another concern is that users may also be confronted with indirectly forced consent mechanisms in the context of IoT, which could lead to low quality or invalid consent. According to GDPR,

> consent of the data subject means any freely given, specific, informed and unambiguous indication of the data subject's wishes by which he or she, by a statement or by a clear affirmative action, signifies agreement to the processing of personal data relating to him or her.

The European Data Protection Supervisor suggests that consent is the legal basis that should:

> be principally relied on in the IoT. In addition to the usual requirements (specific, informed, freely given and freely revocable), end users should be enabled to provide (or withdraw) granular consent: for all data collected by a specific thing; for specific data collected by anything; and for a specific data processing. However, in practice it is difficult to obtain informed consent, because it is difficult to provide sufficient notice in the IoT .

[4]

The European Data Protection Supervisor even suggests that:

> no one shall be denied any functionality of an IoT device (whether use of a device is remunerated or not) on grounds that he or she has not given his or her consent [...] for processing of any data that is not necessary for the functionality requested .

[4]

The European Data Protection Board specifically states that consent withdrawal must be as easy as giving consent, but this does not entail that the exact same procedure should be applied [48, p. 22]. In other words, when consent is obtained via electronic means, for example through a swipe or a mouse-click, the withdrawal of consent should be equally easy and simple [48, p. 22]. Specifically, when consent is obtained through the interface of an IoT device, the data subject must be able to withdraw their consent via the same electronic interface and should not be required to use another interface solely for withdrawing their consent as a different way would entail undue effort [48, p. 22]. Moreover, the withdrawal should not entail detriment for the data subject and must be possible free of charge or without resulting in a lowered service quality [48, p. 22].

It is also important to note that IoT devices constitute "terminal devices" under EU law and thus any storage of information or access to information stored on an IoT device requires the end user's consent in line with the ePrivacy Directive [9, p. 14]. Legitimate interest is unlikely to constitute a legal ground for processing given the seriousness of the intrusion.

1.4.5 Exercising Data Subjects' Rights

Another challenge constitutes the heterogeneity of the data protection policies of the interconnected objects, which may become even more complex depending on the context where the objects are used and the different applicable legal frameworks, even though some data protection and privacy considerations will be similar. Accessing one's data is also a challenge in the IoT context, in particular since the right of individuals extends not only to the displayed data or the requested data (e.g., registration data), but also to the raw data processed in the background. To that end, can be added the inferences and the processing of data for secondary purposes, including the detection of behavior patterns and profiling which could amount to surveillance.

1.4.6 Incidental Collection of Personal Data, Including Processing of Nonusers' Data

The IoT environment is a multiuser context, where devices may have more than one users and data of nonusers may be collected in a way that can be perceived as nearly covert. For instance, when a sensor collects data of persons regularly visiting a building or when a device is used by different members of a family.

The quality of the consent can be also impacted by difficulties in providing information to individuals who are not the end-users of a device, for instance to individuals whose data get incidentally collected.

1.4.7 Mixed Datasets

In real-life application of the IoT, datasets will be mixed, meaning that they will be likely composed of both personal and nonpersonal data. Nonpersonal data are data that are not "personal" as defined in the General Data Protection Regulation. In other words, data that did not originally relate to an identified or identifiable natural person; for example, data relating to the weather collected by building sensors. And data that were originally personal, but they were later made anonymous, which means that the data can no longer be attributed to a particular individual.

The assessment of whether the data are properly anonymized and cannot be re-identified—not even with additional data—must be done on a case-by-case basis, by taking a look at all means reasonably likely to be used by a controller or by another person in order to identify the individual. It is important to remind the reader that anonymized data are nonpersonal data whereas pseudonymized data are personal data [59, p. 5]. Pseudonymization for example would be the case, if personal data are replaced by unique attributes in a dataset and the actual personal data are kept separately from the assigned unique attributes in a secured database.

The assessment demands regular reviews, given the technological progress that can lead to re-identification of personal data. The Commission in its Guidance gives as an example the quality control reports on production lines which can relate to specific employees [59, pp. 6–7]. In some cases, even the data relating to legal entities could be considered personal data, for instance, if the name of the legal entity corresponds to a living natural person [59, pp. 6–7].

If a dataset is composed of both personal and nonpersonal data; in other words, it is a mixed dataset, then the following would apply [59, p. 9]:

- the Free Flow of Non-Personal Data Regulation would apply to the nonpersonal data part of the dataset;
- the General Data Protection Regulation would apply to the personal data part of the dataset;
- "if the non-personal data part and the personal data parts are 'inextricably linked', the data protection rights and obligations stemming from the General Data Protection Regulation [would] fully apply to the whole mixed dataset, also when personal data represent only a small part of the dataset." [59, p. 9] None of the two regulations define the concept of "inextricably linked." In practice, this can refer to a situation where separating the personal from the nonpersonal data would be impossible, not technically feasible or economically inefficient [59, p. 10].

1.4.8 PROFILING AND DISCRIMINATION

Wachter identifies three ways of profiling that could lead to discrimination in IoT systems:

a. *data collection that leads to inferences about the person (e.g., Internet browsing behavior);*
b. *profiling at large through linking IoT datasets (sometimes called 'sensor fusion'); and*
c. *profiling that occurs when data are shared with third parties that combine data with other datasets (e.g., employers, insurers)"* [20, p. 10].

Data Protection and Privacy Issues

IoT devices can only work if they collect data and make inferences, but this can be very intrusive into a person's private life [20, p. 11]. Even data which in a first look seem to be neutral, such as a postcode, when connected to other datasets, can lead to discrimination, based on ethnicity or gender [20, p. 12].

In the IoT ecosystem, a digital identity can be considered:

> a type of profile, made up of all information describing the user that is accessible to a decision-maker, based on observations or prior knowledge (e.g., age, location), or inferences about the user (e.g., behaviours, preferences, predicted future actions) .
>
> [20, p. 7]

The way an individual perceives themselves might be different from the way external entities perceive them, especially given that those entities may only have in their disposal segments of an individual's profile [20, p. 7]. Moreover, the individuals may not be fully aware of those external identities, may lack control or may be not able to assess the validity of those inferences about them [20, p. 7]. Those segmented profiles constitute virtual identities which are rather contextual and may again interfere with a person's right to data protection as well as a person's right to non-discrimination.

1.5 SALIENT USE CASES

1.5.1 DOMOTICS: SMART HOME APPLIANCES

Several studies analyze the so-called consumer dilemma: how consumers choose and use Smart Home appliances and personalized information services over increased privacy concerns [60] and try to introduce contextual privacy norms in the Smart Home.

Studies show that users of Smart Home devices acknowledge the risks and are able to identify threats in relation to smart infrastructure [76, pp. 444–446]. Nevertheless, due to an observed "trust paradox," they do not consider those threats as serious or important enough to take concrete measures except for trying to avoid providing more information than they consider necessary [76, pp. 444–446]. They may for instance avoid to download a particular app or use a device if they do not feel comfortable with the requested data sharing, or not consent to particular information flows related to a feature or further sharing with an advertising partner or social media platforms [77, p. 16].

Moreover, even though users are able to recognize general data flows, they are highly uncertain about specific data practices and wish to have greater awareness and control through user-friendly mechanisms and visual indicators which can assist users to understand and further contribute to the security of their smart devices [76, p. 446].

1.5.2 SMART GLASSES

The European Data Protection Supervisor has outlined the main concerns regarding smart glasses projects.

> Smart glasses are wearable computers with a mobile Internet connection that are worn like glasses or that are mounted on regular glasses. They allow to display information in the user's view field and to capture information from the physical world using, e.g., camera, microphone and GPS receiver, e.g., for augmented-reality (AR) applications. As Internet-enabled devices, they can belong to the Internet of Things (IoT).
>
> [61, p. 4]

Smart glasses have been piloted or are already in use mainly in the service of law enforcement agencies, whereas companies have launched projects for the production of affordable devices for widespread adoption. For instance, in China, use of AI-powered smart glasses have been reported to be used for combatting coronavirus spread [62]. According to EDPS, the considerations are similar as in the previous A29WP opinion on IoT. Even though the value of such technology in some fields has been recognized, the key concern is the capacity to record video and audio in a way that persons would not be aware of the recording [61], with or without the intention of the actual user [61, p. 11].

Thus, smart glasses like other wearable devices may collect data about a data subject who is not the owner or the user of the device. The A29WP highlighted that the application of the EU data protection law does not depend on ownership or use of a device, but on "the processing of the personal data itself, whoever the individual concerned by this data is" [9, p. 13].

1.5.3 CONNECTED VEHICLES AND MOBILITY RELATED APPLICATIONS

As one of the largest vehicle technologies exporters, the EU recognizes the driverless mobility benefits and risks in its approach from 2018 toward connected and automated mobility [63, p. 2], foreseeing a revision of the General Safety Regulation for motor vehicles. Connected vehicles have the potential to generate an enormous amount of data, both nonpersonal and personal. Those data will be vehicle generated data but also customer provided data, in other words data provided by the vehicle keepers, drivers, and passengers. Embedded communication devices will create new and personalized services and products, including roadside assistance and vehicle repair [63, p. 13], as well as a need for new ways ensuring liability such as data recorders to identify who was driving the car during an accident. Data will be increasingly and intensively shared in a wireless fashion with several stakeholders.

Connected vehicles can raise higher concerns in relation to data security, because security incidents can compromise road safety and ultimately physical integrity. For example, cyber-attacks could lead to remote taking control of the vehicle [63, p. 12]. This is even more so because there is no sector-specific approach on vehicle protection toward cyberattacks, provided that connected cars are composed of several components [63, p. 12].

Moreover, the processing of location data can lead to intrusiveness into one's private life and indiscriminate surveillance or misuse of the data [64, p. 10]. The complexity of such systems also lays in the dynamic and multi-actor environments of machine-to-machine (M2M), vehicle-to-vehicle (V2V), and vehicle-to-infrastructure (V2I) communications. The EDPB Opinion prioritizes the legal ground of consent over other legal grounds. This prioritization has been criticized to be a static concept for a very dynamic framework. This concern has been specifically shared by the UNECE World Forum for Harmonization of Vehicle Regulations (WP.29) stating that such concept is "not workable" in the context of the interconnected vehicles.

The European Commission Directorate General for Research and Innovation in its independent expert report on Ethics of Connected and Automated Vehicles provides several recommendations on road safety, privacy, fairness, explainability, and responsibility [65].

1.5.4 Pandemics and Other Emergencies: Lessons Learnt by COVID-19

An emergency could legitimize specific restrictions of freedoms under the condition that those restrictions are proportionate and limited to the emergency period [66, p. 1].[8] The COVID-19 pandemic showed how an *offline virus* could have a big impact on several data processing operations, led either by state or/and private entities. A few areas relevant also in the IoT context were as follows:

1. processing of health data by medical staff, employers, and state authorities, including law enforcement agencies;
2. processing of different types of data for the monitoring of the spread and the containment of the pandemic, through digital apps and wearables, quarantine enforcement mechanisms, data donation apps for scientific purposes, mobility maps, drones, and other data-driven approaches;
3. a wide dependence on digital means for remote working/education, social contact, and entertainment;
4. an increase in data security incidents and a rise in COVID-19 related cyber-threats and cyber-attacks [67].

The European Commission suggested a pan-European and coordinated approach, which included mobile applications as one of the proposed measures in the combat against the pandemic. The European Data Protection Board highlighted that even in these exceptional times, the data controllers and processors must ensure the personal data protection, in particular, any measures must respect the general principles of law and must not be irreversible [68, p. 1]. Specifically, concerning the development and implementation of apps, the Board emphasized that they

> should be made in an accountable way, documenting with a data protection impact assessment all the implemented privacy by design and privacy by default

mechanisms, and the source code should be made publicly available for the widest possible scrutiny by the scientific community .

[68, p. 2]

It continues that location tracking for contact tracing apps is not necessary; location tracking in this context would violate the principle of data minimization and would create major risks for security [68, p. 3]. Concerning data storage in contact tracing apps, EDPB gives priority to a decentralized solution, but it suggests that both a centralized and a decentralized approach could be supported given that the appropriate safeguards have been applied and the relations among the data controllers have been clarified [68, p. 2]. Any data collected through such an emergency system shall be erased or anonymized after they are not needed anymore for the purpose which they were initially collected for [68, p. 3].

In Europe, there are several initiatives for the creation of apps with different functionalities, ranging from information provision to registration of travelers, contact tracing, and quarantine enforcement. Numerous states have been developing and/or have already launched their own applications, like Poland ("selfie app" and "health diary app")[69], Czech Republic ("memory maps") [70], Austria, France, Ireland and Germany (contact tracing Bluetooth-based applications). Specifically, in relation to contact tracing, heated debates took place. The Pan-European Privacy-Preserving Proximity Tracing (PEPP-PT) consortium made its appearance quite early promoting a centralized design for contact tracing apps, whereas another initiative, the Decentralized Privacy-Preserving Proximity Tracing consortium (DP3T), has been promoting a decentralized design, grounded upon an open protocol for Bluetooth Low Energy-based proximity tracing on mobile devices and ensures that personal data will not leave the individual's device [53, p. 10]. In the meantime, other initiatives have also been born which promote hybrid solutions or prioritize one or the other design. Global initiatives, like that of Google and Apple who jointly announced also a broader Bluetooth-based solution by building the functionality into their platforms [71], have also sparked intense discussions and led to the adoption of diverse solutions by the states. A list of contact tracing apps, as of 15 April 2020, can be found in: Health Network, "Mobile applications to support contact tracing in the EU's fight against COVID-19 Common EU Toolbox for Member States Version 1.0" [46, pp. 10–12].

The European Commission has assisted the EU Member States to create a common toolbox for mobile contact tracing and warning apps, by providing a number of minimum requirements for built-in features, Cybersecurity, and data protection standards [46]. Moreover, the Commission has developed specific Guidance on data protection matters, including ten recommendations for a trustful and accountable use of contact tracing apps. Given the sensitivity of the data, the data controllers of those apps should be national health authorities and the individual user must remain in control of their data at all instances [72, p. 3]. The processing must be based on a unambiguous, freely given, specific and informed consent in line with Article 5 ePrivacy Directive and the respective national legislation and the storage and/or access to information in the individual's device is strictly necessary for the requested information society service [72, pp. 4–5]. Moreover, if the processing occurs by national health care authorities, the legal ground to be invoked is Article 6(1)(c) and Article 9(2)(i) GDPR or when such processing is necessary for the performance of a task carried out to further the public interest recognized by EU or Member State law [72, pp. 4–5].

A particular exception is provided in Recital 46 GDPR, which provides:

> The processing of personal data should also be regarded to be lawful where it is necessary to protect an interest which is essential for the life of the data subject or that of another natural person. Processing of personal data based on the vital interest of another natural person should in principle take place only where the processing cannot be manifestly based on another legal basis. Some types of processing may serve both important grounds of public interest and the vital interests of the data subject as, for instance, when processing is necessary for humanitarian purposes, including for monitoring epidemics and their spread or in situations of humanitarian emergencies, in particular in situations of natural and man-made disaster.

Moreover, the Commission reminds in relation to contact tracing and warning apps, which provide automated proximity warnings that subjecting individuals to a decision based solely on automated processing which produces a legal effect or similarly significantly affects him or her is prohibited in line with Article 22 GDPR. Further, it emphasizes that the development of the apps should take into consideration the principle of data minimization based on desired functionalities (a symptom checker app will require the processing of different [personal] data than contact tracing, proximity, and warning apps) [72, p. 6]; limitation of the access to data and subsequent disclosure of data, for example, data of infected persons [72, p. 7]; purpose limitation, expressly stating that *"prevention of further COVID-19 infections"* is not a specific enough purpose [72, p. 8]; storage limitation, based on medical relevance and administrative necessity; [72, p. 8] data security with encryption and pseudonymization [72, p. 8]; data accuracy, relying on technologies that permit a more precise assessment; as well as involvement of the national supervisory authorities and conduct of a data protection impact assessment [72, p. 9].

Several national supervisory authorities have issued their own guidelines and opinions on the processing of health data during the pandemic through data-driven solutions but also the use of digital tools for supporting remote work and education. Even though data security including integrity and confidentiality is one of the obligations explicitly described in GDPR for data controllers, COVID-19 brought to the front security inefficiencies and lack of digital skills, which allow cyber-criminals to exploit old and new vulnerabilities, including COVID-19 related cyber-attacks such as widespread phishing campaigns, DDoS and registration of malicious domain names [73]. With a noticeable dependence on digital services for distance learning and remote work, as well as entertainment and social contact, it became apparent that both employers and providers of services as well as employees or other individuals being at the receiver side of those services, with a big number of IoT devices involved in the process—from smartphones and smart cameras to fit trackers and entertainment platforms—must ensure a high standard of data security by default and by design and users must receive digital literacy trainings, on how to stay safe using all those interconnected devices.

1.6 BEST PRACTICES AND RECOMMENDATIONS

This section aims to provide an overview of the recommendations provided in guidelines and opinions of the European Data Protection Board, its predecessor A29 WP, national supervisory authorities, and stakeholders consultation. In most documents, the recommendations address specifically the different entities involved in the IoT ecosystem, namely (a) the users/consumers and others, (b) the Operating System (OS) and device manufacturers, (c) the third-party application developers, (d) other third parties, (e) the social networking platforms, f) the standardization/certification bodies, and (g) data platforms.

1.6.1 HUMAN-CENTRIC APPROACH THROUGH AWARENESS AND EDUCATION

The European Commission has emphasized that the IoT ecosystem should be human centric, empowering the data subjects to retain control over their data [17]. Some authors suggest to that end a user-centric privacy management [74, p. 385], combined with raising awareness and educating individual users. IoT devices and applications should be accompanied by proper, sufficient, and understandable information, otherwise consent obtained on the basis of incomplete data protection policies may not be considered as informed and thus as nonvalid. For example, smart sex toys that can reveal very intimate information remind users to rethink the information provided to them, the information they choose to share and the security risks relating not only to the directly collected data but also to the inferred or derived [75].

Education must present engineering students and manufacturers of IoT devices with examples of "good privacy designs" and simple design principles to follow when creating smart objects and systems, addressing technical, organizational, and philosophical questions in a co-design process [78]. Re-thinking the design of objects, having in mind "privacy first," could, for instance, simply entail the addition of a switch that could allow consumers to visually understand the data transmissions of their devices and let them select whether the latter shall communicate with the cloud, with the local network, or shall refrain from transmitting any data [79].

This brings us to the argument that, to move from a traditional enterprise-centric approach of security and privacy to a human-centric one, empowering the data subject, requires cross-disciplinary efforts [25, p. 191], including research on philosophical, technical, and societal topics [65, p. 38]. Enabling user's choice beyond *"take-it-or-leave-it"* models would require the re-thinking of how to provide consent alternatively through agile and continuous consent management [65, p. 37].

1.6.2 INCENTIVES AND AUDITS

Many of the documents studied prioritize a necessary mind shift. One way to motivate such change could be through Key Performance Indicators (KPIs) and metrics

Data Protection and Privacy Issues

for security and data privacy, to provide incentives for manufacturers and retailers to consider data protection and privacy by design and by default as *"key selling point of innovative technologies"* [9, p. 21]. For the KPIs to be effective, they should be combined with continuous monitoring by internal and external oversight mechanisms and be reviewed through independent privacy and security audits.

1.6.3 Information by Design for the Sake of Transparency

The data controllers should offer simple opt-outs and/or granular choices on the type of the data being processed and the frequency of the data gathering, whenever applicable [9, p. 21]. They should also ensure that they have the right legal basis for further processing, in the case of repurposing or sharing with other entities and making sure that the appropriate technical and organizational measures are in place [9, p. 21].

The term "privacy policy" could be considered as misleading, as it creates the impression to data subjects that their privacy is protected, whereas in practice privacy policies function more as liability disclaimers [20, p. 16]. The primary aim of privacy policies though is to provide transparency and not to protect companies from litigation. Transparency in the roles and in the processing operations is necessary not only toward the data subject but also toward the multitude of companies receiving data from a single IoT device. In this regard, most recommendations include the provision of a single point of contact when it comes to personal data protection and privacy. Thus, the provision of information should take place in a user-friendly and immediate manner, in what the A29 WP calls *"[to] design devices to inform"* [9]. As discussed earlier trust is essential for the success of the IoT. Transparency and awareness of the possible risks of the use of IoT devices is essential for the users to make an informed decision whether to use a device or not; the opposite, a dishonest stand of hiding the risks would hinder trust [20, p. 21].

The A29 WP has identified as "appropriate measures" for the provision of information for screenless smart technology and the IoT environment, the following:

> icons, QR codes, voice alerts, written details incorporated into paper set-up instructions, videos incorporated into digital set-up instructions, written information on the smart device, messages sent by SMS or email, visible boards containing the information, public signage or public information campaigns.
>
> [52, p. 21]

As Recital 60 GDPR reads, information may be provided in combination with standardized icons—so-called privacy icons [80, p. 358], allowing the data controllers to take a multi-layered transparency approach, by reducing the need for lengthy written data protection/privacy policies [52, p. 25]. Specifically, icons that are presented electronically, which will be often the case for IoT devices or IoT device packaging, must be machine-readable [52, p. 25]. Currently, Personal Information Management Systems are emerging to enable user's control over their personal data, but the need to develop new tools and mechanisms to facilitate the exercise of data subjects' rights remains pressing [81, p. 6].

For instance, Wachter proposes a three-step transparency model, which requests data controllers in the IoT context to describe the possible risks, explain the kind of safeguards in place to limit in particular inaccurate assumptions and discrimination risks, and demonstrate transparent mitigation plans for addressing the risks in case of system compromise [40, pp. 12–13].

1.6.4 DOCUMENTING THE OPERATIONS AND MAPPING THE PARTNERSHIPS

Given the European Commission's intention to render data sharing compulsory under some circumstances and promote the creation of sectoral common data spaces, it is important that there is in place *"a clear limitation on the cross-context use of data,"* by defining the purposes of each data space from the onset, for instance, prohibiting the use of special categories of personal data for purposes others than those initially collected [81, p. 11]. Such limitations may permit a more efficient mapping of partnerships and operations in the IoT space.

1.6.5 DATA PROTECTION IMPACT ASSESSMENT

The GDPR is built upon a risk management approach, which is contextual and harm based. The risk management should take place per given context or sector and ideally user preferences could be taken already into account, by identifying the data contexts and uses and the respective risks and mitigation mechanisms [15, p. 143].

Conducting a data protection/privacy impact assessment before releasing a device or a system, even if this is not mandatory by law, could be considered a best practice in the IoT landscape [9]. The assessment is a living instrument of compliance monitoring. It should start with the conceptualization of a product or a service and continue through design and actual development and market release until the product has ceased to be offered. In other words, this assessment accompanies the product or service throughout its entire life-cycle and aims to satisfy both the need for awareness about a) the regulation and the respective obligations as well as about b) the data processing operations including complex layers of data controllership, by providing hands-on tools, such as data processing matrixes and risk management boards. As a best practice, clear information should be provided to the retailer of the device or service and all other partners, who would need to provide this information to their customers.

There are several templates and recommendations as to how better comply with the requirement for a data protection impact assessment. The d.pia.lab (Brussels Laboratory for Privacy and Data Protection Impact Assessments) has been studying templates and methods to optimize the conduct of data protection impact assessments in the EU and they have developed a template based on Articles 35 and 36 GDPR as well as an outline of impact assessment best practices, which aims to assist with decision making with an optimised use of resources, through an 11-step clearcut process [82].

1.6.6 Appointment of a Data Protection Officer

Not all entities are obliged to appoint a Data Protection Officer (DPO). However, appointment of a DPO is compulsory for public authorities and bodies as well as private entities over a particular workforce size and involved in specific data processing activities. The appointment of a DPO can help demonstrate compliance, in line with the principle of accountability. A DPO must be independent, with proven expertise in data protection and must be provided with adequate resources. Their role is to assist the data controller to monitor compliance, to inform and advise about data protection obligations as well as to assist with the conduct of a data protection impact assessment. The DPO is also the point of contact for data subjects and the supervisory authorities.

1.6.7 Privacy and Data Protection by Design and by Default

The concept of Privacy by Design was first widely presented in the 1990s, incorporating privacy-enhancing technologies (PETs) straight into the design of information technologies and systems. Specifically, *"PET stands for a coherent system of ICT measures that protect privacy by eliminating or reducing personal data or by preventing unnecessary and/or undesired processing of personal data, all without losing the functionality of the information system"* [83].

In the data protection context, the GDPR for the first time addressed Data Protection by Design as a legal obligation for data controllers and processors, referring explicitly to data minimization and the possible use of pseudonymization (Article 25 GDPR). In relation to that, it also introduced the obligation of Data Protection by Default, encouraging engineers to include protection of personal data as a default property of systems and services.

Data protection by Design and by Default are regarded as a many-sided notion, comprising of multiple technological and organizational elements, which integrate privacy and data protection principles in systems, devices, and services. Manufacturers and developers should anonymize the personal data, wherever the latter are not necessary for the purposes of the processing, implement data minimization by conducting reviews on a *"need to know basis,"* apply cryptography (e.g., encryption and hashing), partner with data-protection friendly service providers and storage platforms and make arrangements to enable data subjects to exercise their rights [84]. Futhermore, seeking to certify services and products with Data Protection by Design and by Default seals can create a competitive advantage for manufacturers and data controllers, and enhance data subjects' trust [85, p. 26].

If certification is not available, there may be other guarantees to ensure that a specific product or service complies with the requirements [85, p. 26]. The conduct of a data protection impact assessment, as presented in Articled 35 GDPR, can assist with defining the specific circumstances of processing and enable the decision about which means of processing as well as tools and methods to select and deploy during the conceptualization, design, validation and implementation phase, based on the particular context.

1.6.8 STANDARDIZATION AND CERTIFICATION

Standards are used for *"for developing technical specifications in a specific context of a product type and, provide a framework for security evaluation of product"* [86, p. 8]. Standards aim on the one hand to achieve interoperability and on the other confidence and trust. Concerning the latter, the particularities of the IoT ecosystem with its multifaceted devices, constitute a challenging case for standardization, provided the inherent connectivity and interdependencies, which *"call for holistic solutions"* [86, p. 8].

ENISA has recognized the complexities of the IoT landscape in a series of studies and has published lists of requirements for IoT devices corresponding to existing standards, without however prioritizing a specific solution for the entire IoT sector [87, pp. 12–22]. Specifically, for privacy by design, ENISA suggests the standard ISO 29550 to show that privacy constitutes integral part of the system; the ISO/IEC 27005 for defining and carrying out the data protection/privacy impact assessment; the ETSI TS 103 305, ISO/IEC 27002 and ISO 55000 for establishing and maintaining asset management procedures and configuration controls; the ISO/IEC 15408-1 and -2 for developing a risk analysis defining the scope of security evaluation for identifying the intended use and environment of an IoT device [86, pp. 13–14].

Moreover, clarifying whether web standards can be used for the IoT sector is another recommendation often seen in literature, in combination with an effective monitoring from the national supervisory authorities which should follow closely the development in the sector [55].

Standards or other widely accepted technical specifications form the basis for certification [86, p. 5]. Certification is not defined in the GDPR; albeit, the International Standards Organisation (ISO) defines certification as *"the provision by an independent body of written assurance (a certificate) that the product, service or system in question meets specific requirements"* [88, p. 5]. In line with Articles 42 and 43 GDPR, certification refers to third party attestation concerning personal data processing operations, to show conformity, whereas seals and marks refer to logos or icons proving the successful completion of a certification procedure [88, p. 5].

The EU Cybersecurity Act has laid out the framework for the establishment of European Cybersecurity Certificates for products, services, and processes, explicitly aiming to improve the security of IoT devices, by introducing the principle of security by design and allowing for independent verification of security features. Several certification mechanisms have been developed and certification schemes have been mapped in S. Ziegler (ed.), Internet of Things Security and Data Protection.

1.6.9 INVESTING IN SECURITY BY DESIGN

Several Operators of Essential Services (OES), as defined in the NIS Directive [89], providing products and services in the sectors of energy, transport, water, banking, financial market infrastructures, healthcare and digital infrastructure as well as Digital Service Providers (DSP), including search engines, cloud computing services, and online marketplaces often rely heavily on IoT both for their inside operations and for the provision of services and products to their customers [90, p. 7]. Both

OESs and DSPs will have to adopt appropriate measures to enhance security and to notify serious security incidents to the respective national authorities [90, p. 7].

Security challenges are actually inherent to the networking technologies. Nevertheless, IoT implementations present new risks, mainly because the protection of IoT entails the protection of several diverse systems, including the device, the cloud, the maintenance and the diagnostic tools [90, p. 7]. In the beginning of 2018, an increase in the number and duration of DDoS attacks was observed, which was attributed to IoT botnets; ENISA foresees that *"larger and more destructive attacks"* will occur the next years given that more empowered mobile devices will be added in the IoT ecosystem [91, p. 49].

The range of attacks against IoT infrastructure is great. The Agency has provided a threat taxonomy, recognizing seven categories of threats: nefarious activity/abuse, eavesdropping/interception/hijacking, outages, damage/loss, failures/malfunctions, disaster, physical attacks and includes malware, DDoS, device destruction, Man in the Middle, failure of system, software vulnerabilities, even natural and environmental disasters [92, p. 35]. ENISA highlights the lack of protection mechanisms in low-end IoT devices and services, which creates a pressing call for IoT protection architectures and best practice [91, p. 7]. On top of that, the Agency has identified further gaps which hinder IoT security, including lack of awareness and knowledge, fragmentation in existing security approaches, insecure design and development, lack of interoperability and of incentives, and last but not least, improper product lifecycle management.

Thus, in November 2019, building upon its 2017 work, ENISA published a study on good practices for IoT security inherent to the Software Development Life Cycle (SDLC), underlying the need for *"security by design"* and adopting software development guidelines for secure IoT products and services [90, p. 5]. SDLC consists of different phases which aim to effective and efficient systems based on their design and functional requirements [90, p. 11].

As a manufacturer or a retailer, investing in building or purchasing secure products -except for a legal requirement for launching a product in the EU market and one of the data protection legal obligations for the data controllers - leads to increased consumer trust which could have long term positive effects, providing a competitive advantage [93]. Cybersecurity solutions by design shall also take into consideration data protection principles, including data minimization and data protection by design and by default [94].

The general baseline security measures for IoT include measures that fall in three categories: a) policies, b) organizational, people and process measures and c) technical measures [92, p. 47]. On the other hand, the SDLC recommendations focus on people, processes, and technologies [90, p. 51]. No matter the distinction, the emphasis is given on trainings, the establishment of a security culture, the proper separation of roles and responsibilities, the efficient management and secure design and deployment, as well as the conduct of security reviews, the release of secure code (in-house and third-party), the use of secure communication and access control.

Another recommendation is the data isolation, achieved through the functional separation of datasets and databases. Raw data should be deleted from the device if data have been extracted and no longer needed for the provision of a function or for security reasons [9].

1.6.10 NEW LEGISLATIVE INITIATIVES AND REVIEW OF EXISTING LEGISLATION

In relation to the recently published EU Strategy for data which calls for the development and implementation of IoT solutions, the European Data Protection Supervisor acknowledges that increased deployment and dependence on IoT devices and services will lead to an increase in the number and seriousness of data protection risks, since they enlarge the "attack surface" for cyberattacks and amplify the adverse impacts on individuals [81, p. 6]. This problem should be addressed via new legislative initiatives which may be related to the EU consumer law but also through the review of existing legislation, with particular emphasis on the NIS Directive (the Directive on security of network and information systems) [81, p. 6]. Equally important is the review of the e-Privacy Directive and further adoption of the proposed e-Privacy Regulation aiming to complete the EU legal framework in relation to data protection and privacy [81, p. 6].

Last but not least, in literature, in particular in relation to IoT in conjunction with other innovative technologies, it has been pointed out that new approaches in relation to "group privacy" are necessary, in parallel with individual privacy, especially when profiles are built in relation to so-called reference groups [40, p. 16].

1.7 CONCLUSIONS

In this chapter, the authors discussed the data protection and privacy issues in relation to the IoT in a language that can be understood by a broader audience with no particular data law expertise but with an interest to understand the core notions. With a specific focus on three IoT categories, the authors presented the relevant legal framework, putting the focus on EU law. Exploring the General Data Protection Regulation and the ePrivacy Directive and underlying their in-between relation as a starting point of reference, the authors outlined a common glossary relevant for the IoT framework, including a clarification of the data protection principles and the data subjects' rights. The chapter also encompassed a special reference to data transfers, children, and other vulnerable data subjects as well as special categories of data and inferences.

Further, the chapter analyzed the data protection and privacy challenges raised by the selected IoT technologies, including salient use cases, for instance, the reference to smart glasses projects, connected vehicles, and mobility-related applications as well as use of mobile applications during the COVID-19 pandemic.

Given the objective of this publication, the authors close this chapter with a collection of recommendations and best practices which include: a human-centric approach, incentives, and audits, transparency by design through in-detail documentation of operations and data protection impact assessments, data protection and privacy by design and by default, the appointment of a Data Protection Officer, the implementation of standardization and certification as well as codes of ethics, the appropriate technical measures and investment in security measures.

The authors conclude that the IoT ecosystem with its perplex relations among stakeholders, diverse landscape, emerging security risks and increasingly large market share, will continue to create implications in relation to privacy and data protection. However, having our view to new technological developments, all the stakeholders involved in the IoT sector and its supporting technologies and services shall make sure to follow the highest data protection and privacy standards. It is important to always keep in mind the statement included in the explanatory report of the Convention 108+ that *"[h]uman dignity requires that safeguards be put in place when processing personal data, in order for individuals not to be treated as mere objects"* [95, p. 16]. This is especially true because *"big data enables new, nonobvious, unexpectedly powerful uses of data"* [96, p. 54].

NOTES

1 The latter is also regulated with Regulation (EU) 2018/1807 of the European Parliament and of the Council of 14 November 2018 on a framework for the free flow of non-personal data in the European Union.
2 The Convention 108+ with the OECD Privacy Guidelines of 1980 constitute the 1st generation of privacy standards.
3 Article 4 (14) GDPR: *"Biometric data means personal data resulting from specific technical processing relating to the physical, physiological or behavioural characteristics of a natural person, which allow or confirm the unique identification of that natural person, such as facial images or dactyloscopic data."*
4 Article 4 (15) GDPR: *"Data concerning health means personal data related to the physical or mental health of a natural person, including* the provision of health care services, which reveal information about his or her health status."
5 Data exchanges in the law enforcement sector are governed by the Law Enforcement Directive (article 36 of Directive (EU) 2016/680) and are not covered by the existing adequacy decisions.
6 UN Convention on the Protection of the Child, Article 1: *"[…] a child means every human being below the age of eighteen years, unless under the law applicable to the child, majority is attained earlier."*
7 According to Article 4 (25) GDPR an information society service means a service as defined in point (b) of Article 1(1) of Directive 2015/1535: *"(b) 'service' means any Information Society service, that is to say, any service normally provided for remuneration, at a distance, by electronic means and at the individual request of a recipient of services. For the purposes of this definition: (i) 'at a distance' means that the service is provided without the parties being simultaneously present; (ii) 'by electronic means' means that the service is sent initially and received at its destination by means of electronic equipment for the processing (including digital compression) and storage of data, and entirely transmitted, conveyed and received by wire, by radio, by optical means or by other electromagnetic means; (iii) 'at the individual request of a recipient of services' means that the service is provided through the transmission of data on individual request."*
8 At UN level, the Human Rights Committee in its 'Statement on derogations from the Covenant in connection with the COVID-19 pandemic' from 24 April 2020 acknowledges that "[S]tates parties confronting the threat of widespread contagion may resort, on a temporary basis, to exceptional emergency powers and invoke their right of derogation from the Covenant under article 4, provided this is required to protect the life of the nation. Still, the Committee wishes to remind States parties of the requirements and conditions laid down in article 4 of the Covenant and explained in the Committee's General Comments, most notably in General Comment 29 on States of Emergency (2001), which provides guidance on the following aspects of derogations: (1) official proclamation of a state of emergency; (2) formal notification to the Secretary General of the UN; (3) strict necessity and proportionality of any derogating measure taken; (4) conformity of measures taken

with other international obligations; (5) non-discrimination; and (6) the prohibition on derogating from certain non-derogable rights."

REFERENCES

1. C. Maple, "Security and privacy in the internet of things," *Journal of Cyber Policy*, vol. 2, no. 2, pp. 155–184, May 2017, doi: 10.1080/23738871.2017.1366536.
2. K. Ashton, "That 'Internet of Things' Thing," *RFiD Journal*, vol. 22, no. 7, pp. 97–114, 2009.
3. European Commission, "The European data strategy - shaping Europe's digital future," Brussels, Feb. 2020.
4. European Data Protection Supervisor, "Opinion 6/2017 opinion on the proposal for a regulation on privacy and electronic communications (ePrivacy regulation)," Brussels, Apr. 2017. [Online]. Available: https://edps.europa.eu/sites/edp/files/publication/17-04-24_eprivacy_en.pdf. [Accessed: Apr. 03, 2020].
5. L. A. Remotti et al., *Study on mapping Internet of Things innovation clusters in Europe*. 2019.
6. Alliance of Internet of Things Innovation and 5GIA Infrastructure Association, "Joint 5GIA-AIOTI vision on future networks, services and applications –high societal and economic impact potentials for a collaborative approach in the Horizon Europe Programme," Mar. 2019. [Online]. Available: https://aioti.eu/wp-content/uploads/2019/03/Vision-on-5GIA-AIOTI-partnership-v1.0.pdf. [Accessed: Apr. 01, 2020].
7. Article 29 Data Protection Working Party, "Working document on data protection issues related to RFID technology," Brussels, 10107/05/EN WP 105, Jan. 2006.
8. L. Atzori, A. Iera, and G. Morabito, "The Internet of Things: A survey," *Computer Networks*, vol. 54, no. 15, pp. 2787–2805, Oct. 2010, doi: 10.1016/j.comnet.2010.05.010.
9. Article 29 Data Protection Working Party, "Opinion 8/2014 on the on Recent Developments on the Internet of Things," Brussels, 14/EN WP 223, Sep. 2014.
10. Multistakeholder Expert Group to support the application of Regulation (EU) 2016/679, "Contribution," Brussels, Jun. 2020. [Online]. Available: https://ec.europa.eu/transparency/regexpert/index.cfm?do=groupDetail.groupMeetingDoc&docid=41708. [Accessed: Jun. 22, 2020].
11. Regulation (EU) 2016/679 of the European Parliament and of the Council of 27 April 2016 on the protection of natural persons with regard to the processing of personal data and on the free movement of such data, and repealing Directive 95/46/EC (General Data Protection Regulation), vol. 119. 2016.
12. European Commission, "Staff Working Paper - Impact Assessment Accompanying the document Regulation of the European Parliament and of the Council on the protection of individuals with regard to the processing of personal data and on the free movement of such data (General Data Protection Regulation) and Directive of the European Parliament and of the Council on the protection of individuals with regard to the processing of personal data by competent authorities for the purposes of prevention, investigation, detection or prosecution of criminal offences or the execution of criminal penalties, and the free movement of such data," Brussels, {SEC(2012) 73 final}, Jan. 2012. [Online]. Available: https://eur-lex.europa.eu/legal-content/EN/TXT/PDF/?uri=CELEX:52012SC0072&from=EN. [Accessed: Apr. 01, 2020].
13. G. Buttarelli, "*Keynote speech: Privacy in an age of hyperconnectivity*," presented at the Privacy and Security Conference 2016, Rust am Neusiedler See, Nov. 07, 2016. [Online]. Available: https://edps.europa.eu/sites/edp/files/publication/16-11-07_speech_gb_austria_en.pdf. [Accessed: Apr. 02, 2020].

14. Alliance of Internet of Things Innovation, "Advancing EU IoT research and innovation: AIOTI's position on Horizon Europe and Digital Europe," AIOTI-20180815/01, Aug. 2018. [Online]. Available: https://aioti.eu/wp-content/uploads/2018/09/AIOTI_Position_Paper_HEU_2018_for_publishing.pdf. [Accessed: Apr. 01, 2020].
15. M. Cunningham, "Next generation privacy: The Internet of Things, data exhaust, and reforming regulation by risk of harm," *Groningen Journal of International Law*, vol. 2, no. 2, pp. 115–144, 2014.
16. International Telecommunications Network (ITU), "ITU Internet reports 2005: The Internet of Things - executive summary," *Geneva*, 2005. [Online]. Available: http://www.itu.int/pub/S-POL-IR.IT-2005. [Accessed: Apr. 08, 2020].
17. European Commission, "Report on workshop on security & privacy in IoT," *Brussels*, Jan. 2017. [Online]. Available: https://ec.europa.eu/information_society/newsroom/image/document/2017-15/final_report_20170113_v0_1_clean_778231E0-BC8E-B21F-18089F746A650D4D_44113.pdf. [Accessed: Mar. 19, 2020].
18. A. Rayes and S. Salam, *Internet of Things From Hype to Reality: The Road to Digitization*. Cham: Springer International Publishing, 2019.
19. *AIOTI SPACE*. Available: https://aioti.eu/. [Accessed: Apr. 01, 2020].
20. S. Wachter, "Normative challenges of identification in the Internet of Things: Privacy, profiling, discrimination, and the GDPR," Social Science Research Network, Rochester, NY, SSRN Scholarly Paper ID 3083554, Dec. 2017. doi: 10.2139/ssrn.3083554.
21. European Commission, "Report on the safety and liability implications of Artificial Intelligence, the Internet of Things and robotics," Brussels, COM(2020) 64 final, Feb. 2020. [Online]. Available: https://ec.europa.eu/info/sites/info/files/report-safety-liability-artificial-intelligence-feb2020_en_1.pdf [Accessed: Jun. 22, 2020].
22. "Consumer Internet of Things (CIoT) - what is it and how does it evolve? ," *i-SCOOP*. Available: https://www.i-scoop.eu/internet-of-things-guide/what-is-consumer-internet-of-things-ciot/ [Accessed: Apr. 03, 2020].
23. European Union Agency for Fundamental Rights, *Handbook on European data protection law - 2018 edition*.
24. G. González Fuster, *The Emergence of Personal Data Protection as a Fundamental Right of the EU*. Cham: Springer International Publishing, 2014.
25. S. Ziegler, Ed., *Internet of Things Security and Data Protection*. Cham: Springer International Publishing, 2019.
26. "Report of the Office of the United Nations High Commissioner for Human Rights 'The right to privacy in the digital age'," A/HRC/27/37, Jun. 2014.
27. European Court of Human Rights, "Guide on Article 8 of the European convention on human rights - Right to respect for private and family life, home and correspondence," Council of Europe, Strasbourg, Aug. 2019. [Online]. Available: https://www.echr.coe.int/Documents/Guide_Art_8_ENG.pdf. [Accessed: Apr. 04, 2020].
28. European Data Protection Supervisor, "Guidelines on assessing the proportionality of measures that limit the fundamental rights to privacy and to the protection of personal data," Brussels, Dec. 2019. [Online]. Available: https://edps.europa.eu/sites/edp/files/publication/19-12-19_edps_proportionality_guidelines2_en.pdf [Accessed: Apr. 08, 2020].
29. European Commission, "Data protection rules as a trust-enabler in the EU and beyond – taking stock," Jul. 24, 2019.
30. M. Horten, "The digital single market and Internet of Things," *Open Rights Group*, Nov. 2018. Available: https://www.openrightsgroup.org/policy/virt-eu/iot-and-brexit-briefings/the-digital-single-market-and-internet-of-things. [Accessed: Apr. 03, 2020].

31. Directive 95/46/EC of the European Parliament and of the Council of 24 October 1995 on the protection of individuals with regard to the processing of personal data and on the free movement of such data, vol. OJ L. 1995.
32. European Data Protection Board, "Opinion 5/2019 on the interplay between the ePrivacy Directive and the GDPR, in particular regarding the competence, tasks and powers of data protection authorities," Brussels, Mar. 2019.
33. J. Smith, "Article 29 Working Party," *European Data Protection Board - European Data Protection Board*, Jan. 10, 2018. Available: https://edpb.europa.eu/our-work-tools/article-29-working-party_en. [Accessed: Oct. 01, 2020].
34. "About EDPB," *European Data Protection Board - European Data Protection Board*, Jan. 10, 2018. Available: https://edpb.europa.eu/about-edpb/about-edpb_en. [Accessed: Oct. 01, 2020].
35. "What are Data Protection Authorities (DPAs)?," *European Commission*. Available: https://ec.europa.eu/info/law/law-topic/data-protection/reform/what-are-data-protection-authorities-dpas_en. [Accessed: Oct. 01, 2020].
36. "About the European Data Protection Supervisor," *European Data Protection Supervisor*, Nov. 11, 2016. Available: https://edps.europa.eu/about-edps_en. [Accessed: Oct. 01, 2020].
37. Article 29 Data Protection Working Party, "Opinion 4/2007 on the concept of personal data," Jun. 2007. [Online]. Available: https://www.clinicalstudydatarequest.com/Documents/Privacy-European-guidance.pdf [Accessed: Aug. 11, 2019].
38. R. Mahieu, J. van Hoboken, and H. Asghari, "Responsibility for data protection in a networked world – on the question of the controller, "Effective and complete protection" and its application to data access rights in Europe," *Social Science Research Network*, Rochester, NY, SSRN Scholarly Paper ID 3256743, 2019. [Online]. Available: https://papers.ssrn.com/abstract=3256743. [Accessed: Aug. 11, 2019].
39. "Sheet n°15: Take into account the legal basis in the technical implementation | CNIL." Available: https://www.cnil.fr/en/sheet-ndeg15-take-account-legal-basis-technical-implementation?fbclid=IwAR15k-JcA4U_Q-57lYECsqGM4OnqB3VMrZiuGM-lmu-VepEDLtli8Z5CU5M8. [Accessed Jun. 15, 2020].
40. S. Wachter, "The GDPR and the Internet of Things: A three-step transparency model," *Law, Innovation and Technology*, vol. 10, no. 2, pp. 266–294, Jul. 2018, doi: 10.1080/17579961.2018.1527479.
41. M. Horten, "Open Rights Group - Data Protection (GDPR) and Internet of Things," Nov. 2018. [Online]. Available: https://www.openrightsgroup.org/policy/virt-eu/iot-and-brexit-briefings/data-protection-gdpr-and-internet-of-things. [Accessed: Apr. 03, 2020].
42. "Adequacy decisions," *European Commission*. Available: https://ec.europa.eu/info/law/law-topic/data-protection/international-dimension-data-protection/adequacy-decisions_en. [Accessed Oct. 01, 2020].
43. "Information rights at the end of the transition period - Frequently Asked Questions," Sep. 11, 2020. Available: https://ico.org.uk/for-organisations/data-protection-at-the-end-of-the-transition-period/information-rights-at-the-end-of-the-transition-period-frequently-asked-questions/. [Accessed: Oct. 01, 2020].
44. I. Milkaite and E. Lievens, "Towards a better protection of children's personal data collected by connected toys and devices," *Digital Freedom Fund*, p. 9.
45. I. Milkaite and E. Lievens, "The Internet of Toys: Playing games with children's data?" in G. Mascheroni and D. Holloway (Eds.) *The Internet of Toys*, Cham: Springer International Publishing, pp. 285–305, 2019.

46. Article 29 Data Protection Working Party, "Opinion 2/2009 on the protection of children's personal data (General Guidelines and the special case of schools)," *Brussels, 398/09/EN WP 160*, Feb. 2009. [Online]. Available: https://ec.europa.eu/justice/article-29/documentation/opinion-recommendation/files/2009/wp160_en.pdf. [Accessed: Apr. 02, 2020].
47. G. Malgieri and J. Niklas, "Vulnerable data subjects," *Computer Law and Security Review*, Special Issue on Data Protection and Research, p. 22, Apr. 2020. [Online]. Available: https://papers.ssrn.com/abstract=3569808 [Accessed: Apr. 16, 2020].
48. European Data Protection Board, "Guidelines 05/2020 on consent under Regulation 2016/679," Brussels, May 2020. [Online]. Available: https://edpb.europa.eu/sites/edpb/files/files/file1/edpb_guidelines_202005_consent_en.pdf. [Accessed: May 06, 2020].
49. M. Macenaite and E. Kosta, "Consent for processing children's personal data in the EU: following in US footsteps?" *Information & Communications Technology Law*, vol. 26, no. 2, pp. 146–197, May 2017, doi: 10.1080/13600834.2017.1321096.
50. L. Jasmontaite and P. De Hert, "The EU, children under 13 years, and parental consent: A human rights analysis of a new, age-based bright-line for the protection of children on the Internet," *International Data Privacy Law*, vol. 5, no. 1, pp. 20–33, Feb. 2015, doi: 10.1093/idpl/ipu029.
51. I. Milkaite and E. Lievens, "Child-friendly transparency of data processing in the EU: From legal requirements to platform policies," *Journal of Children and Media*, vol. 14, no. 1, pp. 5–21, Jan. 2020, doi: 10.1080/17482798.2019.1701055.
52. Article 29 Data Protection Working Party, "Guidelines on transparency under Regulation 2016/679," Apr. 11, 2018.
53. eHealth Network, "Mobile applications to support contact tracing in the EU's fight against COVID-19 Common EU Toolbox for Member States Version 1.0," European Commission, Brussels, Apr. 2020. [Online]. Available: https://ec.europa.eu/health/sites/health/files/ehealth/docs/covid-19_apps_en.pdf [Accessed: Apr. 20, 2020].
54. S. Wachter and B. Mittelstadt, "A right to reasonable inferences: Re-thinking data protection law in the age of big data and AI," *Social Science Research Network*, Rochester, NY, SSRN Scholarly Paper ID 3248829, Oct. 2018. [Online]. Available: https://papers.ssrn.com/abstract=3248829. [Accessed: Apr. 03, 2020].
55. "Mauritius Declaration on the Internet of Things," Balaclava, Oct. 2014, p. 2.
56. FTC Staff report, "Internet of Things: Privacy & security in a connected world," Jan. 2015. [Online]. Available: https://www.ftc.gov/system/files/documents/reports/federal-trade-commission-staff-report-november-2013-workshop-entitled-internet-things-privacy/150127iotrpt.pdf. [Accessed: Apr. 29, 2020].
57. A. Josephina and A. Andreas, "Case study the Internet of Things and ethics," *The ORBIT Journal*, vol. 2, no. 2, pp. 1–29, 2019, doi: 10.29297/orbit.v2i2.111.
58. "What does EU regulatory guidance on the Internet of Things mean in practice? Part 1," *Fieldfisher*. Available: https://www.fieldfisher.com/en/services/privacy-security-and-information/privacy-security-and-information-law-blog/what-does-eu-regulatory-guidance-on-the-internet-of-things-mean-in-practice-part-1. [accessed: Apr. 03, 2020].
59. European Commission, "Guidance on the Regulation on a framework for the free flow of non-personal data in the European Union," Brussels, May 2019. [Online]. Available: https://eur-lex.europa.eu/legal-content/EN/TXT/?uri=COM:2019:250:FIN. [Accessed: Apr. 24, 2020].
60. R. K. Chellappa and R. G. Sin, "Personalization versus privacy: An empirical examination of the online consumer's dilemma," *Information Technology and Management*, vol. 6, no. 2, pp. 181–202, Apr. 2005, doi: 10.1007/s10799-005-5879-y.

61. European Data Protection Supervisor, "Technology report no 1: smart glasses and data protection," Brussels, Jan. 2019. [Online]. Available: https://edps.europa.eu/sites/edp/files/publication/19-01-18_edps-tech-report-1-smart_glasses_en.pdf. [Accessed: Apr. 02, 2020].
62. A. Hope, "Chinese authorities employ AI-powered smart glasses for detecting coronavirus infections," *CPO Magazine*, Apr. 02, 2020.
63. European Commission, "On the road to automated mobility: An EU strategy for mobility of the future," Brussels, May 2018. [Online]. Available: https://eur-lex.europa.eu/LexUriServ/LexUriServ.do?uri=COM:2018:0283:FIN:EN:PDF. [Accessed: Apr. 24, 2020]
64. European Data Protection Board, "Guidelines 1/2020 on processing personal data in the context of connected vehicles and mobility related applications," Brussels, Version adopted for public consultation, Jan. 2020. [Online]. Available: https://edpb.europa.eu/sites/edpb/files/consultation/edpb_guidelines_202001_connectedvehicles.pdf. [Accessed: Apr. 02, 2020].
65. Horizon 2020 Commission Expert Group to advise on specific ethical issues raised by driverless mobility (E03659), "Ethics of connected and automated vehicles: recommendations on road safety, privacy, fairness, explainability and responsibility," Publication Office of the European Union:Luxembourg, 2020. [Online]. Available: https://ec.europa.eu/info/sites/info/files/research_and_innovation/ethics_of_connected_and_automated_vehicles_report.pdf. [Accessed: Oct. 01, 2020].
66. European Data Protection Board, "Statement on the processing of personal data in the context of the COVID-19 outbreak," Brussels, Mar. 2020.
67. B. Buntz, "Cybersecurity crisis management during the coronavirus pandemic," *IoT World Today*, Mar. 24, 2020. Available: https://www.iotworldtoday.com/2020/03/24/cybersecurity-crisis-management-during-the-coronavirus-pandemic/. [accessed Apr. 16, 2020].
68. European Data Protection Board, "EDPB letter concerning the European Commission's draft Guidance on apps supporting the fight against the COVID-19 pandemic," Apr. 14, 2020.
69. M. Scott and Z. Wanat, "Poland's coronavirus app offers playbook for other governments – POLITICO," *Politico*, Apr. 02, 2020.
70. S. Stolton, "'Major security and privacy issues' in using location data for COVID-19 apps, Commission says," www.euractiv.com, Apr. 16, 2020. Available: https://www.euractiv.com/section/digital/news/major-security-and-privacy-issues-in-using-location-data-for-covid-19-apps-commission-says/. [Accessed: Apr. 20, 2020].
71. "Apple and Google partner on COVID-19 contact tracing technology," *Apple Newsroom*, Apr. 10, 2020. Available: https://www.apple.com/newsroom/2020/04/apple-and-google-partner-on-covid-19-contact-tracing-technology/. [Accessed: Apr. 20, 2020].
72. European Commission, "Guidance on Apps supporting the fight against COVID 19 pandemic in relation to data protection 2020/C 124 I/01," Brussels, C/2020/2523, Apr. 2020. [Online]. Available: https://eur-lex.europa.eu/legal-content/EN/TXT/?qid=1587141168991&uri=CELEX:52020XC0417(08). [Accessed: Apr. 20, 2020].
73. Europol, "Catching the virus cybercrime, disinformation and the COVID-19 pandemic," Apr. 2020. [Online]. Available: https://www.europol.europa.eu/publications-documents/catching-virus-cybercrime-disinformation-and-covid-19-pandemic. [Accessed: Apr. 21, 2020].
74. I. Torre, F. Koceva, O. R. Sanchez, and G. Adorni, "*A framework for personal data protection in the IoT*," in *2016 11th International Conference for Internet Technology and Secured Transactions (ICITST)*, Barcelona, Spain, Dec. 2016, pp. 384–391, doi: 10.1109/ICITST.2016.7856735.

75. "Picking up good vibrations – data protection and the Internet of Vibrating Things," *CITIP blog*, May 16, 2017. Available: https://www.law.kuleuven.be/citip/blog/picking-up-good-vibrations-data-protection-and-the-internet-of-vibrating-things/. [Accessed: Apr. 03, 2020].
76. M. Tabassum, T. Kosiński, and H. R. Lipford, "'*I don't own the data*': *end user perceptions of smart home device data practices and risks*," in *Proceedings of the Fifteenth USENIX Conference on Usable Privacy and Security*, Santa Clara, CA, USA, Aug. 2019, pp. 435–450. [Accessed: Jun. 23, 2020]. [Online].
77. N. Apthorpe, Y. Shvartzshnaider, A. Mathur, D. Reisman, and N. Feamster, "Discovering smart home Internet of Things privacy norms using contextual integrity," *The Proceedings of the ACM on Interactive, Mobile Wearable Ubiquitous Technol.*, vol. 2, no. 2, pp. 1–23, Jul. 2018, doi: 10.1145/3214262.
78. "Design My Privacy – 8 principles for privacy design – Tijmen Schep." http://www.tijmenschep.com/design-my-privacy/. [Accessed: Sep. 29, 2020].
79. "Candle – the privacy friendly smart home – Tijmen Schep." Available: https://www.tijmenschep.com/create-candle/. [Accessed: Sep. 29, 2020].
80. Z. Efroni, J. Metzger, L. Mischau, and M. Schirmbeck, "Privacy icons," *European Data Protection Law Review*, vol. 5, no. 3, pp. 352–366, 2019, doi: 10.21552/edpl/2019/3/9.
81. European Data Protection Supervisor, "Opinion 3/2020 on the European strategy for data," Brussels, Jun. 2020. [Online]. Available: https://edps.europa.eu/sites/edp/files/publication/20-06-16_opinion_data_strategy_en.pdf. [Accessed: Jun. 22, 2020].
82. K. Dariusz et al., "Data protection impact assessment in the European Union: Developing a template for a report from the assessment process," vol. 2020, no. 1, p. 52, Sep. 2020. [Online]. Available: https://cris.vub.be/en/publications/data-protection-impact-assessment-in-the-european-union-developing-a-template-for-a-report-from-the-assessment-process(2300a8d5-7e5d-4e63-86cb-51288e2eaca4).html. [Accessed: Oct. 01, 2020].
83. "On Promoting Data Protection by Privacy Enhancing Technologies (PETs)," Brussels, COM (2007) 228 final, May 2007. [Online]. Available: http://www.icsd.aegean.gr/website_files/metaptyxiako/997662742.pdf. [Accessed: Sep. 29, 2020].
84. European Commission, *Ethics and Data Protection*. 2018.
85. European Data Protection Board, "Guidelines 4/2019 on Article 25 data protection by design and by default," Brussels, Nov. 2019. [Online]. Available: https://edpb.europa.eu/sites/edpb/files/consultation/edpb_guidelines_201904_dataprotection_by_design_and_by_default.pdf. [Accessed: Apr. 17, 2020].
86. EU Agency for Cybersecurity (ENISA), "IoT security standards gap analysis," Report/Study, Jan. 2019. [Online]. Available: https://www.enisa.europa.eu/publications/iot-security-standards-gap-analysis. [Accessed: May 05, 2020].
87. EU Agency for Cybersecurity (ENISA), "ENISA threat landscape report 2018," Report/Study, Jan. 2019. [Online]. Available: https://www.enisa.europa.eu/publications/enisa-threat-landscape-report-2018. [Accessed: Apr. 22, 2020].
88. European Data Protection Board, "Guidelines 1/2018 on certification and identifying certification criteria in accordance with Articles 42 and 43 of the Regulation 2016/679," Brussels, May 2018. [Online]. Available: https://edpb.europa.eu/sites/edpb/files/consultation/edpb_guidelines_1_2018_certification_en.pdf. [Accessed: May 07, 2020].
89. Directive (EU) 2016/1148 of the European Parliament and of the Council of 6 July 2016 concerning measures for a high common level of security of network and information systems across the Union, vol. OJ L. 2016.
90. EU Agency for Cybersecurity (ENISA), "Good practices for security of IoT - secure software development lifecycle," Athens, Report/Study, Nov. 2019. [Online]. Available: https://www.enisa.europa.eu/publications/good-practices-for-security-of-iot-1. [Accessed: May 04, 2020].

91. EU Agency for Cybersecurity (ENISA), "Threat landscape report 2018," Brussels, Report/Study, Jan. 2019. [Online]. Available: https://www.enisa.europa.eu/publications/enisa-threat-landscape-report-2018. [Accessed: Apr. 08, 2020].
92. EU Agency for Cybersecurity (ENISA), "Baseline security recommendations for IoT," Report/Study, Nov. 2017. [Online]. Available: https://www.enisa.europa.eu/publications/baseline-security-recommendations-for-iot. [Accessed: May 04, 2020].
93. P. Brown, "Making or selling Internet of Things (IoT) devices? Six reasons you need to be thinking about data protection," Mar. 07, 2018. Available: https://ico.org.uk/about-the-ico/news-and-events/blog-making-or-selling-internet-of-things-iot-devices-six-reasons-you-need-to-be-thinking-about-data-protection/. [Accessed: Apr. 03, 2020].
94. O. Gkotsopoulou et al., "*Data Protection by Design for Cybersecurity systems in a Smart Home environment*," *2019 IEEE Conference on Network Softwarization (IEEE NetSoft)*, 24–28 June 2019, Paris, France, p. 10, 2019.
95. Council of Europe, *Convention 108 +Convention for the protection of individuals with regard to the processing of personal data - Explanatory Report*.
96. Executive Office of the President, "Big data: Seizing opportunities, preserving values," May 2014. [Online]. Available: https://www.eff.org/document/2014-presidents-big-data-and-privacy-working-group-report. [Accessed: Apr. 29, 2020].

2 IoT Reference Architectures

V. Kelli, E. G. Sfakianakis, B. Ghita, and P. Sarigiannidis

CONTENTS

2.1	Introduction. What is IoT?	48
2.2	IoT Architecture	50
	2.2.1 Three Layer (Tier) IoT Architecture	50
	2.2.2 Four-Layer Architecture	51
	2.2.3 Five Layer Architecture	52
2.3	Application Layer	53
	2.3.1 MQTT	53
	2.3.2 MQTT-SN	55
	2.3.3 CoAP	56
	2.3.4 DDS	58
	2.3.5 XMPP	59
	2.3.6 AMQP	60
2.4	Novel Application Areas	61
	2.4.1 Smart Cities	62
	2.4.2 Smart Home	62
	2.4.3 Manufacturing	63
	2.4.4 Transport and Automotive	64
	2.4.5 Energy	65
	2.4.6 Healthcare	65
	2.4.7 Assisted Living	66
	2.4.8 Supply and Logistics	66
	2.4.9 Agriculture	66
	2.4.10 Environmental Monitoring	67
	2.4.11 Public Safety	67
2.5	Service Support and Application Support Layer	68
2.6	Networking Layer	68
	2.6.1 Short-Range Wireless	68
	2.6.1.1 Bluetooth Mesh Networking and Bluetooth Low Energy	68
	2.6.1.2 ZigBee	69

        2.6.1.3   Z-Wave 70
        2.6.1.4   Wi-Fi 70
        2.6.1.5   Light Fidelity (Li-Fi) 71
    2.6.2   Cellular Technologies 72
        2.6.2.1   Extended Coverage-GSM-IoT (EC-GSM-IoT) 72
        2.6.2.2   Narrow-Band IoT (NB-IoT) 72
        2.6.2.3   LTE-Machine Type Communication (LTE-MTC) and
                  Enhanced Machine Type Communication (eMTC) 73
    2.6.3   Long Range Wireless Technologies 73
        2.6.3.1   LoRa (LoRa PHY and LoRaWAN) 73
        2.6.3.2   Sigfox 74
        2.6.3.3   Weightless 74
        2.6.3.4   Wi-Max 75
        2.6.3.5   Satellite 75
        2.6.3.6   Infrared Communication 76
    2.6.4   Thread Protocol 78
    2.6.5   Insteon 79
2.7   Perception Layer 80
    2.7.1   Pressure Sensors 81
    2.7.2   Temperature Sensors 82
    2.7.3   Touch Sensors 82
    2.7.4   IR/Light Sensor 83
    2.7.5   Humidity Sensor 83
    2.7.6   Accelerometer 84
    2.7.7   Proximity Sensor 84
    2.7.8   Actuators 86
    2.7.9   Level Sensors 87
    2.7.10 Gas Sensors 87
    2.7.11 Water Quality Sensor 88
2.8   Ubiquitous Computing 88
2.9   IoT Security Issues 90
2.10 Conclusions 93
References 94

2.1   INTRODUCTION. WHAT IS IOT?

Internet of Things (IoT) means "things" or "objects" that are connected to the Internet and each other. This could be almost anything such as computers, tablets or smartphones, cameras, medical instruments, vehicles, buildings, animals, all connected to the Internet, communicating, and sharing information. These interconnected objects provide data that are collected and analyzed. IoT technologies are expanding constantly in different sectors like smart homes and cities, agriculture, health, assisted living, public safety, environmental monitoring, transport, and industry. By the end of 2020 deployment of IoT technologies, it is estimated to connect over 50 billion devices to the network impacting nearly every aspect of our daily lives. The IoT is a

IoT Reference Architectures

concept that has its roots in various network, sensing, and information processing approaches. To provide value-added capabilities, advanced technology must be used in sensing/actuating, connectivity, edge computing, machine learning, networked systems, security, and privacy, as seen in Figure 2.1. Significant efforts have been made to present a framework that supports the convergence of IoT architecture approaches.

In this chapter, the three most predominant IoT architectural models will be presented, and the corresponding application and network protocols will be further addressed. Hardware utilized for IoT technology, such as sensors, will be further analyzed, in order to cover a wide range of devices. Finally, an introduction to Ubiquitous computing, followed by security issues IoT systems may encounter is closing this chapter.

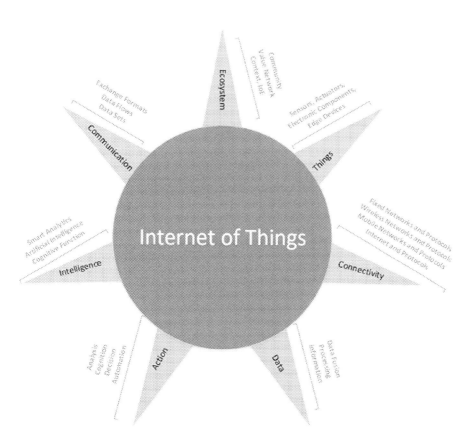

FIGURE 2.1 Internet of Things. Redrawn from [1]

2.2 IOT ARCHITECTURE

The domain of the IoT will enclose a wide range of technologies. Hence, single-reference architecture cannot be used as a road map for all possible IoT implementations and there is no single unified IoT architecture that is agreed by standardization organizations, industrial or academic. There are multiple models under development such as the ITU-T model, the NIST model for Smart Grid, the M2M model from ETSI or the Architectural Reference Model from the EU IoT-A project and related work in other international standardization developing organizations (SDOs) as the IETF, W3C, etc [2]. While a reference model can probably be identified, it is likely that several reference architectures will coexist in the IoT and should be taken into consideration for future development. Figure 2.2 displays the three IoT architectural models, which will be analyzed further below.

2.2.1 THREE LAYER (TIER) IoT ARCHITECTURE

The three-level architecture is a simplified reference model for the IoT Architecture, which has been designed and implemented in a number of systems. In this stack, you can see three different layers: Perception layer, Network layer, and Application layer, as displayed in Figure 2.3 [2].

1. Perception or Sensor layer—Sensors, actuators, and edge devices that interact with and collect data from the environment. Sensors are communicating with the network and with an application that is going to manage that object as a networking object

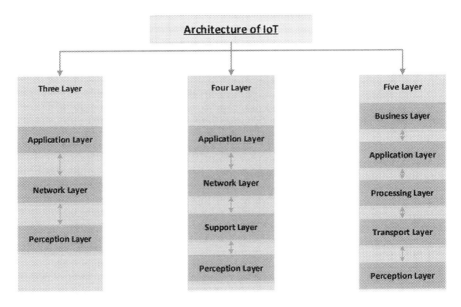

FIGURE 2.2 The 3 IoT architectures

IoT Reference Architectures

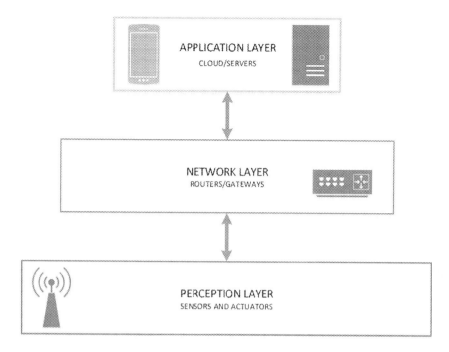

FIGURE 2.3 IoT architecture in three layers. Redrawn from [2]

2. Network Layer—Discovers, connects, and translates devices over a network and in coordination with the application layer. This layer is also called to as "transport layer." Its main function is the transmission and processing of information received from the perception layer. The transmission medium can be wired or wireless using a 3G, Wi-Fi, Bluetooth, Infrared or ZigBee protocol, etc.
3. Application Layer—Manages the sensor or device and Performs data processing and storage with specialized services and functionality for users.

2.2.2 Four-Layer Architecture

The three-layer architecture is a simplified architecture. Because of the continuous development and deployment of IoT, it could not satisfy all the requirements. With so many applications of IoT, multiple organizations have defined models of how IoT network could be organized. One of the main models presented by the International Telecommunication Union (ITU) which introduces a four-layer architecture for IoT in 2012 with the ITU-T Y 2060 recommendation. The ITU-T model provides universally common understanding of the essential functions and capabilities of the IoT architecture. It helps to reduce the implementation complexity and promotes interoperability among various IoT applications and communication technologies. The ITU-T-layered reference model consists of four horizontal layers and the common management and security capabilities related with all layers. Figure 2.4 presents the ITU-T-layered architecture.

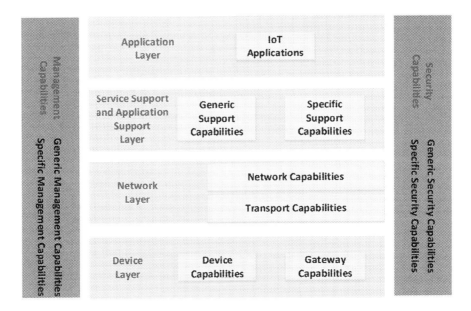

FIGURE 2.4 ITU-T reference model. Redrawn from [60]

The top layer is Application layer, the second layer is the Service support and Application support layer, the third layer is the network layer, and the bottom layer is the Device or Perception layer. The support capabilities are divided into generic and application-specific capabilities. Management and security capabilities are also categorized into generic and specific capabilities and cross over all layers. The generic management capabilities include device management functions such as remote activation, status monitoring and control, software update, network topology management, and traffic and congestion control. The generic security capabilities include access control, authorization and authentication, privacy protection, and integrity protection

2.2.3 FIVE LAYER ARCHITECTURE

The five-layer architecture, as seen in Figure 2.5, is composed of three of the four-layer architecture's components, adding the business layer as a top level, and an additional layer named processing layer. All the layers presented in this architecture are, from top to bottom, namely, business layer, application layer, processing layer, transport layer, and perception layer. Specifically, the business layer is placed at the top level, in contrast with the three- and four-layer architectural models. This layer's main responsibility is to manage the IoT system as a whole. Applications, business, and profit models are controlled through this layer. Additionally, it is responsible for providing privacy to the user. The transport layer interacts with the processing layer, also known as middleware layer. The processing layer's main purpose is to remove any irrelevant information, by analyzing the data received from the transport layer [3], therefore minimizing the massive amounts of data to be passed up to the rest of

IoT Reference Architectures

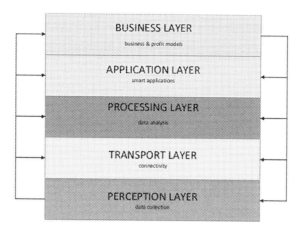

FIGURE 2.5 5-layer architecture

the layers. The transport layer's goal is to transfer data received by the perception layer to the processing layer and it operates similarly to the network layer introduced in the previous architectural schemes.

2.3 APPLICATION LAYER

The Application layer is considered the top tier layer of both the three-layer and four-layer IoT architecture. This layer handles the communication between IoT devices that use the same application layer protocol. Essentially, this layer's responsibility lies in creating the data to be communicated, according to the user's needs. IoT does not benefit from regular internet application layer protocols, due to the resource-constrained nature of IoT devices; therefore, new protocols have emerged, specifically catering to IoT needs. In this section, the main IoT application layer protocols will be presented and analyzed.

2.3.1 MQTT

Message Queuing Telemetry Transport (MQTT) is a lightweight, topic-based publish-subscribe application layer messaging protocol used for machine-to-machine (M2M) communication. Every data exchange involves two network entities, a broker, and one or more client(s).

The broker acts as an intermediate, by routing published messages to the subscribed recipients. A client may act as a publisher, a subscriber, or both. Client interaction involves communication strictly with the broker, and never with each other. Specifically, a publisher uploads, or publishes, sensor data to the broker addressed for a specific topic. The broker, in turn, handles the distribution of the aforementioned data to the clients subscribed to this topic. A simple example can be seen in Figure 2.6. There is no limit to the number of topics a client can be subscribed to, or publish their data. Additionally, publishers are not aware whether there are any

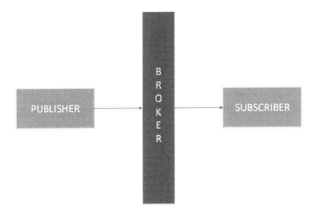

FIGURE 2.6 Simple MQTT interaction

subscribing nodes for the topic they publish; however, messages are published, nonetheless.

Topics are represented in string format and may be composed by many levels, separated by the slash character. For instance, the topic "home/living_room/humidity" has three levels, pointing to data regarding the humidity levels in the living room. A humidity sensor in the living room will publish their readings to this topic, and the broker will handle the distribution of this data only to the clients who are subscribed to this topic. A client may be subscribed to all published data to a topic level, for instance, the topic "home/living_room/#" will allow the subscriber to receive updates from all sensors located in the living room. An example can be observed in Figure 2.7.

MQTT is designed for use in TCP/IP networks. MQTT uses the connection-oriented Transmission Control Protocol (TCP) as a transport layer protocol. Port 1883 is the standard MQTT port, while port 8883 can be used for secure MQTT message exchange, with Transport Layer Security/Secure Socket Layer (TLS/SSL) encryption.

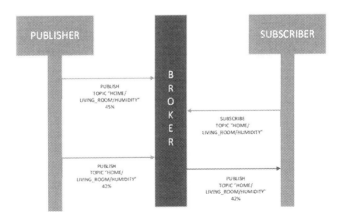

FIGURE 2.7 MQTT Publish and Subscribe Exchange

Finally, MQTT features three Quality of Service (QoS) levels, enabling the communicating parties to prioritize traffic and ensure the delivery of the message. The higher the QoS, the higher the priority of the message.

- **QoS 0**: This is the lowest QoS available. It ensures delivery at most once, meaning that there is no acknowledgment by the recipient of the message that the message has been indeed received.
- **QoS 1**: This QoS level ensures delivery at least once, requiring acknowledgment from the receiver that the message has been received. If no acknowledgment gets provided, the sender retransmits the message.
- **QoS 2**: The highest QoS level that MQTT supports, enables for message delivery exactly one time. This involves two acknowledgments from the receiver of the message. This is the most reliable QoS level as message deliver is ensured; however, high QoS can result in delays with processing other network traffic.

2.3.2 MQTT-SN

Carrying many similarities with MQTT although not as used, Message Queuing Telemetry Transport for Sensor Networks (MQTT-SN) handles messaging for wireless environments with heavily constrained entities, such as battery-operated embedded devices with limited processing capabilities [65]. MQTT-SN can therefore be considered as an MQTT version for environments requiring wireless communication.

To accommodate to these networks, MQTT-SN attempts to reduce payload size. To achieve that, MQTT string topic names are replaced with a shorter topic name or a 2-byte topic identifier field for message length reduction. Initially, clients register the topic names they are interested in with the gateway, in order to obtain the topic ID, and later on refer to that topic with its corresponding ID. Shorter, 2-byte topic names and topic IDs are known in advance by both the device and gateway, so that the registering process may be avoided. Additionally, battery-operated devices benefit from the offline keep-alive procedure introduced with this MQTT variation, which allows such devices to go to a sleeping state, and messages destined to them will be delivered when they wake up.

MQTT-SN clients do not communicate directly with an MQTT broker entity, but interact with the gateway instead. The gateway acts as an intermediate between the clients and the broker. However, the gateway may be integrated with a broker. In any case, the communication between clients and gateways is achieved with MQTT-SN, and in the case that the gateway is not integrated with a broker, MQTT is utilized for communication between the gateway and the broker. The gateway's main function is to translate MQTT-SN to MQTT for the broker, and act as a broker for the clients. A new entity is introduced, called MQTT-SN forwarder, whose main purpose is to forward traffic between the clients and the gateway, without changing the data. The example of MQTT-SN communications can be observed in Figure 2.8.

In contrast with the standard MQTT protocol, MQTT-SN does not utilize TCP as a transport layer protocol as TCP is considered to consume more resources for devices with limited capabilities. For this reason, MQTT-SN was originally designed

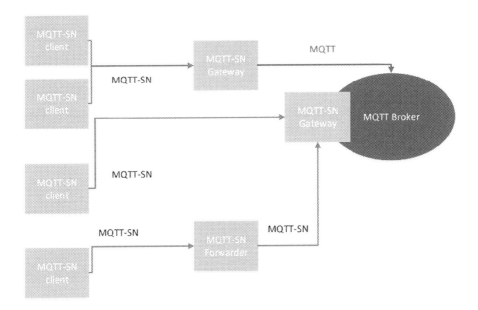

FIGURE 2.8 MQTT-SN

to run on top of ZigBee; however, it can be used with the connectionless User Datagram Protocol (UDP) or Bluetooth for transportation purposes.

2.3.3 CoAP

Constrained Application Protocol (CoAP) is a lightweight web transfer protocol used by constrained devices for communication purposes. CoAP is easy to proxy both to and from HyperText Transfer Protocol (HTTP) and therefore can be integrated with the existing web so that CoAP clients can access HTTP content while accommodating to the constrained environment's needs with a low, 4-byte message overhead.

Similar to HTTP, CoAP supports requests for resources and responses between clients and servers, respectively, to enable M2M communication. A simple graph of client/server communications can be observed in Figure 2.9. CoAP is based upon the Representational State Transfer (REST) model, meaning resources are available under a certain URI, instead of a topic or topic ID.

CoAP header carries the request method or the response code for client requests and server responses, respectively. CoAP messages are related to a specific message ID, mentioned in the header as well for acknowledgment purposes. Request messages are divided to confirmable and nonconfirmable, with confirmable messages being the most reliable option, as an acknowledgment message with the same message ID is required by the recipient. If no acknowledgment is sent after a default timeout threshold, the message gets retransmitted. On the other hand, nonconfirmable messages do not require any sort of acknowledgment that the message has been received. Finally, reset messages are sent in the case that a confirmable message is indeed received; however, the node cannot process the message. An example of a

IoT Reference Architectures 57

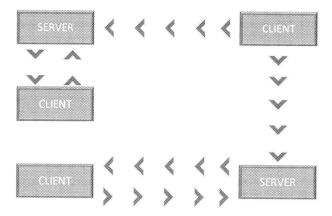

FIGURE 2.9 CoAP Client/Server Communications

confirmable message and nonconfirmable transaction between a client and a server is

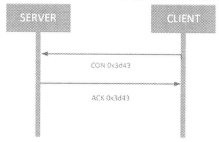

FIGURE 2.10 CoAP reliable message transaction

pictured in detail in Figures 2.10 and 2.11, respectively.

CoAP uses UDP for transportation purposes. The CoAP protocol stack is displayed in Figure 2.12.

Four security modes are embedded in CoAP, employing Datagram Transport Layer Security (DTLS) [4].

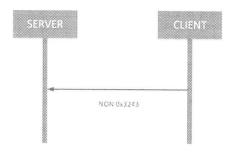

FIGURE 2.11 CoAP unreliable message transaction

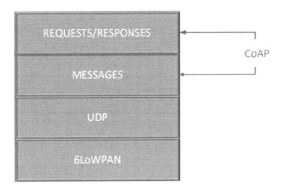

FIGURE 2.12 CoAP protocol Stack

- **NoSec**: DTLS is not activated.
- **PreSharedKey**: DTLS is activated. There are a number of pre-shared encryption keys and the nodes they are matched with.
- **RawPublicKey**: DTLS is activated. Devices possess a pair of asymmetric encryption keys. Each device has an identity calculated from the public key, and a list of other devices' identities that it can communicate with.
- **Certificate**: DTLS is activated. An asymmetric encryption key pair is used, in conjunction with an X.509 certificate. The certificate is signed by a common trust root.

2.3.4 DDS

Data Distribution Service (DDS) is a messaging middleware that was designed by Object Management Group (OMG). DDS follows the Data-Centric Publish-Subscribe model (DCPS). Similar to MQTT, DDS performs a publishing procedure, posting data under a specific topic, and a subscription procedure, in which exist the subscribers who are interested in specific published data. Data are categorized under a topic and a type. A topic has similar function to MQTT topics, while types are used for informing the middleware about the security or the handling method required for the specific data. DDS aims to deliver data in real time, without the need for a broker to handle the routing process.

In DDS according to DCPS, publishers publish their data, and a data writer is used to send this data under a specific topic. Data writers are associated with a single topic; therefore, if a publisher wants to publish to many topics, that can be accomplished by owning many data writers; however, each data writer can only be owned by one publisher. Similar to this, subscribers, the actual recipient of the data, own data readers, who are associated to a single topic. Just as the data writers, a data reader can be owned by a single subscriber, but a subscriber can own many data readers if they are subscribed to multiple topics. Subscribers receive and process published data and proceed with sending the data over to the data reader to store. An example is displayed in Figure 2.13.

IoT Reference Architectures

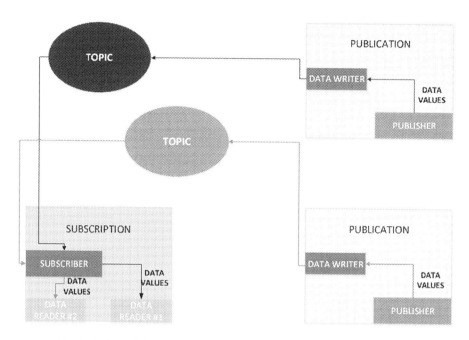

FIGURE 2.13 DDS Publication and Subscription Example

Essentially, in DDS there is no need for a central broker to route the published messages to the suitable recipients, as DDS handles distributing the message from the publisher to the subscriber.

2.3.5 XMPP

Extensible Messaging and Presence Protocol (XMPP) is a communication protocol based on Extensible Markup Language (XML). This protocol supports near real-time data exchange between two network entities. XMPP utilizes a distributed client-server architecture for communications. This specifies that a client will have to connect to a server in order to exchange any information. As far as the distributed aspect of XMPP, clients can exchange data between each other, regardless of the server they are associated with.

Distributed communication takes place in a client-to-server-to-server-to-client manner, where servers handle routing the messages from the sender clients to the recipient clients. Therefore, in the case of remote clients, i.e., clients associated with different servers, servers manage to route data from the sending client to the receiving client's server, who in turn delivers the data to the aforementioned recipient. A simple figure of XMPP communications is displayed in Figure 2.14.

Data exchanged with XMPP, known as XML "stanzas" [5], can be divided into three categories, presented below.

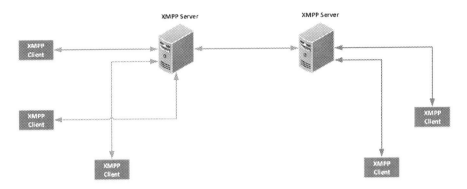

FIGURE 2.14 XMPP communications

- **Message**: The message stanza refers to pushing information from one entity to another, in which the recipient's address must be mentioned. When a server receives a message stanza, it is attempted to route the message to the addressed receiving client.
- **Presence**: The presence stanza refers to a publish/subscribe broadcast mechanism, in which the client/publisher sends data to the server without necessarily specifying the address of the recipient(s). The server broadcasts the messages to the entities that are subscribed to the specific publishing client, in contrast with MQTT, in which the data published are organized in topic areas. However, if a recipient address is specified by the publishing client, the server handles routing the message to the addressed client.
- **Info/Query (IQ)**: Similar to HTTP, IQ is a request-response mechanism, according to which an entity can seek information from another entity with a request. The requested entity's reaction will be a response.

Regarding the means of transportation, XMPP utilizes TCP. Consequently, clients and servers must establish a TCP connection to communicate with each other and exchange data.

2.3.6 AMQP

Advanced Message Queuing Protocol (AMQP) is a communication protocol used for message transportation between clients and messaging middleware servers, also known as brokers.

In particular, the broker receives data from a data-producing node, who also provided all the routing information for the message; the data are received by a broker message routing agent called the "exchange." There are two exchange types, namely direct exchange that routes based on a routing key, and topic exchange, which routes based on a routing pattern. The routing key is an address that the exchange may apply to route the message. For publish/subscribe routing, the routing key is the topic. The routing pattern follows the same principles as the routing key. Therefore, the exchange

IoT Reference Architectures 61

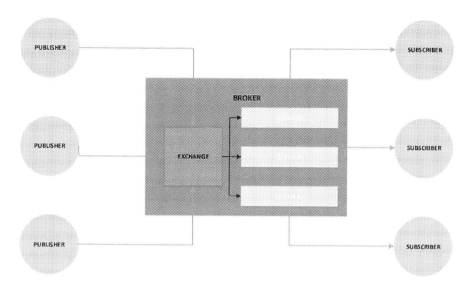

FIGURE 2.15 AMQP example

decides how to route the message, and appends it to one or multiple First-In-First-Out (FIFO) buffers, or queues This can be observed in Figure 2.15.

Specifically, the queue's function is to store the received messages from the data producers and when in turn, as fast as possible, forward them to the receiving nodes, usually a consumer application. Essentially, AMQP implements message deliveries for subscriptions as queues. Queues can be created by consumer applications, or reading clients, and may possess the following attributes:

- **Name**: The name of the queue to be created.
- **Exclusive**: If the client enables this option, the queue is exclusive to the particular connection, consequently, if the connection gets teared-down the queue gets deleted.
- **Durable**: The queue remains active.

AMQP assumes a reliable network transport layer; therefore, TCP may be used as a transport protocol, to ensure connection reliability. For security enhancement purposes, it is proposed that an encryption scheme is implemented, such as TLS.

2.4 NOVEL APPLICATION AREAS

The strength of IoT comes from its ability to be integrated in virtually all organizational, business, and social activities, as it enables or enhances data collection, physical world interaction, and network connectivity. Given these three benefits, there are a number of areas that are fully utilizing IoT and hence are likely to see the most benefits. This section will provide a brief outline of these areas, highlighting the impact made by IoT.

2.4.1 SMART CITIES

The concept of smart cities acts as an umbrella for data collection and connectivity integration and reporting at all levels in a metropolitan environment. Given its coverage, applications that are part of smart cities also intersect with some of the other areas listed later in the chapter, such as transport and healthcare, and assisted living. Three main areas are the major beneficiaries of IoT in this context: utilities, transportation, and social services.

To begin with the utilities, IoT enables full activity reporting throughout an urban area and can redistribute and optimize the level of allocated resources. The first steps in this area were taken more than a decade ago with the introduction of smart meters and smart lighting, albeit with dedicated, limited, or no network connectivity to take full advantage of them. More recently, the range was expanded with waste collection, which provides the data to adapt the service to the needs of the population. These types of interaction allow loose connectivity and limited dynamic adjustment of the provision. The next step was achieved with the utility infrastructure monitoring, which provides better integration with utility companies to ensure demand is met also to the benefit of the recipients rather than the providers.

More interesting and challenging is the monitoring and reporting of traffic and public transport; this connects the utilities and the transportation applications. At a basic level, as vehicles fully report speed and position, the users are informed of current availability and transport duration, from information about next bus to estimates about the overall duration of their journey. The combined data of road congestion and positioning and utilization of public transport also allows the municipality to redirect or redistribute traffic to fully use the infrastructure, alleviate congestion, and ultimately respond to demand. Similar to the flow of data traffic in a computer network environment, the data collection and interfacing with the system is only one of the challenges, the most significant one being the creation of intelligent, adaptable algorithms to provide the optimal solution as the environment constantly changes.

Beyond public transport, live information about traffic also allows enhancing of ancillary services, including smart parking and video surveillance, replacing human monitoring, and enforcing with automated measures.

Some of the above applications link to the third aspect of smart cities, which are the resident services, providing a seamless, automated service to residents. The aspect of surveillance is often a contentious one in the context of smart cities; discussing its drawbacks on privacy has been acknowledged repeatedly in the legal and ethical domains and is beyond the scope of this book. What is worth pointing out from a technological perspective is the enhancement of public safety monitoring, from traffic violations to crime. Moving across to public services, medical and emergency assistance have also seen a boost with the advent of IoT, particularly in the areas of elderly long-term care and ambulance response.

2.4.2 SMART HOME

We are entering the era of automation, composed by context-aware devices which possess sensing capabilities and communicate with each other wirelessly to provide life-easing services to humans. Smart home devices aim to gather data and exchange information between each other in order to replace common daily tasks with automated procedures, without the need for humans to control them. The "smart" aspect of such devices, refers to their ability to be activated when certain conditions are met.

Devices that compose the smart home environment could be smart security systems, smart thermostats, smart fridges, smart lighting systems, windows, etc. Smart lighting systems can be used in order to detect the presence of people in the room and therefore be activated or deactivated accordingly. Smart thermostats are currently controlled remotely by the user; in the near future, they could be activated based on the smart home's occupant's schedule, in order to provide heat at a minimum level if there is no one at home, set the heat according to environmental variables or sense the user's activity and therefore adjust heating to an optimal temperature. Thus, such devices, apart from contributing to living an effortless life, can also be applied for monitoring energy consumption and for energy-saving purposes for the smart household.

Due to the automated and intelligent nature of smart home environments, they can be implemented in order to aid people who require constant monitoring and assistance. Therefore, such infrastructures can be utilized for assisted living purposes, explained in detail a few sections below.

The incorporation of intelligent devices in smart homes should be performed in a seamless manner. It is essential that the user does not feel constantly monitored, rather elegantly assisted. Such systems should be produced so that they still feel like homes, as it is essential for the occupants to feel like they are safe and in control of their lives, and not left at the mercy of technology.

2.4.3 MANUFACTURING

The smart cities, as highlighted in the previous section, do have a major impact on welfare and therefore are the most apparent for the general public. However, IoT does have an increasing impact on the manufacturing sector, with recent studies such as the one from IoT analytics [6] indicating that industrial applications are currently occupying the top spot in terms of growth, interest, and current impact. The benefits brought in are obvious and natural, as data collection and environment interaction are automated and the human factor is replaced with IoT equivalents.

The manufacturing applications extend across several domains throughout the manufacturing process, starting with data collection using smart sensors, automated processing using actuators, and reporting results and communicating with other systems through network connectivity. Building a parallel with the smart cities' applications, IoT-enabled manufacturing converts physical or human resources into IoT that perform the respective tasks. A number of concepts have been introduced to support this paradigm, in addition to IoT. The conversion is reflected through cyber-physical systems, that exchange data with other entities in order to perform their tasks. The service components are typically part of cloud infrastructures, to allow full relocating of resources and services and optimize expenditure. The decision-making process is driven by big data analytics, which replace experts and expert systems to provide an intelligent adaptive solution using the historical data and predicted development. The interaction is underpinned by the information and communication technology, which ubiquitously connects the smart elements. The paradigm itself is summarized by Industry 4.0, an industrial revolution that is driven by information and intelligent processing, with data and IoT at its core [7].

2.4.4 Transport and Automotive

The benefits of IoT in the context of transport and automotive industry have already been introduced in the previous two sections, in the context of traffic information and manufacturing revolution.

The impact of IoT for transport extends beyond the urban realm, as traffic, proximity, and environmental sensors can be placed throughout the road network. Significant progress has been achieved in terms of the road information on two fronts—sensors placed throughout the infrastructure, reporting on traffic volumes, and, at application layer, users providing crowd measurements through their mobile devices, which can also be aggregated for a holistic view of the network. A number of options have been subsequently identified, such as automatic traffic diversions, either signposted on the motorway or transmitted to the mobile applications to communicate to users, or variable availability of the network (including opening and closing the emergency lanes for extra capacity and additional toll charging).

One area that prevails in terms of IoT benefits is the automotive industry and products. Given its peculiarities, it is reported here separately from manufacturing, because its main benefits come from the data generated. Data collection and usage coupled with controlling features on smart cars led to the concept of Softwarization linked to cars and car services. The value of data has been long acknowledged in the automotive industry. Going back to its beginnings, the insurance industry introduced the driving data collection black boxes, aiming to monitor (or even police or enforce if necessary) the behavior of drivers. In parallel, the engine control module has also been expanded over the years to collect and store engine parameters that would support the car performance monitoring and diagnostics. On a third direction, manufacturers also introduced a level of in-car intelligence through the in-car entertainment [and management] systems. While all these systems did have loose connectivity, it was really the IoT hardware, software, and communication development that allowed the provision of a full solution, collecting across all aspects of the car and driving, and reporting them either for internal consumption or externally. The benefits are two-fold—the users have increased information and control about their own car and services and, equally important, the manufacturers can fully study the evolution of each car. Indeed, telemetry crowd measurements provide a very reach feature data set that, among other things, can be used by manufacturers to increase their after-sales service and understanding of performance. The continuous availability of a communication line between the car and the manufacturer also allows for the control of services provided, from adjusting engine parameters and efficiency to extra comfort services, such as entertainment or heating, with each aspect available through ancillary payments.

One step further, already reached but developing slower, is *car-as-a-service*, whereby the usage of a car is a service rather than the result of ownership. Given

IoT Reference Architectures 65

sufficient control of access and resources, potential beneficiaries would not need to buy a car, they could simply access and use a car, similar to a cab sourcing/hailing app.

Car-based services, both data collection and provisioning to other partners have raised a number of questions related to the principle of owning a car, which should allow full privacy and control of existing features, rather than be followed by a number of software additions.

2.4.5 Energy

As it is the case with the previous examples, some aspects of energy applications can be found as part of smart cities. The most significant range of applications however stems from providing full monitoring of the generation, distribution, and consumption of energy via IoT sensors, meters, and applications, named smart grid. The smart grid provides a number of benefits to all the stakeholders: producers, distributors, and consumers. To begin with, in an ideal scenario, energy producers would be required to generate a constant amount of energy throughout the daily cycle; constant energy demand is convenient to fulfill as there are a number of cheap, conventional fossil fuel-based sources. In reality, demand fluctuates and the producer must use more expensive, on-demand sources to respond; the higher and unpredictable the variations are, the more expensive is the required energy. This also affects the energy distribution network, which must maintain parameters (frequency in particular) above certain thresholds to ensure appliances can still be powered. Finally, the consumer is typically provided with a flat rate, calculated as a result of the combined energy prices from the various power generation methods employed (and the traditional process of collecting consumer usage, through meter reading, is a rather expensive one). All these limitations can be removed by the introduction of the smart grid. A network of sensors throughout the distribution grid allows optimization based on demand and supply; the smart meters provide full granularity measurements for energy consumption and allow propagation of up to date information about usage to providers and suppliers. Finally, collating all the current and historical information, in combination with big data statistics allows suppliers to make more accurate energy predictions and reduce the cost, wastage, and outage of energy.

2.4.6 Healthcare

Healthcare is intrinsically expensive due to the required environment, devices used for data collection, ensuring the data flow between patient and the various provisioning entities, and the required decisional expertise. IoT aims to reduce either directly or indirectly several of these factors through the introduction of intelligent patient monitoring. At a minimum, having connected sensors that collect patient vitals and pass them to monitoring systems and doctors eliminates additional expenditure and delays due to recording and handling of data; if implemented remotely, this further eliminates the need for dedicated facilities and support and is beneficial for all stakeholders.

The healthcare domain is by far the most sensitive in terms of the sensitivity of data handling. Although all areas carry a level of threat and are subject to confidentiality, healthcare requires additional measures put in place, from security the data collection and transport to particularly access and processing.

2.4.7 ASSISTED LIVING

As examined in the previous application areas and especially in the healthcare sector, IoT can improve the overall quality of human life, ensuring the constant monitoring of patient status. Assisted living, may be considered as a derivative of the healthcare application domain, as IoT applications will support and provide continuous assistance for people in need, such as the elderly or people with general disabilities, allowing them to live an independent and safe life.

In conjunction with constant vital sign monitoring, assisted living can provide instant emergency aid such as immediate ambulance response if an incident occurs; therefore, real-time event observation is of essence. Data gathered from the user's environment, or from the user's vitals, can be viewed and processed in real time, allowing caregivers to identify any abnormal or dangerous behavior and provide assistance. Assisted living benefits people with disabilities by providing visual or hearing aid, when required.

Security is of paramount importance in assisted living, as health data represent the current status of a person that can potentially require immediate assistance. Consequently, any service disruptions or privacy breaches impose direct hazard for human life. Therefore, data should be handled with maximum levels of security and privacy.

2.4.8 SUPPLY AND LOGISTICS

As supply chains and shipping are becoming increasingly transnational, with companies typically operating across multiple geographical areas, the infrastructure must also minimize delay and maintain constant granular information about stocks. The introduction of IoT benefits greatly this sector as it allows monitoring of merchandise down to individual items throughout their travel. Following the use of barcodes, which allowed partial automation, RFID tags in particular and active tags or marking of items led to the seamless flow of information between the entities providing, transporting, delivering, and receiving goods.

2.4.9 AGRICULTURE

Agriculture is one of the areas that could be perceived as most remote from IT infrastructures, with the entities and processes employed requiring typically human monitoring, analysis, and intervention. In recent years, IoT applications did however have a significant impact on agriculture, particularly with regards to monitoring and farming automation. Sensor readings such as soil quality, humidity, and environmental temperature can provide farmers with all the necessary knowledge and can consequently contribute to better crop quality.

2.4.10 ENVIRONMENTAL MONITORING

Environmental pollution, such as air, water, and soil pollution, can have severe impacts on human health. According to the World Health Organization (WHO) [8], air pollution-related fatalities are estimated at about 4.2 million people at an annual level. Industrial emissions, as well as fumes produced by gas-based means of transportation can contribute to the rise of air pollution levels [9]. For this purpose, IoT can be integrated, offering constant monitoring of environmental parameters. By having a clear view of current environment status, authorities can make informed decisions to mitigate potential dangers.

This application area includes wildfire, tsunami, and earthquake monitoring with sensors. Public safety mechanisms can be activated and early warnings can be issued if abnormalities are observed.

2.4.11 PUBLIC SAFETY

Major technological advancements such as the IoT can bring in great benefits by getting integrated in public safety application domains, such as military, defense, law enforcement, and rescue.

In military context, IoT has already been employed for various reasons. Automation has led to many changes in this domain, with examples being unmanned aircraft and autonomous combat vehicles. Apart from commercial applications, IoT has already been successfully integrated for military and defense purposes. Devices can communicate with each other to accomplish various goals in the warfare, such as reconnaissance. For instance, sensor and IoT technology can bring in real-time, real-world data and therefore benefit this sector by providing constant situational awareness with real-time surveillance and intelligence, with no need for human involvement and consequently minimizing the risk for loss of human life. Therefore, decisions can be taken in a more confident way.

As far as IoT integration for public safety contribution in smart cities, law enforcement can be informed about incidents in real time from sensors and cameras and therefore react faster when needed. Emergency response personnel, in the case of a natural disaster or a rescue mission, can be fully informed about the situation in the field, thus minimizing any risks for further loss of life. Wearable IoT technology can provide full context about the whereabouts and vital readings of rescue crew, providing in this way a sense of safety and security while on the field.

There is indeed an element of hype in IoT industry, fueled by businesses and organizations identifying opportunities to expand or strengthen using IoT, faced by the challenges of implementing the necessary steps to engender the change. Nevertheless, the number of IoT devices is growing at a significant and steady pace of 10–20% per year [10]. In 2017, Mark Huang, Gartner Research Vice President, was estimating that, by 2020, there will be 20 billion Internet-connected devices that will contribute to the IoT success [11]. Two years, later, Gartner expanded on the estimates and identified that the enterprise and automotive segment alone was going to total 5.8 billion endpoints by 2020. In parallel, IoT analytics was estimating in 2018 a slightly lower figure of 7 billion devices, with a prediction of almost 10 billion IoT devices connected by 2020 [12].

2.5 SERVICE SUPPORT AND APPLICATION SUPPORT LAYER

The Service and Application support layer includes generic support capabilities as well as application-specific support capabilities. As the name indicates, the generic-support capabilities are common capabilities applicable to many applications, whereas the application-specific capabilities serve a particular application's requirements.

This layer analyses, stores, and processes large amount of data. A general combination of data collection, data purification (removing unwanted, irrelevant data), and analysis takes place in this layer in order to deliver data to the user in understandable form. Might use databases, cloud computing, big data processing resources. Considering the distributed nature of IoT services and applications, there exists generic functionality such as data processing and storage as well as specialized functionality per application and service

A general combination of data collection, data purification (removing unwanted, irrelevant data), analysis, refining the data in order to deliver it to the user in understandable form, takes place in this layer. We can call this the brain of IoT.

2.6 NETWORKING LAYER

The third layer is the network layer, which includes the networking and transport capabilities. They include functions for access control, routing, mobility management, resource allocation, authentication, authorization and accounting (AAA), etc. The transport capabilities include functions for transporting IoT application data as well as control and management instructions. The networking layer of IoT will enable universal connectivity. We use various network protocols to connect the smart devices to network and we will present the most outstanding in the following sections.

2.6.1 SHORT-RANGE WIRELESS

Short-range wireless communication refers to information traveling between two or more endpoints in close proximity. The distance short-range wireless can support, must be no longer than a few centimeters or a few meters.

2.6.1.1 Bluetooth Mesh Networking and Bluetooth Low Energy

Bluetooth mesh networking is a protocol based upon Bluetooth Low Energy that allows for many-to-many communication over Bluetooth radio. Bluetooth Low Energy (Bluetooth LE, or BLE, formerly marketed as Bluetooth Smart) is based on mesh networking principles. The mesh network operation is designed to:

- enable messages to be sent from one element to one or more elements;
- allow messages to be relayed via other nodes to extend the range of communication;

- secure messages against known security attacks, including eavesdropping attacks, man-in-the-middle attacks, replay attacks, trash-can attacks, brute-force key attacks, and possible additional security attacks not documented here;
- work on existing devices in the market today;
- deliver messages in a timely manner;
- continue to work when one or more devices are moved or stop operating; and
- have built-in forward compatibility to support future versions of the Mesh Profile specification BLE technology operates in the same spectrum range (2.400–2.4835 GHz ISM band) as classic Bluetooth technology but uses a different set of channels. It can serve IoT devices in cells with a radius over 100 m and supports data rates in the range 1–2 Mbps.

2.6.1.2 ZigBee

The ZigBee or ZigBee/IEEE 802.15.4 protocol is a specification created for wireless networking. It includes hardware and software standard design for WSN (Wireless sensor network) requiring high reliability, low cost, low power, scalability, and low data rate. ZigBee-style self-organizing ad-hoc digital radio networks were conceived in the 1990s, but the IEEE 802.15.4-2003/ZigBee specification was ratified on December 14, 2004. Half year later the ZigBee Alliance [13] announced availability of Specification 1.0 (on June 13, 2005).

ZigBee PRO and ZigBee Remote Control (RF4CE), among other available ZigBee profiles, are also based on the IEEE802.15.4 protocol, operating at 2.4 GHz targeting applications that require relatively infrequent data exchanges at low data-rates over a restricted area and within a 100 m range such as in a home or building.

ZigBee, like Bluetooth, has a large installed base of operation, although perhaps traditionally more in industrial settings.

ZigBee/RF4CE due to its characteristics is well positioned to take advantage of wireless control and sensor networks in M2M and IoT applications.

The latest version of ZigBee is the recently launched 3.0, which is essentially the unification of the various ZigBee wireless standards into a single standard.

There are several advantages of using the ZigBee protocol when comparing with other protocols for WSN.

- One of the main advantages is that ZigBee is standardized at all layers, this ensures that products from different manufacturers are compatible with each other.
- Another advantage supported by Zigbee, is the ability of mesh networking. In mesh networking, there isn't a single route to a node, instead, due to the interconnection of nodes with each other, multiple routes to a single node exist. That makes every node of the network reachable from every other node, also, self-healing is provided, if the preferable path to a node fails there are other paths to reach the node. The more devices you have the more reliable the network is.
- Low consumption of energy and working in the network even without the necessity of a battery (Green Power). Energy-harvesting devices lack batteries,

getting it by extracting the energy they need from the environment (by tapping into motion, light, piezo/pressure, or the Peltier effect). This is especially effective for devices that are only sometimes on the network (when they have power) and lets these devices securely go on and off the network, so they can be off most of the time and not need any energy.
- And high scalability, ZigBee networks can to thousands of devices and they will communicate with each other using the best available path.

2.6.1.3 Z-Wave

Z-Wave is a low-power RF communications technology that is primarily designed for home automation for products such as lamp controllers and sensors among many others. Optimized for reliable and low-latency communication of small data packets with data rates up to 100 kbps, it operates in the sub-1 GHz band.

Z-Wave has a fairly simple addressing mechanism compared to Bluetooth and Zigbee protocols. The addressing scheme is kept simple because of the effort to minimize traffic and preserve power.

2.6.1.4 Wi-Fi

Wi-Fi, short name for wireless fidelity, is the name of a wireless networking technology that uses radio waves to provide wireless high-speed Internet and network connections. Is based on the IEEE 802.11 family of standards which are commonly used for local area networking of devices and Internet access [14]. A way of getting broadband internet to a device using wireless transmitters and radio signals. Once a wireless transmitter receives data from the internet, it converts the data into a radio signal that can be received and read by Wi-Fi-enabled devices. Information is then exchanged between the transmitter and the device

It remains the most widespread and generally known wireless communications protocol. *It ranges approximately 50 m and includes IoT cloud infrastructure access.* Wi-Fi frequency bands are frequency ranges within the wireless spectrum that are designated to carry Wi-Fi: 2.4 GHz and 5 GHz. Wi-Fi Certified products by the Wi-Fi Alliance organization are interoperable with each other, even if they are from different manufacturers since they fulfill *the Institute of Electrical and Electronics Engineers' (IEEE) 802.11 standards*

Its broad usage across the IoT world is mainly limited by higher-than-average power consumption resulting from the need of retaining high signal strength and fast data transfer for better connectivity and reliability.

As 5G is to cellular, Wi-Fi 6 (or 802.11ax) is the most recent standard for wireless network transmission and like 5G, it promises to be faster, broader, and smarter than previous generations [15, 16].

IPv6 Low-power wireless Personal Area Network (6LowPAN)

Low-Power Wide-Area Network (LPWAN) is an emerging network technology for IoT which offers long-range and wide-area communication at low-power. It thus overcomes the range limits and scalability challenges associated with traditional short-range wireless sensor networks. Due to their escalating demand, LPWANs are gaining momentum, with multiple competing technologies currently being

IoT Reference Architectures

developed. Despite their promise, existing LPWAN technologies raise a number of challenges in terms of spectrum limitation, coexistence, mobility, scalability, coverage, security, and application-specific requirements which make their adoption challenging.

2.6.1.5 Light Fidelity (Li-Fi)

Li-Fi stands for Light Fidelity and is a Visible Light Communications (VLC) system which runs wireless communications that travel at very high speeds. With Li-Fi, a light bulb can be used router. It can utilize a common household LED light bulb to enable data transfer, boasting speeds of up to 224 gigabits per second. The term Li-Fi was coined by University of Edinburgh Professor Harald Haas during a TED Talk in 2011.

ITU has proposed the standard G.9991, to join ITU's series of standards for home networks connecting in-premises devices and interfacing with the outside world. The standard details the system architecture, physical layer, and data link layer specification for "high-speed indoor VLC transceivers," the VLC access points within lightbulbs.

The IEEE 802.15.7 standard defines the physical layer (PHY) and media access control (MAC) layer.

In 2016, it was announced that an extension of standard Wi-Fi was to be launched called Wi-Fi HaLow.

Every LED lamp should be powered through an LED driver; this LED driver will get information from the Internet server and the data will be encoded in the driver. Based on this encoded data, the LED lamp will flicker at a very high speed that cannot be noticed by the human eyes. But the Photo Detector on the other end will be able to read all the flickering and this data will be decoded after Amplification and Processing. An example is displayed in Figure 2.16.

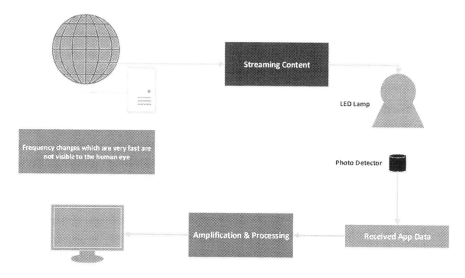

FIGURE 2.16 Li-Fi. Redrawn from [61]

Philips Lighting has tested LiFi in France. PureLifi also completed the world's first Skype call on a Lifi-enabled mobile phone back in 2018 [17]. The call was completed using a large case around the phone to conduct the signal. Patents have been filed by several companies for LiFi application in the automobile sector, meaning that the technology could be used to increase road safety or communicate road conditions to traffic lights. But these are just patents and we may be a long way away from practical use.

The Li-Fi market was projected to have a compound annual growth rate of 82% from 2013 to 2018 and to be worth over $6 billion per year by 2018 [18]. However, the market has not developed as such and Li-Fi remains with a niche market, mainly for technology evaluation.

2.6.2 Cellular Technologies

Cellular technologies refer to wireless communications that cover a significantly larger geographical area than short-range wireless. Information is distributed to and from a cell tower, correspondent to each cell.

2.6.2.1 Extended Coverage-GSM-IoT (EC-GSM-IoT)

Extended Coverage-GSM-IoT is 3GPP standard-based LPWAN technology. EC-GSM-IoT is based on enhanced GPRS (eGPRS), designed to support long-range, low-power, and high-capacity communication. EC-GSM-IoT is backward compatible with existing GSM technologies. Hence, it can be added to the existing cellular network as a software upgrade, reducing the cost of infrastructure and deployment.

EC-GSM-IoT extends the coverage of GPRS by 20dB. To support various application requirements, EC-GSM-IoT provides two modulation options, Gaussian Minimum Shift Keying (GMSK) and 8-ary Phase Shift Keying (8PSK). Using these two modulations, it achieves peak data rate of 10 kbps and 240 kbps, respectively. Additionally, EC-GSM-IoT improves battery lifetime by using extended Discontinued Reception (eDRX) technique, which allows the device to choose the number of inactivity periods depending on the application requirements. EC-GSM-IoT can support up to 50k devices using a single BS.

Hence, EC-GSM-IoT can provide coverage for M2M devices in locations with challenging radio coverage conditions.

2.6.2.2 Narrow-Band IoT (NB-IoT)

NB-IoT (Narrowband IoT) is a 3rd Generation Partnership Project (3GPP) LPWAN technology. It is a 3GPP Release 13 feature that reuses various principles and building blocks of the LTE physical layer and higher protocol layers. NB-IoT is offering flexibility of deployment by allowing the use of a small portion of the available spectrum. It supports up to 50k devices per cell and requires minimum 180 kHz of bandwidth to establish communication. It can be deployed as a standalone carrier with available spectrum exceeding 180 kHz, in-band within an LTE physical resource block, or in the guard-band inside an LTE carrier. NB-IoT uses resource mapping to preserve the orthogonality of LTE signals by avoiding mapping signals to resources currently used by LTE signals.

IoT Reference Architectures

2.6.2.3 LTE-Machine Type Communication (LTE-MTC) and Enhanced Machine Type Communication (eMTC)

LTE for Machine Type Communication (LTE-MTC), similarly to NB-IoT, is a 3GPP low-power wide area network, used for various cellular devices and services, particularly for IoT or M2M applications. LTE-MTC optimizes the production of next-generation IoT devices with low costs, longer battery life, and better coverage. This is achieved by supporting data rates of up to 10 Mbps for LTE-MTC devices, which will further decrease the production cost, ensuring however that LTE will effectively compete against long-range wireless technologies [19].

eMTC is a modified version of the LTE-MTC standard, supporting a bandwidth of 1.08 MHz within an existing LTE deployment, and 1.4 MHz in standalone deployment. The key difference between LTE-MTC EMTC is that the latter supports data rates of up to 1 Mbps, since it is limited to a 1.08 MHz channel width [20].

The summary of LPWAN technologies can be seen in Table 2.1.

2.6.3 LONG RANGE WIRELESS TECHNOLOGIES

Long-Range Wireless Communications enables information distribution through larger geographical areas, which can reach a few kilometers in size.

2.6.3.1 LoRa (LoRa PHY and LoRaWAN)

LoRa stands for Long Range and it is a digital wireless data communication technology which uses license-free sub-gigahertz radio frequency bands like 169 MHz, 433 MHz, 868 MHz (Europe), and 915 MHz (North America). LoRa enables very-long-range transmissions (above 10 km in rural areas) with low battery.

LoRaWAN defines the communication protocol and system architecture for the network, while the LoRa physical layer enables the long-range communication link. LoRaWAN is also responsible for managing the communication frequencies, data rate, and power for all devices. LoRa and LoRaWAN permit inexpensive, long-range connectivity for the IoT devices in rural, remote, and offshore industries. Typical uses of LoRa products can be found in the following industries: mining,

---

TABLE 2.1
Accelerometer Sensors Suitability for Different Applications (Adapted from [64])

Application	Piezoelectric	Capacitive MEMS	Piezoresistive
Static Acceleration (gravity)		x	x
G- Force (aircraft)		x	x
Seismic (earthquake)	x		
Low Frequency Vibration(human motion)		x	x
General Vibration (electric motor)	x	x	
High Frequency Vibration (gear noise analysis)	x		
General Shock (general testing)	x	x	x
High Impact Shock (drop testing)	x		x
Extreme Shock (vehicle crash testing)	x		x

natural resource management, renewable energy, transcontinental logistics, and supply chain management. Fleet Space Technologies [21] uses LoRaWAN to provide massive connectivity to IoT sensors and devices in rural, remote, and offshore areas.

2.6.3.2 Sigfox

An alternative long-range technology is Sigfox, which in terms of range comes between Wi-Fi and cellular. It uses the ISM bands, which are free to use without the need to acquire licenses, to transmit data over a very narrow spectrum to and from connected objects. The idea for Sigfox is that for many M2M applications that run on a small battery and only require low levels of data transfer, then Wi-Fi's range is too short while cellular is too expensive and also consumes too much power. Sigfox uses a technology called Ultra Narrow Band (UNB) and is only designed to handle low data-transfer speeds of 0.01 to 1 kbps. It consumes only 50 microwatts compared to 5000 microwatts for cellular communication or can deliver a typical standby time 20 years with a 2.5 Ah battery while it is only 0.2 years for cellular.

Already deployed in tens of thousands of connected objects, the network is currently being rolled out in major cities across Europe. The network offers a robust, power-efficient, and scalable network that can communicate with millions of battery-operated devices across areas of several square kilometers, making it suitable for various M2M applications that are expected to include smart meters, patient monitors, security devices, street lighting, and environmental sensors.

2.6.3.3 Weightless

Weightless is a proposed proprietary open wireless technology standard for exchanging data between a base station and thousands of end-user devices around it (using wavelength radio transmissions in unoccupied TV transmission channels) with high levels of security. As of 2018 weightless devices are operating in license-exempt sub-GHz frequency bands (e.g., 138 MHz, 433 MHz, 470 MHz, 780 MHz, 868 MHz, 915 MHz, and 923 MHz).

The defining characteristics of Weightless are the following: 100% bidirectional, fully acknowledged communication for reliability; optimized for a large number of low-complexity end devices with asynchronous uplink-dominated communication with short payload sizes (typically < 48 bytes); optimized for ultra-low-power consumption (at the expense of latency and throughput compared to cellular technologies). The weightless standard data rates vary from 0.625 kbps to 100 kbps. Typical End Device transmit power of 14 dBm (up to 30 dBm) while the Base Station transmit power of 27 dBm (up to 30 dBm).

A typical Weightless network is composed of the following elements:

1. End Devices (ED): the user nodes in the network which are of low-complexity, the low-cost.
2. Base Stations (BS): the central node in each cell, with which all EDs communicate via a star topology.

Base Station Network (BSN): interconnects all Base Stations of a single network to manage the radio resource allocation and scheduling across the network, and handle authentication, roaming, and scheduling.

Weightless Special Interest Group (Weightless-SIG) [22] proposed Weightless, an open standard offering LPWAN connectivity. Weightless-N (nWave) is similar to SigFox. Only supporting unidirectional communication for end-devices to the BS [23]. It achieves communication range of up to 3 km with a maximum data rate of 100 kbps. The MAC protocol of Weightless-N is based on slotted ALOHA.

Additionally, Weightless-P is the latest standard introduced by Weightless-SIG. Unlike Weightless-N, it offers bi-directional communication with support for acknowledgments. It achieves data rate around 100 kbps. Compared to Weightless-N, Weightless-P has shorter communication range (2 km) and shorter battery lifetime.

2.6.3.4 Wi-Max

Wi-Max, short for Wireless Interoperability for Microwave Access, is a long-range wireless network protocol based on the IEEE 802.16 set of standards. Wi-Max was originally designed to support 30–40 Mbps transmission rate, but it is expected to support rates of up to 1 Gbps [24]. Frequencies are on the lower end, ranging from 2 GHz to 11 GHz; thus, transmissions are not easily interrupted by physical obstacles. Wi-Max was designed to deliver information from device to device over large geographical distances. Specifically, Wi-Max allows coverage of up to 50 km [25].

Wi-Max is utilized in order to deliver communications with QoS. Since Wi-Max is designed to operate similarly to Wi-Fi, it handles sending data from one endpoint to another via radio signals [26], but with larger geographical coverage and a larger number of users supported. Wi-Max requires two parts to operate: a Wi-Max base station, and a Wi-Max receiver, integrated within the receiving device. A base station operates similarly to a cellular tower, and allows immediate connection to the internet using wired links. Wi-Max is a technology which can be used in order to connect buildings, houses [27], etc., with each other in order to exchange information; thus, the integration of Wi-Max for wireless communications in sensor network environments such as smart homes, smart cities, healthcare or even the industrial sector can be proven to be beneficial.

It is possible for Wi-Fi and Wi-Max to co-exist in a collaborative manner [28]. Such a scenario would include the utilization of Wi-Max in order to interconnect Wi-Fi hotspots. Therefore, Wi-Fi will be able to expand the area in which it is accessible.

2.6.3.5 Satellite

As observed in the previous sections, a sufficient amount of IoT device networking options exist in order to interconnect devices such as sensors over vast amounts of geographical areas. However, the possibility that terrestrial infrastructures such as cellular services are not enough to cover reliably remote areas still remains. Smart objects may be distributed over a large amount of areas, a portion of which may be unreachable through the aforementioned wireless technologies. Therefore, satellite IoT networking can be utilized in order to transmit various sensor readings and provide all the benefits of IoT technology even in distant areas, making IoT truly global.

Satellite infrastructures for IoT communications is a relatively novel approach to networking; however, it is considered a very promising candidate; as of 2025, the market for satellite technology for the interconnection of smart devices is expected to reach 5.9 billion USD, while the number of satellite-supporting smart devices is expected to be around 30.3 million [29] at that time. One of the many benefits of utilizing satellites is the constant and global coverage such technology allows. Additionally, satellites support broadcast, multicast, and geocast operations, i.e., send information to all or multiple nodes, or to all nodes in a particular geographical area [30], respectively.

Satellite communications use frequencies ranging from 1.53 GHz to 31 GHz [31], with data rates in the 10 Mbps to 1 Gbps interval. Given the fact that satellite communications are more reliable than their terrestrial equivalents, they can be utilized in various critical IoT application areas that require constant and reliable communications, as any disruptions could lead to severe consequences; such application areas are the healthcare and public safety domain and especially emergency responses and the industrial domain. Therefore, in the case that terrestrial communications fail to respond during an emergency, such as natural disasters, satellite technology can be utilized.

2.6.3.6 Infrared Communication

Infrared radiation (IR), or infrared light, is a type of electromagnetic radiation that's invisible to human eyes but that we can feel as heat produced when atoms absorb and then release 1.3 visible light and just below red visible light, hence the name "infrared." IR has frequencies from **430 THz** to **300 GHz** [66] and a wavelength from 780 nm to 1 mm [32], as pictured in Figure 2.17.

Wireless infrared communication refers to the use of free-space propagation of light waves in the near-infrared band as a transmission medium for communication.

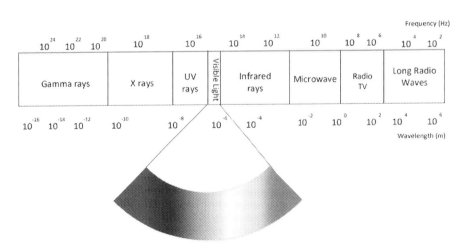

FIGURE 2.17 Electromagnetic spectrum

IoT Reference Architectures

The two most popular mediums in the wireless arena are Infrared (IR) and Radio Frequency (RF). Infrared Data Association, or IrDA in short [33], is a group of device manufacturers that developed a standard for transmitting data via infrared (IR) light waves. It provides specifications for the complete set of protocols for wireless IR communication.

The IrDA Data protocol defines a standard for an interoperable two-way wireless IR transmission data port. The protocol stack, which is displayed in Figure 2.18, consists of four layers. The **IrPHY** is the physical layer of the IrDA specifications and there are six different data rates that can be supported SIR standard low speed infrared, MIR medium speed infrared, FIR very fast infrared, *UFIR* ultrafast *infrared* and Giga-IR with data rate up to 1 Gbit/s. The Link Access Protocol **IrLAP** is responsible for establishing and preserving a link between two devices which includes device discovery, connection, and negotiation. The Link Management Protocol **IRLMP** provides multiplexing of the IrLAP layer as well as service registry and discovery mechanisms. The Information Access Service **IAS** is responsible for detecting services on other devices, and for supplying the appropriate mechanisms for advertising to remote devices.

The Tiny Transport Protocol TinyTP enables the flow control at IrLMP level and the segmentation and reassembly (SAR). Part of the Session and Application layers of IrDA protocol include IrOBEX, IrCOMM, IrLAN, IrSimple, IrSS, and IrFM. The Object Exchange IrOBEX provides communication and data exchange between devices. The Infrared Communications Protocol IrCOMM enables the infrared device to act like as a serial or parallel port.

The IrDA standards provide the base for the creation of low cost and low power transceivers that can be used in numerous portable devices like laptops, printers, PDAs, wireless computer mice and keyboards, and wireless door keys for home or vehicle access [34]. The IrDA offers safe, secure, and high rate low distance communication and it is universally adopted and implemented.

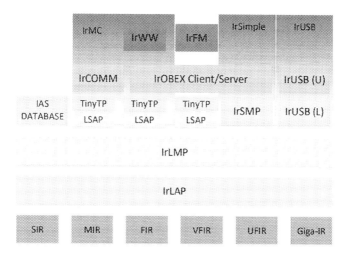

FIGURE 2.18 IrDA protocol stack

2.6.4 THREAD PROTOCOL

Smart homes and buildings filled with connected products are loaded with possibilities to make our lives easier and more comfortable. But except from that smart home IoT devices can help reduce costs and conserve energy.

To address the needs of IoT network such as energy efficiency, speed, security, scalability, resiliency, and reliability a new connectivity technology was needed. Thread is an IP based, low power, secure, and mesh networking protocol for IoT devices.

Although the thread is new technology its design utilizes many existing protocols. At the radio layer it uses IEEE 802.15.4 protocol as ZigBee. At the network layer, it uses the IPv6 internet protocol *6LowPAN, as seen in* Figures 2.19 and 2.20. By using an existing widely deployed technology, low-cost implementations are achieved which are fast available to the market.

The Thread group is responsible for defining and maintaining the Thread protocol specification

By using the Thread protocol, different IoT devices can be integrated easily and become part of same secure mesh network.

A single application layer communicating across multiple link technologies such as Wi-Fi and Cellular. When a device is attempting to participate in a Thread Network has to go through the phases of discovery, commissioning, and finally attaching before it can join in the network. All joining is user-initiated in Thread Networks. Once joined, a device is fully participating in the Thread Network and can exchange application layer information with other devices and services within and beyond the Thread Network. The basic thread topology can be viewed in Figure 2.21. [35, 36].

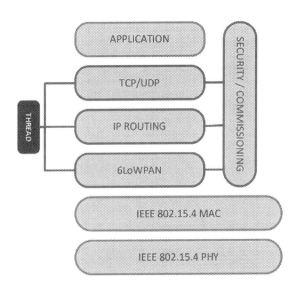

FIGURE 2.19 Thread: An IoT protocol

IoT Reference Architectures

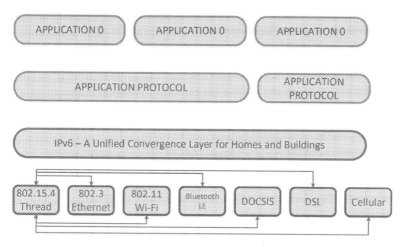

FIGURE 2.20 Thread: An IoT protocol

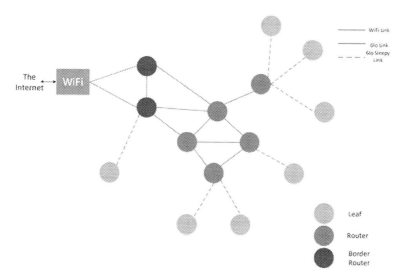

FIGURE 2.21 Basic Thread Network Topology and Devices

2.6.5 INSTEON

Insteon technology, developed by SmartLabs in 2005, is designed to accommodate specifically to the needs of smart homes. Information is transmitted from device to device through power lines, radio frequencies, or both. Additionally, the data rate is around 13 Kbps, with frequencies in the 131.65 kHz to 915 MHz interval, depending on the usage of power lines or radio frequencies. If no obstacles are present, Insteon can cover ranges of up to 45 meters. Moreover, Insteon networking is defined by a dual-mesh topology; thus, devices are allowed to directly address each other and connect to as many nodes needed, for cooperation purposes.

Devices behave as peers; therefore, a device may relay messages to another device, if the source was not powerful enough to reach the target. Specifically, data can be transmitted using the multi-hop approach. All devices retransmit data they receive, unless they are the recipient of the message. However, messages can be relayed three times only; thus, only three hops are permitted. Therefore, the more devices are added to the environment, the better reliability is provided.

Every message undergoes error detection for security enhancement purposes, specifically, with 8-bits checksums. Finally, every Insteon-compatible device possesses its own unique ID, equivalent to a MAC address.

2.7 PERCEPTION LAYER

As outlined in Application layer section, the IoT applications require basic tasks relating to inputs (data collection) or outputs (data sharing or interaction with another system). The defining feature of an IoT device is its presence in a physically remote, relatively unconnected environment, which requires either/both a data collection task or a type of interaction. To achieve that, an IoT entity requires its perception layer, which represents its interface with the surrounding environment. To enable the respective interaction, the IoT environment critically relies on its sensors and actuators to collect data and/or undertake a specific action.

Sensors provide the IoT domain with the ability to sense the environments; from a functionality perspective, they use a transducer to convert the observed variable into an electrical signal that is then processed and used by the IoT device. Among the targeted observations, the most used types of sensors replicate the human senses and collect data on temperature, pressure, movement, sound, imaging, or proximity. Beyond human senses, there are variations that focus specifically on industrial processes (such as on/off, voltage, positioning, flow, fluid, level, or humidity) or light (infrared/UV) sensors.

From a strict terminology perspective, there are multiple classifications available [37]. Depending on their requirement to have an excitation or power signal, sensors can be divided into *active* (requiring an excitation signal) and *passive* (not requiring an excitation signal). Related to the nature of the inputs and outputs, there are *analog* or *digital* sensors, depending on whether the acquired and generated signals are continuous or discrete. Beyond the nature of the signal, a more relevant classification is derived from the means of detection, including categories such as electric, biological, physical, chemical sensing, or the nature of the input, such as photo- or thermoelectric, electromagnetic or electrochemical, optical, and so on.

Given their role, sensors include as their defining characteristics a combination of parameters from the electronics world and physical accuracy properties. To begin with, the sensors are evaluated by their ability to measure the respective variable, which includes accuracy, range, resolution, and sensitivity, accompanied by noise, filtering, and hysteresis. The names of the parameters are self-defining; what is worth pointing out is that some of them lead to a compromise, such as range versus resolution, sensitivity versus filtering, or accuracy versus noise.

Beyond their measurement abilities, IoT sensors are part of lightweight systems; therefore, they are severely bound by two other characteristics: their size and their

IoT Reference Architectures

power consumption. Indeed, from a manufacturing perspective, more accurate or sensitive devices are typically larger in size and may also utilize more power to acquire the data. To support their minimization, a significant amount of recent research focused on offloading their duties towards the application and processing abilities of the IoT hardware. This allowed using less accurate or sensitive devices, subsequently lighter and less power-demanding, to provide their raw data to be pre-processed and lead to similarly accurate results when compared to larger and more power-demanding sensors. An example of this is [Ghita and Thomas]…

The following sections will provide an outline of individual types of sensors.

2.7.1 Pressure Sensors [38]

Pressure sensors consist of a pressure-sensitive component that measures the pressure of a liquid or gas against a diaphragm and converts it to an electrical signal. Depending on the type of measure against a reference level, there are four types of pressure sensors:

- Absolute pressure, which measures the pressure versus pure vacuum. The areas that require such sensors are environment measurements, either for weather prediction or for applications that operate in a variable air pressure environment (such as aerospace)
- Gauge pressure, which refers to pressure relative to the ambient pressure. This is particularly useful for systems that are working against the open environment, such as tire pressure
- Differential pressure, which allows comparative measurements between two different environments, relevant in a number of applications, from air conditioning and exhaust ducts to medical devices

A graphical overview of these above types of pressure is provided in Figure 2.22.

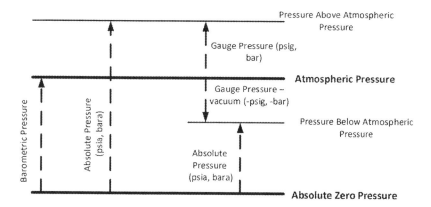

FIGURE 2.22 Graphical overview of different pressure types. Redrawn from [62]

2.7.2 Temperature Sensors [39]

Temperature sensors can be divided into four main categories: thermocouples, thermistors, resistance temperature detectors (RTDs), and semiconductor-based.

Thermocouples are the most used type of temperature sensor across all domains, as they are passive, operate over a wide temperature range and are mostly linear. They are based on the Seebeck effect, which creates a voltage difference between two conductors joined together and exposed to the same temperature. As this difference is linearly dependent on the temperature of exposure, it can be electrically measured and converted to a reading. The differences between existing sensors in terms of measured temperature and sensitivity come from the two materials joined together.

RTDs and thermistors are similar in terms of sensing principle. They both rely on the fact that the resistance of a material changes with the temperature that is exposed to. RTDs represent the more accurate and stable alternative, providing a linear measurement response of the variations in temperature but also being the more expensive option. Thermistors are easier to integrate due to size and price, but they have a nonlinear relationship between temperature and resistance and therefore require a correction circuit, typically a resistor and a voltage divider, to convert their reading.

Semiconductor-based temperature sensors are the most suited for embedded design, given they are rather mechanically delicate in comparison with the other categories and addition of mechanical housing makes them less accurate. The fundamental concept for this category is that the properties of a semiconductor vary with the temperature that it is exposed to; by connecting the semiconductor into an integrated circuit, the variance can be converted to an electrical reading. There are five types of such sensors, depending on the measured characteristic: voltage, current, digital, silicon resistance, or diode. The first two categories are self-explanatory as part of an amplifier, the digital sensor also includes an ADC converter, while the last two exploit the variations in resistance or conductivity of the silicone material. In all cases, the sensors are linear and have a narrower range compared to the other categories.

2.7.3 Touch Sensors

Touch sensors are critical when it comes to user interaction, as they provide the necessary interface with the system. The existing types of sensors convert and electromagnetic field, acoustic field, mechanical pressure, or interference with a light source into a signal. The most frequently used types of such sensors are the wire resitive sensors, which put in contact metallic conductors when applying pressure in an area, surface capacitive, which allows the surface to react to the static electrical capacity of the body (or an electrically charged tool, such as a stylus). A variation of the second category is the projective capacity sensor, which builds up an electromagnetic field pattern that can be interacted with, requiring an even lower variation in electrical capacity. The other two options for touch sensors are surface acoustic wave (SAW) and infrared. SAW sensors use piezoelectric pairs of transducers and receivers to establish an acoustic field on the surface, which will be affected when the surface is touched. Infrared sensors achieve a similar effect by creating an infrared grid above the surface of the screen, also affected when interrupted by touch.

2.7.4 IR/Light Sensor

Light sensors are passive sensors that measure the radiant light energy within a frequency band by converting the light energy into electrical energy. There are several classes of light sensors: photo resistors, photovoltaic materials, photoemitters, semiconductors, and thermal. While they are all relevant for various applications, photovoltaic materials and semiconductors are dominating the IoT sensor market due to their ability to integrate as part of circuitry. Photovoltaic materials are the preferred choice when the application requires not only responding to light but also actually powering the device, as they can generate sufficient voltage for running low-powered devices. Photoresistors and photodiodes change their properties when illuminated. While photoresistors are passive devices and will only vary their resistance depending on the level of incidental light, photodiodes have an active response, either photovoltaic or photoconductive. In photovoltaic mode, the photodiode will generate a variable low current, which will have its intensity controlled by the same mechanism as the photovoltaic materials. In photoconductive mode, the element will change its capacitance and can be integrated in a circuit. The standard traditional application for light sensors is as an on-off switch for intelligent lightning. More recently though, they have been integrated into IR and proximity sensors, allowing to detect the presence of a person or measure temperature of ambient objects depending on their emission of IR.

2.7.5 Humidity Sensor

A humidity sensor measures and reports moisture and, optionally, air temperature.

The typical observation variable in this context is relative humidity, which is the ratio between the current moisture in the air and the highest amount of moisture at a particular air temperature, which is a relevant variable both in industrial processes as well as in personal comfort. Depending on the employed mechanism, there are three classes of humidity sensors: capacitive, resistive, and thermal. To begin with, the capacitive sensors are based on oxidation; they read the level of oxidation of a strip of metal placed between two electrodes. Given the employed mechanism is a nonlinear chemical one, this category requires significant processing and normalization to ensure a linear reading. The resistive sensors rely on the presence of ions in a salt medium present between two electrodes; changes in humidity will also the concentration of ions and, subsequently, the resistance of the electrodes. Finally, the thermal sensors observe the variation in resistance between two electrodes, one in dry nitrogen and one at ambient temperature.

The humidity sensors have a number of defining parameters, from the accuracy and linearity to reliability. In addition, their drift is measured through repeatability and, given they are based on a relatively slow chemical process, the response time once the stimulus is applied.

While the importance of humidity is less obvious from a human perspective, apart from the level of comfort, measuring it is critical for industrial and agricultural applications. A few particular areas to highlight are the textile industry, as humidity changes the fabric properties, and agriculture, to determine the level of moisture in the air and predict the quality of these. [40].

2.7.6 Accelerometer [41]

An accelerometer is a type of sensor that detects moving acceleration in terms of amplitude and frequency. Traditionally accelerometers have been used in relation to monitoring human movement, from observing possible movement issues (from motor recovery to fall detection) to detecting and improving posture and gait. With the advent of wearable devices, accelerometers have been used in a range of monitoring and classification of activities, from sitting or standing to running or cycling.

Accelerometers are conceptually based on a damp mass on a spring. Given an external acceleration on the direction of the spring, the mass will expand or contract the spring proportionally with the strength of the acceleration. Depending on the type of response, there are two types of accelerometers—DC response and AC response. The first category measures static accelerations, such as gravity, that do not vary in time, hence have zero variation. AC-response accelerometers can observe variation in acceleration (such as the one produced by a hand movement) or vibration.

Implementation-wise, there are three types of accelerometers: capacitive Micro-Electro-Mechanical Systems (MEMS) [42], piezoresistive, or piezoelectric. As their name indicates, MEMS are a combination of electrical and mechanical component which transforms the capacitance of a mass that is subject to acceleration. Due to their small factor, MEMS are very suitable for integration into surface-mount devices or printed circuit boards; in terms of the above classification, they are DC-coupled. The miniature size does impact on their accuracy, as measurements may be rather noisy and allow a rather narrow measurement range. Piezoresistive accelerometers are also DC-coupled, but rely on the changes in resistance when acceleration is applied to the sensor mass. They benefit from a wide measurement range, so can measure high-impact events, such as shocks, but have a low sensitivity, hence should not be relied upon for accurate measurements. Because of their construction characteristics, piezoresistive sensors are the most expensive category of accelerometer. Finally, piezoelectric sensors rely on the property of lead zirconium titanate to produce an electrical charge when acceleration is applied to it. This category of sensors benefits from high sensitivity, low noise levels, but the range of applications is slightly limited due to their AC-coupled nature and can saturate if subjected to an acceleration outside their measurement interval. The accelerometer sensor suitability for different applications is presented in Table 2.2.

2.7.7 Proximity Sensor [43]

One of the requirements of IoT is to allow more natural, human-oriented interaction. One option to achieve this is to augment sensing from switch-, action-based to observational. A basic example is sensing the presence of a user rather than requiring them to interact with a system, undertake a task, or press a switch. Proximity sensors are one of the distinctive categories fulfilling this requirement, as they convert the presence of a subject into an electrical signal. From a physics perspective, there are four underlying mechanisms: inductive, electrical capacitive, magnetic, and ultrasonic. To start, the first category monitors the eddy currents generated in a metallic sensing object due to electromagnetic induction; second type of sensors observe variations in

TABLE 2.2
Summary of LPWAN Technologies

	NB-IoT	EC-GSM-IoT	LTE Cat M1	LoRa	SigFox	IQRF	RPMA	Telensa	DASH7	Weightless-N	Weightless-P	SNOW
Modulation	QPSK, OFDMA (UL), SC-FDMA (DL)	GMSK, 8PSK	QPSK	CSS	DBPSK, GFSK	GFSK	DSSS, CDMA	FSK	GFSK	DBPSK	GMSK, OQPSK	BPSK
Band	Licensed, Sub-GHz	Licensed, Sub-GHz	Licensed, Sub-GHz	Unlicensed, Sub-GHz	Unlicensed, Sub-GHz	Unlicensed, Sub-GHz	Unlicensed, 2.4 GHz	Unlicensed, Sub-GHz	Unlicensed, Sub-GHz	Unlicensed, Sub-GHz	Licensed, Sub-GHz	Unlicensed, TV white spaces
Max Range (Km)	15	15	15	15	10	0-5	15	1-10	0-5	0-3	0-2	5
Peak Data Rate (kpbs)	250 (UL) 170(DL)	10	375	27	1	20	80	65	9.6, 5.66, 166.766	100	100	50
Security	Yes	Yes	Yes	Yes	Yes	Yes	Yes	Yes	Yes	Yes	Yes	N/A
Indoor	Yes	Yes	Yes	Yes	No	Yes	Yes	No	No	No	Yes	Yes
Link Budget (dB)	164	164	164	164	N/A	N/A	177	N/A	N/A	N/A	N/A	N/A
Mobility	No	Yes	Yes	Yes	No	Yes	Limited	No	N/A	No	No	N/A
Battery lifetime (years)	10	10	10	10	5	N/A	15	10	N/A	N/A	N/A	N/A

electrical capacity when approaching the sensing object, the third category, as per its name, use magnets and reed switches, and the last one observes the reflecting sonic waves as they bounce off a detected object. Due to the respective physical phenomenon observed, the three categories also react to different stimuli. The inductive sensors will detect the presence of a metallic object, the capacitive sensors will detect the presence of a wider range of materials starting with metallic objects down to resins or powders, the magnetic sensors will only detect magnets, while the ultrasonic sensors can be used for distance measurement and monitoring. Given their sensitivity to variation in electrical charge or current, the first two are very susceptible to electrical noise, from power lines in their vicinity down to long cables between the sensing element and the interpreter within the sensor.

2.7.8 Actuators [44]

Actuators provide, where necessary, the output to support the IoT functionality, by converting an electrical power input into a type of motion. They can be split into several categories, depending on the type of motion generated and the transmission medium used for converting the electrical input. The aim of this section is to provide a high-level taxonomy of actuators, together with indication of the likely applications to encounter. Given the nature of the interaction with an IoT system and the fact that IoT inherited some of the components from industrial processes that it allowed replacing, the range of output actions is significantly wider than the sensed inputs and hence not suitable for an exhaustive review.

There are two main types of output movement: linear or rotary. The linear actuators will produce movement along a direction or may act as a push-pull mechanism within a system. The rotary ones will produce a rotational movement, typically through the use of a motor.

Based on the type of system that they require interaction with, actuators can be split into two five categories: electrical, magnetic, mechanical, hydraulic, or pneumatic. Each of these comes with a rather separate range of applications, given the physical characteristics of the employed mechanism. Given the stimulus in the IoT domain is invariably electrical, only the first two categories (electrical and magnetic) are relevant in this context.

Electrical actuators are perhaps the most versatile ones, as they allow interacting with a variety of other systems, typically through an electromagnetic switch/latch or motor. Due to the relatively high level of control, such applications will provide significant accuracy in terms of movement; the typical example for accurate movement is a step motor, which allows rotational movement for a specific angle, allowing a precise positioning of a device in a rotational plan. A rotational electric actuator can also be used as an (equally precise) linear actuator through the use of a lead screw within a shaft or drive nut.

The magnetic actuators include either distant magnetic interaction or magnetically controlled active materials [45]. The distant interaction requires either a moving coil, a moving magnet, or a moving iron. A mobile coil will start moving in a static magnetic field when electricity is applied to it; similarly, a mobile magnet placed between two magnet poles can be switched between poles through an electrical coil stimulation. The moving iron actuators are based on the tendency of a soft magnet placed into a coil to move so that the resulting magnetic energy is minimal. The magnetically controlled active materials include magnetorestrictive materials, which deform when affected by a magnetic field, and MRF (magnetorheological fluid), which solidify when subjected to a magnetic field.

Beyond their taxonomy, it is worth pointing out some of the typical usage of actuators as part of the IoT applications. There are three areas that have seen a significant upsurge due to the advent of IoT actuators: switches and latches, robotic controls, and accurate positioning.

2.7.9 Level Sensors

Level sensors are devices used for taking level measurements for liquid substances. Such substances may include water, petroleum, fuel, etc. Level sensors can be integrated for IoT applications, such as environmental monitoring, for sensing changes in sea level, providing early warnings for floods, in the manufacturing sector for monitoring the production process with tank liquid level readings and even in the smart city application area, in smart home appliances such as fridges. Level sensor capabilities may include monitoring the level of nonliquid substances, which however behave in a liquid-like manner. Such substances may include, for instance, powders.

Level sensors convert analog data, which is the level of the substance measured, in a digital format. These sensors can be classified into two main categories:

- Point level sensor: These level sensors are used to provide information about whether the substance has reached a certain threshold sensing point or not.
- Continuous level sensor: Contrary to point level sensors, continuous sensors constantly provide information regarding the amount of substance in the container.

It is possible to classify level sensors even further, based on the means they take the measurements. For instance, if the sensor must be in contact with the liquid in question in order to take the level measurements, then this sensor is classified as an invasive sensor. On the other hand, sensors that employ ultrasonic or microwave technology in order to perform their tasks, are called noncontact sensors.

2.7.10 Gas Sensors

Gas sensors are used in order to sense various qualities in the air, or detect the concentration of different gases, such as carbon monoxide, carbon dioxide, oxygen, etc. Their integration in IoT can be observed in many application areas. For instance, such sensors are be applied in smart cities and smart homes for carbon monoxide and other toxic gasses detection. Additionally, such technology can benefit the environment, by constantly monitoring air pollution levels and therefore aid in preventative measures to be rendered. Oxygen level recordings can be proven to be beneficial for patients in the healthcare and assisted living application areas. Additionally, in the

Public Safety application area, oxygen monitoring sensors can be employed for vital checks in rescue crew.

Gas sensors may be divided into potable and fixed-type sensors. Portable sensors are usually battery-powered wearables, which can provide feedback to the user if the gas amount to be detected exceeds some levels. On the other hand, fixed-type sensors are integrated within the environment.

2.7.11 Water Quality Sensor

Water quality sensors can be used to measure the concentration of different elements in the water, such as pH, temperature, chlorine concentration, etc. Such sensors can be widely used for multiple IoT application domains; however, integration would be highly beneficial in the environmental area. With such technology, constant water pollution monitoring would be possible; therefore, it would be easy to detect timely water contamination. Proper measures can be taken in order to avoid further damage. Additionally, water quality sensors can be used for quality control purposes in the agriculture application area.

Most commonly used water quality sensors include chlorine sensors, which can monitor chlorine percentage in water, pH sensors, which can indicate the acidity of water, etc.

2.8 UBIQUITOUS COMPUTING

Computing has evolved over the years and the evolution is accelerating. In the early 1970's one mainframe computer was serving many people. In the 1980s, the mainframe computer was replaced by personal computer based on the idea of one computer for one person. Through the ages the technology has dramatically transformed and nowadays we are in ubiquitous computing era (ubicomp). The whole concept of IoT and Ubiquitous computing can be seen in Figure 2.23.

Mark Weiser the father of ubiquitous computing invented the term ubiquitous to describe smart computing devices or IoT devices appearing everywhere in any location and in any format. According to the famous quote from his article *The Computer of the 21st Century* published in Scientific American in 1991 "The most profound technologies are those that disappear. They weave themselves into the fabric of everyday life until they are indistinguishable from it." [46] The IoT refers to a type of network to connect anything with the Internet based on specific protocols through devices that collect data, process, and analyze that data and then return it back to the user. The goal of the IoT is to enable things to be connected anytime, anyplace with anything and anyone ideally using any path/network and any service. Mark Weiser's visions can be drawn from his position towards three dominant trends in computer science at his time virtual reality, artificial intelligence, and user agents.

Modern processors are made from tiny chips cheaper and generally better from the past. As the technology improves permits to place a wireless transceiver on things we use in everyday life. Giving a digital identity to the objects by using an electronic code such as "ucode" could be one of the keys for the Ubiquitous IoT Network.

IoT Reference Architectures

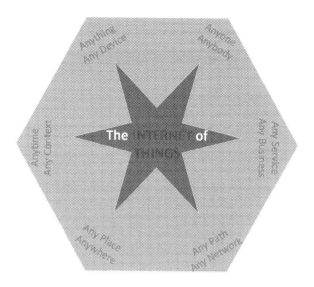

FIGURE 2.23 Internet of Things. Redrawn from [63]

The IERC definition states that IoT is "A dynamic global network infrastructure with self-configuring capabilities based on standard and interoperable communication protocols where physical and virtual "things" have identities, physical attributes, and virtual personalities and use intelligent interfaces, and are seamlessly integrated into the information network" [47]

Ubiquitous computing can offer various advantages, improving the quality of life in many aspects. Internet-connected things, such as wearable IoT technology or sensors placed in the user's environment, are communicating with each other, exchanging information and therefore are able to provide insight about situations and individuals that would otherwise be difficult to obtain. Hence, ubiquitous computing has become a big trend nowadays.

Furthermore, machine learning solutions can provide an extra layer of applicability for IoT devices, by adding the personalization factor. Pattern and activity recognition can be proven to be beneficial in multiple IoT application domains. For instance, the authors in [48] presented a Human Activity Recognition System, which utilizes a wearable sensor and deep learning techniques in order to recognize the most common daily activities of the users. Such personalized systems can be used in order to learn the user's behavioral patterns and assess the user's well-being as well as predict any possible anomalies threatening the wearer [49]. Thus, the predictive capabilities of machine learning can be successfully integrated in the IoT healthcare and assisted living domain. Additionally, by analyzing and learning from human behaviors, predictive models can be developed, aiming to detect in a timely manner behavioral irregularities, for instance in the mental health sector. With machine learning solutions, services can be tailored to the needs of smart city inhabitants [50]. Essentially, by monitoring a user's behavior, personalized recommendations can be proposed to the citizen, with the help of artificial intelligence.

The purpose of Ubiquitous computing is the presence of computing in every aspect of human life, with seamless integration. Personalized systems utilizing machine learning techniques can further benefit both commercial and noncommercial users. Thus, ubiquitous computing has great potential to be the technology of the future.

2.9 IOT SECURITY ISSUES

It is estimated that as of the current year (2020), there are about 31 billion IoT devices [51] connected on the internet, a number that corresponds to roughly 4.1 devices for each human. It is believed that about 127 new devices are connected to the internet every single second [52]. Hence, highly personal, and sensitive data are produced and exchanged in very large volumes every day. This has inevitably contributed to data privacy leakages and service disruptions, as a result of cyber-attacks targeting various IoT application domains. However, as recent events indicate, cyber-attacks against IoT environments can be proven to cause more severe consequences than just the aforementioned ones.

Since little to no security solutions are implemented in order to protect IoT devices, there are many examples pointing to the hazards of having smart devices connected to the internet. For instance, smart traffic lights aim to route traffic in an intelligent way, by using sensor and camera technology. An adversary could easily manipulate unsecure, internet-connected traffic lights and therefore induce accidents that could possibly be fatal. Possibly the most well-known malware targeting IoT device is the Mirai malware. Mirai targeted unsecure IoT devices, by scanning for vulnerabilities. If the default log-in credentials were not changed, Mirai used them to log-in, infect the device and take control of the device, turning it into a bot, therefore forming the first ever known botnet that was composed by IoT devices. The botnet was believed to be used against US's most used Domain Name System (DNS) provider, in order to successfully cause a distributed denial-of-service (DDoS) attack. As a result of the attack, the provider lost about 8% of its customers following the incident, therefore causing a significant loss in revenue [53]. Another impactful cyber-attack targeting the industrial sector was the 2015 Ukrainian power grid attack, which resulted in power supply disruptions for about 225,000 people. As discussed in the previous sections, the healthcare is one of the most user-sensitive application domain, with data privacy and security being the biggest considerations. However, the implantation of smart pacemakers, without any communication security measures poses significant threats against the patient's life, if a malicious adversary wanted to manipulate the device.

As noticed, many security issues exist and need to be addressed in order to make IoT secure and safe for everyone. This section's purpose is to provide a detailed description of IoT-related security problems and present possible countermeasures as well.

- Denial of Service (DoS) and Distributed Denial of Service (DDoS) are possibly the most well-known and applied attacks. DoS and DDoS aim to exhaust the device's resources by producing an enormous amount of network traffic,

also known as flooding the target. As a result of this attack, there are disruptions in the device's operation, and possibly the total failure of the device to operate for the duration of the attack. Therefore, the availability aspect is heavily impacted. As IoT devices are by default resource-constrained, it is very easy to for a DoS attack to be effectively accomplished.
- Botnets, as explained in the Mirai case, are composed by infected vulnerable devices that can be used to perform DDoS attacks. The infected devices are called bots or zombies; they show no sign of exploitation as they continue to operate normally. The device's security is breached, usually with a malware, in order to give full control privileges to the bot master, who can then issue arbitrary commands for the bots to execute against another target.
- Jamming attacks target the network layer of the IoT architectures. These attacks aim to tamper with the communication channel in various ways, in order to cause disruptions in information exchange between IoT devices. For instance, the attacker may produce a radio signal that interferes with the communication channel in a continuous way; thus, IoT devices are unable to communicate. Such as a jamming attack, is known as constant jamming [54]. An attacker may attempt to send constantly packets via the communication channel in order to effectively jam it; thus, deceiving the receiving IoT devices into believing there is more traffic to receive and therefore stop communicating. Additionally, an attacker may choose the jamming attack implementation based on the communications that are taking place in the channel. For instance, they may choose to not initiate the attack in case the channel is not utilized at the specific moment, and wait until devices start to communicate.
- Spoofing attacks aim to represent a malicious adversary as a legal entity by fabricating data, in order to perform malicious actions. This could potentially involve an addition of an IoT device, or an IoT device-behaving software to the network that behaves in a normal manner, which however produces false data. In addition, spoofing attacks can be used as a means to perform further attacks, such as Man In The Middle attacks, the concept of which is explained below.
- A Man-In-The-Middle (MITM) attack involves a malicious adversary who manages in an illegal way, to position himself between two communicating legal entities, therefore eavesdropping all traffic exchanged between them. Apart from simply monitoring traffic that is not intended for him, the attacker could potentially drop all traffic between the nodes, thus inducing a DoS attack as the communication is disrupted, or modify the context of the traffic in order to achieve a malicious goal [54]. There are many means to accomplish a MITM attack, with the most applied one being Address Resolution Protocol (ARP) spoofing.
- Sybil attack affects mostly peer-to-peer networks and aims to create an illusion of multiple entities by having one or more malicious nodes forging new identities, therefore misleading other nodes and compromising the overall effectiveness of the system [55].
- A wormhole attack refers to the creation of a low-latency communication channel between two possibly compromised legitimate network entities. The purpose of this tunnel is to record and forward traffic from one entity to another,

replaying traffic [56] from there. It is also possible that traffic is dropped from the recording node's end, therefore causing disruptions in the network [57].
- Traffic Analysis, although not an actual attack, aims to gather intelligence about specific devices by monitoring their network traffic. As a result, specific attacks can be orchestrated and vulnerabilities can be discovered by analyzing network packets in terms of protocols, addressing, context of exchange, etc. Many packet-analyzing tools are available currently, such as Wireshark [58] or Tcpdump [59], supporting a wide range of protocols.
- Application-layer protocol-specific attacks aim to exploit the way the protocol operates by identifying key vulnerability points in protocol specifications. Since application-layer IoT protocols are not designed with security in mind, this is very easy to accomplish. For instance, in MQTT, it is possible for an adversary to subscribe to all topics with the "#" character in an unsecure MQTT broker, therefore eavesdropping all traffic produced by legal nodes. Additionally, messages are not encrypted, which means that a potential MITM attack against two network entities could result in the total loss of data confidentiality and integrity, if the exchanged messages are modified. In CoAP, the risk of amplification is described, which could result in DDoS'ing the targets. Since response packets can be quite larger than request packets, an attacker could take advantage of that by using vulnerable CoAP nodes to create amplified packets to send to a target.

As presented above, there are many security challenges IoT has to effectively mitigate in order for the services to be provided in an undisrupted, secure, and optimal way. For this reason, many solutions have been proposed, with the most predominant one being an implementation of an Intrusion Detection System (IDS). Intrusion Detection Systems aim to detect in a timely manner possible attacks against the system. There are currently two main categories of Intrusion Detection Systems, namely Anomaly Detection Systems and Signature-Based IDS.

Anomaly Detection Systems usually monitor network traffic and locate possible anomalies. Such systems are currently developed mainly with machine learning solutions. Machine learning for anomaly detection can involve training algorithms such as autoencoders, decision trees, random forest, naïve bayes, deep neural networks, etc. The training phase includes the introduction of a usually normal, nonmalicious dataset, aiming to train the resulting model to fully recognize such traffic as normal; therefore, any deviation from normal behavior will be classified as an anomaly. Machine learning algorithms can also be trained to recognize various behaviors other than normal, by introducing labeled datasets with various attack scenarios to the algorithm. As a result, the trained model can be integrated for the creation of an IDS, in which all traffic will be passed through, in order to determine if the traffic is normal or not, or in the case of training with multiple attack scenarios, determine if a specific attack is under execution, or if the traffic is considered normal. Such intrusion detection systems can recognize attacks that have not yet taken place or are not known by the IDS, thus being a very attractive solution for IoT systems. A drawback of anomaly detection systems is the possibility of training with insufficient normal datasets; thus, the final model will classify normal behavior as anomalous.

On the other hand, signature-based intrusion detection systems are based on the creation of rules, similar to a database, in the form of attack signatures. Attack signatures are characteristics of known attacks that have already taken place. Any traffic that does not match an existing signature will be considered as normal. Signature-based IDS are not usually developed with machine learning techniques. The major drawback of this technique is that, in order for the system to be able to effectively recognize a plethora of attacks, a sufficient amount of rules will have to be established, which can be difficult. Any attacks that are not registered in the IDS's knowledge base in the form of a signature, will be classified as normal. Additionally, this technique is not able to detect unknown attacks, contrary to anomaly-based IDS.

It is possible to create an IDS that integrates a response mechanism, which may mitigate the impact an attack may have against IoT. Such IDSs, are classified as Intrusion Detection and Prevention Systems (IDPS). A response to an attack could be discarding any traffic that was deemed to be malicious, notifying a handler in case an attack was detected, or blocking any malicious IPs.

2.10 CONCLUSIONS

IoT is an emerging technology which however has been incorporated into many aspects of human life. Enabling the performance of tasks with greater ease, assisting people in need and providing major benefits to noncommercial users, IoT is a very attractive solution, accommodating the contemporary lifestyle. This chapter's purpose was to address the three major IoT architectures, with the four-layer architecture being the main focus. All layers of the aforementioned architectural model were analyzed further; therefore, the major IoT application areas were discussed, such as smart homes and smart cities and various others as well, in order to present to the reader the many areas that IoT is applicable to in a technologically advanced society. Additionally, the most predominant application-layer protocols were presented in an in-depth analysis, namely MQTT, MQTT-SN, CoAP, DDS, XMPP, and AMQP. Network layer protocols for IoT environments were presented and categorized based on the range covered, offering a wide informational view. Additionally, the IoT protocols for smart home and smart city device intercommunications, Thread and Insteon, were introduced.

Furthermore, emphasis was given on IoT hardware components, where a variety of sensors were presented, composing the perception layer. Since the deployment purpose of IoT is to be seamlessly integrated within the environment, an introduction to Ubiquitous computing followed. Finally, a brief summary of IoT security issues was provided, with attacks targeting all layers of the four-layer architectural model, and possible mitigation techniques in the form of Intrusion Detection and Prevention Systems were proposed.

REFERENCES

1. O. Vermesan, M. Eisenhauer, H. Sundmaeker, P. Guillemin, M. Serrano, E. Z. Tragos, J. Valino, A. van der Wees, A. Gluhak and R. Bahr, "Internet of Things Cognitive Transformation Technology Research Trends and Applications," in *Cognitive Hyperconnected Digital Transformation : Internet of Things Intelligence Evolution*, River Publishers, 2017, pp. 17–95.
2. [Online]. https://www.netburner.com/learn/architectural-frameworks-in-the-iot-civilization/.
3. R. Kumar, T. W. Au, and W. S. Haji Suhaili, "Exploring data security and privacy issues in Internet of Things based on five-layer architecture," *International Journal of Communication Networks and Information Security*, vol. 12, pp. 108–121, 2020.
4. [Online]. https://tools.ietf.org/html/rfc7252#section-9.
5. [Online]. https://xmpp.org/rfcs/rfc6120.html#stanzas-semantics-message.
6. [Online]. https://iot-analytics.com/top-10-iot-applications-in-2020/.
7. [Online]. https://www.sciencedirect.com/science/article/pii/S2095809917307130.
8. [Online]. https://www.who.int/health-topics/air-pollution#tab=tab_1.
9. K. K. Patel, S. M. Patel, P.G. Scholar, and C. Salazar, "Internet of Things-IOT: Definition, characteristics, architecture, enabling technologies, application & future challenges," *International Journal of Engineering Science and Computing*, vol. 6, pp. 6122–6131, 2016.
10. [Online]. https://iot-analytics.com/state-of-the-iot-update-q1-q2-2018-number-of-iot-devices-now-7b/.
11. [Online]. https://www.gartner.com/imagesrv/books/iot/iotEbook_digital.pdf.
12. [Online]. https://iot-analytics.com/state-of-the-iot-update-q1-q2-2018-number-of-iot-devices-now-7b/.
13. [Online]. https://zigbeealliance.org.
14. [Online]. https://en.wikipedia.org/wiki/Wi-Fi.
15. [Online]. https://www.wi-fi.org/beacon/dave-chen/wi-fi-6-in-the-enterprise-we-re-in-a-new-wireless-era.
16. [Online]. https://www.wi-fi.org/discover-wi-fi.
17. [Online]. https://www.dailydot.com/debug/what-is-lifi/.
18. [Online]. https://en.wikipedia.org/wiki/Li-Fi.
19. [Online]. https://sg.element14.com/trends-cellular-iot-part1.
20. [Online]. https://halberdbastion.com/technology/iot/iot-protocols/emtc-lte-cat-m1.
21. [Online]. https://www.fleet.space/.
22. [Online]. http://www.weightless.org.
23. [Online]. https://www.link-labs.com.
24. [Online]. https://en.wikipedia.org/wiki/WiMAX.
25. H. Zemrane, A. N. Abbou, Y. Baddi and A. Hasbi, "*Wireless Sensor Networks as part of IOT: Performance study of WiMax - Mobil protocol*," 2018 4th International Conference on Cloud Computing Technologies and Applications (Cloudtech), Brussels, Belgium, 2018, pp. 1–8, doi: 10.1109/CloudTech.2018.8713351.
26. [Online]. https://computer.howstuffworks.com/wimax.htm.
27. S. Banerji and R. Singha Chowdhury, "Wi-Fi & WiMAX: A comparative study".
28. S. Song and B. Issac, "Analysis of wifi and wimax and wireless network coexistence," 2014.
29. [Online]. http://satellitemarkets.com/satellite-iot-game-changer-industry.
30. M. De Sanctis, E. Cianca, G. Araniti, I. Bisio and R. Prasad, "Satellite Communications Supporting Internet of Remote Things,"*IEEE Internet of Things Journal*, vol. 3, no. 1, pp. 113–123, Feb. 2016, doi: 10.1109/JIOT.2015.2487046.

31. I. Jawhar, N. Mohamed, and J. Al-Jaroodi, "Networking architectures and protocols for smart city systems," *Journal of Internet Services and Applications*, vol. 9, pp. 1–16, 2018.
32. [Online]. https://en.wikipedia.org/wiki/Infrared.
33. [Online]. https://en.wikipedia.org/wiki/Infrared_Data_Association.
34. J. B. Carruthers, "Wireless infrared communications,"*Wiley Encyclopedia of Telecommunications*, 2003.
35. [Online]. https://www.threadgroup.org/What-is-Thread.
36. [Online]. https://www.threadgroup.org/Portals/0/documents/support/Thread%201.2%20Base%20Features.pdf.
37. [Online]. https://www.electronicshub.org/different-types-sensors.
38. [Online]. https://www.thomasnet.com/articles/instruments-controls/pressure-sensors/.
39. [Online].https://www.digikey.com/en/blog/types-of-temperature-sensors.
40. [Online].https://www.electronicsforu.com/resources/electronics-components/humidity-sensor-basic-usage-parameter.
41. [Online]. https://www.sciencedirect.com/topics/materials-science/accelerometer.
42. [Online].https://blog.endaq.com/accelerometer-selection.
43. [Online]. http://www.ia.omron.com/support/guide/41/introduction.html.
44. [Online]. https://www.progressiveautomations.com/pages/actuators.
45. [Online]. https://www.cedrat-technologies.com/en/technologies/actuators/magnetic-actuators-motors.html.
46. M. Weiser, "The computer for the 21st century," *SIGMOBILE Mobile Computing and Communications Review*, vol. 3, pp. 3–11, 1999.
47. [Online]. http://www.internet-of-things-research.eu/pdf/IERC_Cluster_Book_2012_WEB.pdf
48. V. Bianchi et al., "IoT wearable sensor and deep learning: An integrated approach for personalized human activity recognition in a smart home environment," *IEEE Internet of Things Journal*, vol. PP, pp. 1, 2019.
49. D. Scott Kehler et al., "A systematic review of the association between sedentary behaviors with frailty," 2018.
50. J. Chin, V. Callaghan, and I. Lam, "*Understanding and personalising smart city services using machine learning, the Internet-of-Things and big data*," 2017 IEEE 26th International Symposium on Industrial Electronics (ISIE), Edinburgh, 2017, pp. 2050–2055, doi: 10.1109/ISIE.2017.8001570.
51. [Online]. https://securitytoday.com/Articles/2020/01/13/The-IoT-Rundown-for-2020.aspx?Page=2.
52. [Online].https://www.vxchnge.com/blog/iot-statistics.
53. N. Neshenko, E. Bou-Harb, J. Crichigno, G. Kaddoum, and N. Ghani, "Demystifying IoT security: An exhaustive survey on IoT vulnerabilities and a first empirical look on Internet-scale IoT exploitations,"*IEEE Communications Surveys & Tutorials*, vol. 21, no. 3, pp. 2702–2733, thirdquarter 2019, doi: 10.1109/COMST.2019.2910750.
54. P. I. Radoglou Grammatikis, P. G. Sarigiannidis, and I. D. Moscholios, "Securing the Internet of Things: Challenges, threats and solutions," *Internet of Things*, vol. 5, pp. 41–70, 2019.
55. K. Zhang, X. Liang, R. Lu, and X. S. Shen, "Sybil attacks and their defenses in the Internet of Things," *IEEE Internet of Things Journal*, vol. 1, pp. 372–383, 2014.
56. R. Pandey, "Wormhole attack in wireless sensor network," 2014.
57. D. Buch and D. Jinwala, "Prevention of wormhole attack in wireless sensor network," 2011.
58. [Online]. https://www.wireshark.org/.

59. [Online]. https://www.tcpdump.org/manpages/tcpdump.1.html.
60. International Telecommunications Union, ITU Recommendation Y.4000/Y.2060 (06/12) Overview of the Internet of things.
61. [Online]. https://circuitdigest.com/article/what-is-lifi-how-it-works/.
62. [Online]. https://www.engineeringtoolbox.com/pressure-d_587.html.
63. K. K. Patel, S. M. Patel, P.G. Scholar, and C. Salazar "Internet of Things-IOT: Definition, characteristics, architecture, enabling technologies, application & future challenges," *International Journal of Engineering Science and Computing*, vol. 6, pp. 6122–6131, 2016.
64. [Online]. https://blog.endaq.com/accelerometer-selection.
65. [Online]. https://www.oasis-open.org/committees/download.php/66091/MQTT-SN_spec_v1.2.pdf.
66. [Online]. https://www.elprocus.com/communication-using-infrared-technology/.

3 Threats in Critical Infrastructures

A. Peratikou, S. Shiaeles, and S. Stavrou

CONTENTS

3.1 Introduction 98
    3.1.1 Definition of a Critical Infrastructure 99
    3.1.2 Critical Infrastructure Domains 99
        3.1.2.1 Agriculture 99
        3.1.2.2 Commercial Facilities 100
        3.1.2.3 Financial Services 100
        3.1.2.4 Information and Communication Services......................100
        3.1.2.5 Government Facilities and Administrative Services 100
        3.1.2.6 Healthcare Services and Emergency 101
        3.1.2.7 Water Services..101
        3.1.2.8 Manufacturing 101
    3.1.3 Energy Services and Materials..101
        3.1.3.1 Chemical Industries..102
        3.1.3.2 Transportation Systems and Logistics Services...................102
3.2 Key Security Goals of CIIs Application Domains 102
    3.2.1 Application Domain: Logistics, Tracking, Fleet Management 102
        3.2.1.1 Key Security Goals...103
    3.2.2 Application Domain: Smart Meters 104
        3.2.2.1 Key Security Goals...104
    3.2.3 Application Domain: Capillary Networks 105
        3.2.3.1 Key Security Goals...105
    3.2.4 Critical Infrastructures Incidents...106
        3.2.4.1 Revenge Attack from an Ex-Employee—2000 106
        3.2.4.2 Energy Companies Cyberattacks
                "Night Dragon" 2009–2010 107
        3.2.4.3 Stuxnet—2010..107
        3.2.4.4 German Steel Mill Attacks -2014.......................................108
        3.2.4.5 Ukraine Cyberattacks and Blackout—2015 and 2016 108
        3.2.4.6 WannaCry Ransomware-British Health Exposed-2017 109
        3.2.4.7 Latin America Machete Attack—2019...............................110
        3.2.4.8 Saudi Arabia Drone Attacks at Oil Processing
                Facilities—2019 110

3.3 Relevant Deployment Architectures and threats on CIIs 110
    3.3.1 Relevant Deployment Architectures 110
        3.3.1.1 Fleet Management 110
        3.3.1.2 Smart Grid 111
        3.3.1.3 Capillary Network 112
    3.3.2 Threats on CIIs 113
        3.3.2.1 Out of Date Systems 113
        3.3.2.2 Outdated Hardware 113
        3.3.2.3 Lack of Training and Awareness 113
        3.3.2.4 Lack of Talent 113
        3.3.2.5 Social Engineering 113
        3.3.2.6 Network Attacks 115
        3.3.2.7 COTS Drones 120
    3.3.3 Attacks on Smart Grid CIIs 129
        3.3.3.1 Attacks on the Infrastructure Grid 129
        3.3.3.2 Attacks to Critical Infrastructure Components to
                Acquire Private Sensitive Information 130
        3.3.3.3 Compromise of Interactions of AMI
                Components with the Infrastructure 130
        3.3.3.4 Hijacking Connections Between Meters and
                Demand Response Systems 130
        3.3.3.5 Altering of Meter Data When Transmitted
                Between Meter and Gateway, Gateway and
                Consumer or Gateway and External Entities 131
        3.3.3.6 Attacker Alters Meter Data, Gateway Configuration Data,
                Meter Configuration Data, CLS Configuration Data or a
                Firmware Update in the WAN 131
        3.3.3.7 Compromise of an Existing Data Concentrator 131
        3.3.3.8 False Data is Injected by an Attacker in the
                Smart Grid Traffic 132
        3.3.3.9 Injection of Realistic False Data 132
        3.3.3.10 Load Redistribution Attack (Injection of
                Realistic False Data) 132
        3.3.3.11 Monitoring Data of Other Customers 132
        3.3.3.12 Time Modification of the Gateway 132
3.4 Conclusions 133
References 133

3.1 INTRODUCTION

The threats to Cybersecurity are considered by the World Economic Forum as one of the top five global risks encountered by all nations. Constantly increasing in a logarithmic manner is the financial loss caused by Cybersecurity breaches and theft of

Threats in Critical Infrastructures 99

intellectual property. The core functions of all economies in nations around the world have become an ongoing target of cyber threats, as well as governments on all levels such as local regional and national. Cyber-attacks potential to disturb all critical services of both the private enterprise and nongovernmental agencies is growing at an exponential rate.

Protecting national security assets demands cautious cyber intelligence competence, not only to locate cyber threats but also above all to be able to avert or defend against cyber warfare. The increasing number of both people and IoT devices getting connected greatly impacts specific portions of critical infrastructures. Those critical infrastructures most immediately impacted are electrical grid systems, transportation, and telecommunications. As societies become so interconnected to both their devices and the critical services they require, this increasing dependency raises the vulnerability to disruptions of critical infrastructures.

The importance of critical infrastructures for the proper functioning of the nation and social actions makes them prone to cyber-attacks, threats, and risks. A fact that is demonstrated by various attacks described below.

3.1.1 Definition of a Critical Infrastructure

Critical infrastructures are defined as assets, systems, or subsystems necessary for the maintenance of the vital functions of society, health, physical protection, security, economic, and social well-being of citizens. The term Critical infrastructure indicates all such systems required to maintain governmental operations, both physical and nonphysical along with online resources. Critical infrastructures can be separated into private and public sectors. The public sector consists of all government-owned or controlled operations, while the private sector consists of all entities that are controlled by the state such as private companies, private banks, and any organizations that are not governmental that contribute to the nation's economy.

3.1.2 Critical Infrastructure Domains

According to various European Union, Governments, and the Department of US Homeland Security, Critical infrastructures include the following key resources. The majority of the key resources are held by the private sector in most governments [1]. NIS Directive and GDPR generated a framework for a cross-sectoral approach [2]. In this framework, there is no separation between sectors and the grade of accountability is based on the criticality of services provided.

3.1.2.1 Agriculture

The agriculture sector even though is under private ownership in all countries; it is considered a high-risk sector due to its interdependencies with various governmental resources such as energy, transportation, and water. In cases that there is a high dependence of agriculture on technology, it becomes evident that a malfunction or breakdown of any device or system (e.g., in an IoT sensor) may cause serious damages. For example, food safety may be compromised if some relevant technological resources are not functioning due to a smart sensors' downgrading. Therefore, reliability of the overall system is a key goal, as in any other case of sensor networks. However, as in any other case of connected systems, several security risks are also in place regarding the potential of cyberattacks, such as installing malware on the system or compromising the data confidentiality and/or integrity. Often these systems run on unmonitored networks, which provides greater flexibility to an attacker. It should be also pointed out that this sector is traditionally not Cybersecurity aware and, thus, security design is not incorporated into the solution requirements [3]. As a result, hackers may easily gain access to control systems and manipulate them maliciously or spread malware. Moreover, these internet-connected systems can be used to gain access to other connected third-party systems and ultimately become part of a botnet [3]. In an FBI industry note in 2016, the Bureau said that increased adoption of "precision farming" technology threatens to expose the nation's agriculture sector to the risk of hacking and data theft. Therefore, similarly to any other networking system, the main security goals rest with confidentiality, integrity, and availability of data/services.

3.1.2.2 Commercial Facilities
The commercial facilities include various ranges of retail and entertainment sites that attract a broad crowd of people such as casinos, real estate, and lodging. Due to the ease of access, this sector poses a substantial risk.

3.1.2.3 Financial Services
The financial services sector includes insurance companies, investment firms, and products along with other financing organizations. This sector is one of the most attacked sectors; attackers use various types of threats to extort this sector for financial gain. Illustrated in a recent data breach in Equifax in the US where millions of records were stolen.

3.1.2.4 Information and Communication Services
The communication sector plays a vital role in the economy of the governments as it regulates the operations of all organizations and businesses. The information technology sector provides all control systems and services; thus, organizations, governments, and residents are subservient. The importance of this sector lies in its interdependency with other sectors such as the energy sector to supply power for the cell towers. The financial services sector depends on the communication sector to transmit transactions. The logistic sector relies on telecommunication for tracking requirements.

3.1.2.5 Government Facilities and Administrative Services
The sector includes a range of buildings and facilities that are owned by the government. The criticality of this sector derives from the fact that most of the facilities are open to the public for business activities and the administrative services for

transactions. This sector also includes structures that are not open to public, along with elements for the protection of assets such as CCTV systems.

3.1.2.6 Healthcare Services and Emergency

The composition of this sector includes hospitals, emergency services, and healthcare providing institutions. Its role is of great importance in events of disaster when it comes to response and recovery in all the other sectors. As an example, there will be a disturbance in the financial sector, in case the financial officers cannot be recovered by the healthcare or emergency services. The interdependencies of this sector are transportation, agriculture, communication, energy, and water. A recent ransomware attack of 2017, named "WannaCry," caused chaos in healthcare services in the UK.

3.1.2.7 Water Services

Water is crucial in the health of the citizens and prevention of illnesses. It has been a target of attacks since 3000 BC; thus, protecting this resource from physical and cyber threats is of crucial importance.

3.1.2.8 Manufacturing

The manufacturing sector is critical to the economic prosperity of each government. An attack to the components of manufacturing could result in disturbance of critical functions across various sectors of the critical infrastructure. The application of IoT to the manufacturing industry is called the Industrial Internet of Things (IIoT). IIoT comprises a network of intelligent computers, devices, and objects that collect and share huge amounts of data relevant to a manufacturing/industry process. Following the IoT generic pattern, collected data are sent to a central Cloud-based service where it is aggregated with other data and then shared with end users in a helpful way.

IIoT is expected to radically change manufacturing by enabling the acquisition and accessibility of far greater amounts of data, at far greater speeds, and far more efficiently than before. Several innovative companies have started to implement the IoT by leveraging intelligent, connected devices in their factories [4]. Security is recognized as one of the two most outstanding concerns for IIoT. With more sensors and other smart, connected devices being deployed, the number of security vulnerabilities has considerably increased. This fact, along with the critical nature of IIoT applications, has led to the creation of a security framework for the IIoT (Industrial Internet Consortium, 2016).

3.1.3 Energy Services and Materials

The energy sector is defined by the European Commission [2] as the sector build by the three subsectors: electricity, oil, and gas. The energy supply plays a vital role in the welfare of the citizens; without a steady energy supply all countries cannot function and economy will stay at a stall state. It is considered one of the most critical as it includes nuclear reactors. If this sector gets compromised by attackers, they can gain full control to the system and cause catastrophic results that go far beyond economic destructions.

3.1.3.1 Chemical Industries

Chemical industries represent an important critical infrastructure which in many aspects differ from other utility (e.g., electric, water, gas) critical infrastructures. This differentiation arises from the fact that chemical industries need to also secure their control systems, manufacturing operations but also their intellectual property. A leaked or stolen pharmaceutical formula could worth billions in the black market or most importantly pharmaceutical/chemical "recipes" may be weaponized against population. Similarly, transportation of hazardous or other sensitive chemical materials has to be monitored and protected accordingly.

3.1.3.2 Transportation Systems and Logistics Services

Transportation and logistics services consist of all the services involved in the transportation of people or goods through the surface of the earth, whether is via airplane, via sea, or via land. The subsectors or modes included in this sector are aviation, motor carrier, maritime transportation, mass transit and rail, pipeline systems, freight rail, and postal and shipping services.

3.2 KEY SECURITY GOALS OF CIIS APPLICATION DOMAINS

In this section, we map threats in the context of CII example domains. Each application domain can help in identifying the security goals and requirements, as well as relevant deployment architectures. Deployment architectures are important, since the characteristics of architectures can greatly affect the attack surface and the applicability of threat mitigation measures. Regarding CIIs, the application domain is complex vast [5] where sectors can be subdivided into numerous subsectors. While there may not be a universal architecture for all potential information systems realizing the desired set of services in all sectors/subsectors, there is a generic consensus on the generic characteristics of an architecture for CIIs, where appropriate network segregation and different levels of defense are used. As such, we will examine certain architectures that encompass these characteristics. There is a great variance regarding the security-related characteristics under which different deployments operate. These differences may pertain to the technological measures deployed (e.g., existence or lack of firewalls), established procedures (e.g., taking backups), or human aspects (e.g., security awareness) and transcend all IoT application domains and CIIs. These differences are important to identify, since they are highly relevant to the degree that a deployment is vulnerable to certain threats or to the impact that a data breach may have.

3.2.1 APPLICATION DOMAIN: LOGISTICS, TRACKING, FLEET MANAGEMENT

CIIs can be exploited in the area of logistics, tracking, and fleet management in various and diverse ways [6–10]. Asset tracking is an already widely adopted application domain: location transmitters are installed on pallets, parcels, fleet, returnable

Threats in Critical Infrastructures

containers, trolleys, etc., allowing for real-time knowledge of asset location. Besides location, sensors can report data such as temperature, humidity, tilt, providing full information on the transport conditions and improving safety of goods (extensively used in food transport). IoT can also enhance the security in the logistics domain, by providing the means for monitoring of intrusion or theft indications and reliably transmitting these signals to appropriate control centers. Moreover, the status and load of warehouses can be effectively monitored resulting to their usage being optimized. Similarly, the status of traffic, information for possible congestions, and parking space availability can be used to optimize vehicle movement and transports. With respect to fleet management, (i) fleet operations can be optimized by streamlining logistics using real-time data and alerts to optimize delivery routes, monitor performance, and quickly respond to delays or issues as they happen, (ii) vehicle performance can be maintained through predicting and monitoring maintenance needs, driver status and behavior, fixing of potential issues, etc. Overall, clear visibility of all assets and the movement and status of goods at all stages of the logistics journey can be gained. In all cases, sensors are deployed on assets or information collection points, and data are collected from them and subsequently processed. Actuators may also be present and driven accordingly, to perform physical actions that affect the environment.

The introduction of IoT technology in the area of logistics, tracking, and fleet management has increased the volume and scope of data exchanged, the active services, and the dependence of business processes on IT infrastructure. All these aspects pose security risks, and relevant security goals and policies must be established, followed by threat mitigation actions.

3.2.1.1 Key Security Goals

According to [11], the main threat repercussions in the supply chain pertain to physical harm (attackers may damage equipment of facilities), data corruption (false data can be sent or sending of legitimate data can be blocked), and espionage (data can be snooped by adversaries). [12] reviews threats related to the application of commercial tags for RFID-Based IoT applications, many of which are directly applicable to the domain of logistics, tracking, and fleet management: these are classified under the generic categories of security risks, where adversaries can damage, block, or take advantage from a service in a malicious way, and privacy/confidentiality risks, where attackers may gain access to confidential data. Attacks may be performed at a physical level (including but not limited to physical destruction, removal, displacement, cloning), software level (remote switch off, command injection, attacks to web servers, etc.), and channel attacks (signal interception, relaying, replaying or amplification, jamming and so forth). Taking the above into account, it can be concluded that in the context of the application of IoT technology in the domain of logistics, tracking, and fleet management, operations reliability and continuity, resilience, and maintenance of data integrity and confidentiality are the key high-level security goals, and any threat that jeopardizes the aforementioned goals should be assessed and treated accordingly. It is worth noting that in the domain of logistics, tracking, and fleet management, given that IoT components are inherently geographically dispersed, the attack surface is greater as compared to other setups where components may be confined to physically secured locations; thus, attackers have more opportunities to exploit vulnerabilities.

3.2.2 Application Domain: Smart Meters

The applications of IoT in smart meters are an imperative need, since energy sources are limited, and one of the core ideas behind smart meters is to minimize operational loses. Smart meters record electric energy consumption in a specific time frame in order to display the results as well as to communicate, in real time, these results to the electricity provider. Thus, billing is becoming automated, the energy provider can estimate the needs in electricity more accurately in order to raise/lower the cost accordantly and reduce power outages and energy theft [13].

A typical smart-grid metering and control system consists of a collection of meters/sensors and controllers/actuators that communicate with a substation/data-concentrator, a consumer or technician, and various third-party entities. The communication among different network entities is realized by high-speed wired or wireless links or a combination thereof. A smart-grid metering and control system has a layered network structure through which it collects data and controls the delivery of electricity.

To keep safe, the smart-meter system and the sensitive components we consider some security goals [14–16]:

3.2.2.1 Key Security Goals

The information recorded and transmitted by smart meters is of high importance, since energy providers use this to (i) regulate energy production and energy flow within the distribution network and (ii) charge consumers for the energy they use. Energy consumers may also use these data for optimizing their energy use. It must also be noted that energy consumption data can be analyzed to infer patterns, which can disclose life schedules, personal habits, and events. Therefore, the following security dimensions are associated with smart meters:

1. *Safety*, i.e., system or the devices should operate without causing any risk to technological services, public services, humans, or even to the environment. In the context of smart meters, safety refers not only to the services but also to the smart meter infrastructure.
2. *Security*, i.e., the protection of the system from unintended or unauthorized access, change/disruption, or destruction (e.g., malware, remote attacks)
3. *Reliability*, i.e., the ability the smart meter to perform its required functions under stated conditions.
4. *Resilience*, i.e., the ability to withstand and operate as normal as possible while being under major disruption.
5. *Privacy*, i.e., who is the owner of the collected data. Smart meter collects data that could be analyzed in order to identify the time people are in home or not, among other things.
6. Accuracy, i.e., the term of accuracy referred to system's correct calculation of energy and at the accurate and efficient distribution of information.

7. Availability of resources at any given time. Both the energy provider as well as the consumer must have access to the respective information, e.g., billing information, control messages.
8. Integrity, i.e., the ability of the system to prevent any changes of the collected data as well as control commands.

Taking the above into account, we can conclude that in the context of smart meters, *protection from harm, protection of the environment, resilience, operations reliability and continuity and maintenance of data integrity and confidentiality/privacy* are the key high-level security goals, and any threat that jeopardizes the aforementioned goals should be assessed and treated accordingly.

3.2.3 APPLICATION DOMAIN: CAPILLARY NETWORKS

A Capillary Network is a local network that uses short-range radio access technologies to provide connectivity to a big number of devices. More specifically, by leveraging the key capabilities of cellular networks, ubiquity integrated security, network management and advanced backhaul connectivity, capillary networks can greatly enhance the IoT. It is important to note that the use of short-range links as compared to long-range links enables the reduction of the transmission power, thus improving energy efficiency and reducing interference. Short-range radio technologies provide efficient connectivity to devices within a specific local area. These capillary networks need to be connected to the edge of a communication infrastructure in order to reach service functions that are hosted on the internet or in a cloud server [17, 18]. Furthermore, today's vehicles are equipped with multiple communication capabilities. A car can communicate with the driver, with infrastructure regarding the highway system (vehicle-to-infrastructure (V2I)), with other vehicles (vehicle to vehicle (V2V)), and with cloud infrastructures (Vehicle-to-cloud (V2C)). It is important to note that extensive research is being carried out in the field of connected vehicles in order to encompass communication capabilities with pedestrians (Vehicle-to-Pedestrian (V2P)) and to everything (Vehicle-to-X (V2X)) [19, 20]. For an IoT ecosystem, capillary networks can provide local wireless sensor networks the ability to connect to and efficiently use the capabilities of cellular networks through gateways. As a result, a vast range of constrained devices equipped with only short-range radio can utilize the cellular network capabilities to gain global connectivity, supported with the security, management, and virtualization services of the cellular network [21].

3.2.3.1 Key Security Goals

The information transmitted through capillary networks can be of high importance, e.g., to regulate traffic in a road network so as to reduce delays and energy consumption or to avoid accidents, or in smart cities to leverage urban sensing and use these data to optimize city operation and promote user-centric services [22]. Furthermore, this information may include personal data, such as the position of a car (and its passengers). To keep safe, the capillary network and its sensitive components we take into account the following security goals:

1. Safety, i.e., into the capillary networks system providing management devices and expert individuals which they could handle the system and its data. The capillary networks should operate without the risk of breaking down.
2. Security, i.e., the devices deployed in capillary networks are likely to vary significantly in terms of computational resources, power consumption and energy source (e.g., sensors, connected vehicles). Thus, implementation of appropriate security measures is challenging, and it cannot follow a one-size-fits-all model.
3. Reliability, i.e., the ability of a system or component to perform its required functions under stated conditions for a specified period. Under an internet-integrated deployment, adversaries will be able to exploit security-related systematic failures reliably once those vulnerabilities have been discovered.
4. Privacy, i.e., the right of an individual or group to control or influence what information related to them may be collected, processed, and stored, by whom, and to whom that information may be disclosed. In capillary network, a risk probably is a network, device, connection intrusion into the system.
5. Accuracy, i.e., capillary network system must work accurate, without failures, since all devices are operating under the same network.
6. Network Management, i.e., range of tasks, such as ensuring automatic configuration and connectivity—for devices connected through a capillary network—are fulfilled by network management. In addition, network management needs to establish access control restrictions and data treatment rules for QoS based on SLAs, subscriptions, and security policies. In addition, a service provider should be able to use the management function to adapt service policies and add or remove devices. Considering the above it can be concluded that in the context of capillary networks, protection from harm, protection of the environment, resilience, operations reliability, and continuity and maintenance of data integrity and confidentiality/privacy are the key high-level security goals, and any threat that jeopardizes the aforementioned goals should be assessed and treated accordingly.

3.2.4 CRITICAL INFRASTRUCTURES INCIDENTS

The targets of the attacks were mostly related to the infrastructure that served economic, political, social, or military.

3.2.4.1 Revenge Attack from an Ex-Employee—2000

In 2000 in Australia's Maroochy county [23], an ex-employee who used to work as a supervisor of the control system of a water company illegally accessed the county's drainage control system as a revenge for firing him from the company. To control the 142 sewage pumping stations, an SCADA system was used, each station had a system competent in receiving commands from the master station and transmitting signals back. The communication was via a two-way radio system that operated over repeater stations. The attacker used a laptop that had the control software installed along with a radio modem, connected the laptop to each pumping station resulting in wastewater being spilled into parks and rivers and inoperability of four pumping stations.

3.2.4.2 Energy Companies Cyberattacks "Night Dragon" 2009–2010

Hackers based in China have attacked the systems of five multinational oil and gas companies in an attempt to steal confidential information. The attack was highlighted by McAfee's online security company, which gave it the "Night Dragon" name. McAfee did not release the names of the five companies, simply saying they were "well known" and announced that another seven had become targets. Night Dragon involved a set of attacks that started on 2009 directed at global energy, oil and petrochemical companies. The vice president of McAfee [24] threat research division said the attacks were not particularly sophisticated but achieved their goals, which illustrates how vulnerable critical infrastructure systems are.

The hackers infiltrated either through public websites or through infected emails that were sent to company executives. They designed and executed a various phase attack system. The first phase was infiltrating the extranet web servers via SQL injections, then hacker tools available online were uploaded to those servers which allowed them to gain access to organizations internal network. In the next phase, the attackers used password cracking tools to get the usernames and passwords. Using the compromised web servers, the attackers disabled Internet Explorer proxy settings which allowed direct communication of the infected machines to the Internet. The last phase involved the RAT malware to connect to other machines to obtain sensitive information. The timeline of "Night Dragon" attack is presented in Figure 3.1.

3.2.4.3 Stuxnet—2010

Stuxnet targeted industrial control systems (PLCs) and was the first computer virus designed to damage not only digitally but also to cause physical disruption. It jeopardized Iranian nuclear plants in 2010 [25]. It is believed that at least 14 industrial sites in Iran have been infected by the computer worm named "Stuxnet." The worm was created in 2008 in cooperation with the US–Israel (CIA–Mossad) and with the NSA in order to deprive Iran of the possibility of developing its nuclear program. Stuxnet, code-named "Olympic Games," seemed a less bloodthirsty practice to

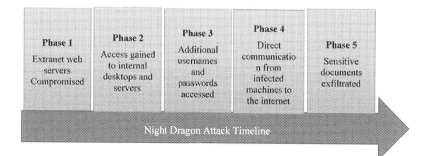

FIGURE 3.1 Timeline of Night Dragon Attack

achieve this. The main target of the Stuxnet virus was Iran's uranium enrichment plants—the virus was able to infiltrate the computers that control centrifuges at Natanz's nuclear plants, causing not only malfunctions but also material damage. The attack of "Stuxnet" was performed in three phases. It was spread via a USB and exploited unknown windows vulnerabilities. The first phase was attacking windows machines and networks, replicating itself on a continuous basis. On the second phase, it infected windows-based siemens step 7, a software that programs industrial control systems. On the last phase, it jeopardized programmable logic controllers.

3.2.4.4 German Steel Mill Attacks -2014

A German steel mill was targeted in 2014, where the attackers managed to take full control of the production software. The attack resulted in serious physical material damage to the site. It is rumored that the process used by the attackers was spear phishing in which they sent fraudulent emails that seemed to be coming from reliable sources; the email had an attachment that injected the malware to the sale software of the site then it moved through the network damaging a number of industrial automation components and systems. This attack is one of the most widely known attacks that caused significant physical damage to an industrial site.

3.2.4.5 Ukraine Cyberattacks and Blackout—2015 and 2016

In 2015, a massive cyberattack in Ukraine left almost a quarter of a million people without electricity. The incident has since been the clearest example of what hackers attacking critical infrastructure can do. When there were anomalies in the Ukrainian electricity grid again in December 2016, the first suspicion was another cyberattack. Research appears to have confirmed this suspicion, raising questions about what this means for Cybersecurity. While investigators believe the same team is responsible for both incidents, there are some significant differences between the cases of 2015 and 2016. Most importantly, the first incident took much longer, the latest attack took place at midnight of December 17 and lasted more than an hour. The attacks were of a different type: the 2016 attack was against a transport facility while the 2015 attack allegedly affected infrastructure facilities. The 2015 incident involved many different critical infrastructure components in a concerted action to cut off electricity, disable administrators' access to basic systems, and "shut down" telephone networks. The attacks targeted the regional distribution level as illustrated in Figure 3.2.

The attackers utilized a range of competences, including spear phishing emails, variants of the known malware BlackEnergy 3, and handling of the Office documents that enclosed the malware in order to gain access into the Information Technology networks of the electricity companies. They harvest credentials and information to gain access to the ICS network. Furthermore, the attackers illustrated competence not only in network infrastructure, such as Uninterruptable Power Supplies (UPSs), but also in operating the ICSs through supervisory control system, Human–Machine Interface (HMI). They implemented two SCADA hijacking techniques: one gnostic and one custom and operated them across different categories of SCADA/DMS at three companies [26]. They also targeted field devices and wrote firmware that was malicious which made the devices inoperable.

Threats in Critical Infrastructures

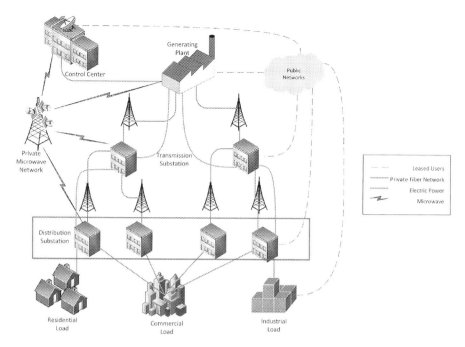

FIGURE 3.2 Overview of Electric System

3.2.4.6 WannaCry Ransomware-British Health Exposed-2017

An attack that was launched on May 12, 2017 had hit more than 75,000 hospitals and organization machines in more than 99 countries in its first day and had increased to over 130,000 in 104 countries by the next day [27]. The attack was accomplished via a ransomware named "WannaCry." Although most of the targets were in Russia and Asia in general, a significant proportion were in Europe, with the main targets being England's NHS system. Based on a weakness of Microsoft's Windows Server system, which has been "patched" by the company since March 2017, it was able to do such a damage because most of the machines that had been hit had not made the necessary security updates. The ransomware was spread by a group of hackers—the so-called Shadow Brokers—who found it in a (not so secret as it seems) US NSA (National Security Agency) server. The attack paralyzed the operations of 16 hospitals in the UK. The exact method that the malicious software had infiltrated the computers is not yet known, but it is believed that it was through a phishing email that contained the wann a cry, that later on spread to the rest of the computers in the network. After installing it, the files got encrypted and the user was asked to pay the ransom in exchange for regaining access to the files. In fact, there were detailed instructions and two accounts on which payment could be made. Unfortunately, the solution to end the attack, which came days after the British Marcus Hatchins came up with the kill switch, failed to prevent all affected users from succumbing to blackmail. Overall, it is estimated that the profits of the promoters of the attack exceeded £958,000.

3.2.4.7 Latin America Machete Attack—2019

A cyber-espionage against Venezuela institutions that stayed under the attack for years. The "Machete" is a malware that was discovered in 2010, documented by Kaspersky in 2014, that continues to expand throughout the years. The malware was designed by a Spanish speaking group in an attempt to steal sensitive information. The virus was spread through phishing and is still alive infecting equipment's of Latin American Government. As of 2019 75% of the computers that have been spied upon with the use of machete belonged to Venezuelan government [28].

3.2.4.8 Saudi Arabia Drone Attacks at Oil Processing Facilities—2019

In 2019, an attack was launched at two state-owned Saudi Aramco oil processing facilities at Saudi Arabia. Drones were used to attack the facilities causing them to be shut down for repairs, which led to cutting down Saudi Arabia's oil production by about half. It is estimated that these two facilities were responsible for 5% of global oil production and the obstruction of their normal operation caused turbulence to the global financial markets.

3.3 RELEVANT DEPLOYMENT ARCHITECTURES AND THREATS ON CIIS

3.3.1 Relevant Deployment Architectures

3.3.1.1 Fleet Management

[29] outlines the architecture of fleet management as illustrated in Figure 3.3. It can be observed that vehicle sensors and actuators (left side) typically communicate through wide-range wireless networks, while for sensors and actuators attached to stationary objects (right side), more communication options (including wired networking) are available. The information is collected into a command center, where it is stored and processed and—when necessary—commands are issued to actuators. In this context, cloud services can be used.

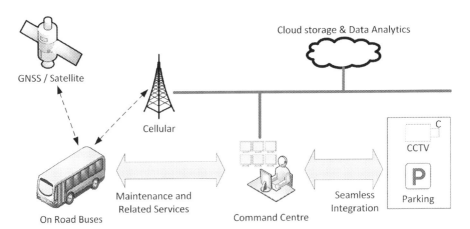

FIGURE 3.3 The fleet management architecture

Threats in Critical Infrastructures 111

Regarding logistics and tracking, sensors attached to merchandise and goods under transport will effectively employ wireless technologies to communicate, while warehouses fall into the category of stationary objects and—similarly to above—have more communication options. It should be noted however that IoT devices attached to goods (either directly or to palettes, containers, etc.) are typically close to vehicles that transport.

3.3.1.2 Smart Grid

Figure 3.4 depicts the generic architecture of a smart-grid system [30]. The main functionalities of each component in a smart-grid metering and control system are as follows.

Utility company: it connects to the substation network through the wide-area network (WAN) interface and the communication channel might be Wi-Fi, satellite, 4G-LTE, Wi-Max, etc. The utility company is responsible for processing alarms and alerts, managing the meter data, and generating bills. Moreover, it may also provide a web portal that allows customers to view their monthly energy consumption and bills.

- Substation/data-concentrator network: it consists of several smart meters in a certain area as well as a data collector. The connection between smart meters and the data collector might through Wi-Fi, ZigBee, power line carrier (PLC), etc. Typically, the smart meters form a wireless mesh network and forward the meter readings to the data collector through multi-hop communications. The data collector then transmits the accumulated data to the utility company.
- Home area network (HAN): it provides the consumer access points to control and monitor the real-time power consumption. The HAN contains a home gateway that receives the power-consumption data from the smart meter and displays it on householder's devices (e.g., laptop, tablet, smartphone). Furthermore, the home gateway may send the power consumption data to a third party for other value-added services (e.g., efficiency advice, supplier selection). The HAN also includes a controller that enables householders to remotely control the status of their home appliances.

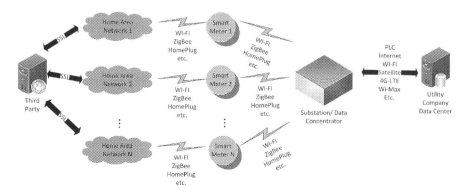

FIGURE 3.4 Architecture of a typical smart-grid system

- Smart meter: it is composed of a microcontroller, a metrology board, and a communication board. Under the control of the microcontroller, the metrology board measures the real-time power consumption, and the meter data is transmitted to both the substation network as well as the home area network through the communication board. The connection between the smart meter and home appliances may be through Wi-Fi, ZigBee, Ethernet, HomePlug, Wireless M-Bus, etc. The smart meter may also contain a disconnection function that (if enabled) allows utility companies or customers to remotely connect or disconnect the home appliances and services.
- Third party: it relies on accurate meter readings to provide value-added services for householders, including power efficiency advice, supplier selection, etc. Those services will help householders to manage their power usage in a cost-effective manner.

3.3.1.3 Capillary Network

The architecture presented in Figure 3.5 comprises three domains: the capillary connectivity domain, the cellular connectivity domain, and the data domain. The first two domains span the nodes that provide connectivity in the capillary network and in the cellular network, respectively. The data domain spans the nodes that provide data processing functionality for a desired service. These nodes are primarily the connected devices themselves, as they generate and use service data though an intermediate node, which, like a capillary gateway, would also be included in the data domain if it provides data processing functionality (for example, if it acts as a CoAP mirror server).

- Capillary Connectivity Domain: domains span the nodes that provide connectivity in the capillary network. When deploying a capillary network, a significant number of capillary gateways need to be installed to provide a satisfactory level of local connectivity.
- Cellular Connectivity Domain: domains span the nodes that provide connectivity in the cellular network.

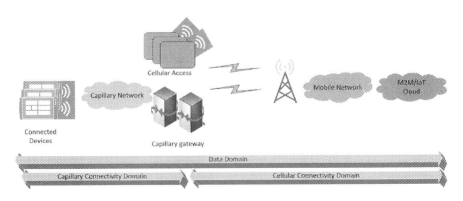

FIGURE 3.5 Capillary network

- Data Domain: domains span the nodes that provide data processing functionality for a desired service.
- Capillary gateway ideally, any service provider should be able to deploy a capillary network, including device and gateway configuration. For this to be possible, deployment needs to be simple and use basic rules—circumventing the need for in-depth network planning.
- Connected devices are primarily the nodes, as they generate and use service data though an intermediate node, which like a capillary gateway, would also be included in the data domain if it provides data processing functionality (for example, if it acts as a CoAP mirror server)
- Capillary network connected to the global communication infrastructure can be achieved through a cellular network, which can be a wide-area network or an indoor cellular solution.
- Cellular access is the selective restriction of access to system or capillary network resource

3.3.2 THREATS ON CIIs

3.3.2.1 Out of Date Systems
Critical Infrastructure may still use outdated not supported operating systems or outdated software applications that may also connect to PLC and SCADA systems.

3.3.2.2 Outdated Hardware
Outdated hardware which may range from outdated not responsive PCs, to custom-specific developed hardware.

3.3.2.3 Lack of Training and Awareness
It is well documented that many breaches are due to the lack of basic security precautions that arise from the fact that operators or employees of a critical facility do not even receive a basic training in Cybersecurity awareness. This lack of basic Cybersecurity culture is exploited by malicious users to pass to the unsuspected employees malicious USB devices that compromise the critical infrastructure. Employees also fell victims of phishing by either divulging their passwords or by executing malicious files send to them, leading to the well-known moto, "people are the weakest link."

3.3.2.4 Lack of Talent
Currently one of the biggest Cybersecurity challenges/threats is the lack of skilled security professionals. Without skilled Cybersecurity personnel, critical infrastructure organizations don't have the capability to develop and deploy the right processes and mechanisms to protect their assets from cyberattacks.

3.3.2.5 Social Engineering
Social Engineering can be considered as the art of gaining access to buildings, systems, and data by exploiting initially human psychology. A social engineer might call an employee and pose an IT support person, trying to trick the employee into

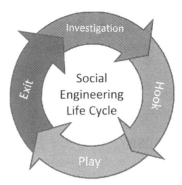

FIGURE 3.6 Social Engineering Attack Lifecycle

divulging his password, or launch a phishing campaign to unsuspecting victims after profiling them. If the social engineer succeeds into tricking a user to click on a malicious link or file, or accept a malicious USB, the critical's infrastructure firewall may not mean much. The social engineering attack lifecycle usually includes 4 phases, as presented in Figure 3.6.

These include investigation, hook, play, and exit. During the investigation phase, the social engineer prepares the ground for an attack and aims to build a successful hook. During this phase he gathers background information, identifies the victims, the plan to engage, and the levers. It also selects the attack methods. During the hook phase, he deceives the victim to gain a foothold. He typically engages the target, spins a story, builds a level of trust, and takes control of the interaction. During the play phase, he reaches the attack objectives by expanding the foothold/strengthens the control of relationship, executes the attack, extracts information, and meets attack objectives. During the last phase, exit, he closes the interaction, ideally without arousing suspicion, brings the charade to natural end, removes all traces of malware, and covers its tracks.

Social Engineering takes multiple forms,

- Phishing—Users of interest are tricked to click/download a malicious link.
- Spray and Pray—Users of interest are mass targeted.
- Spear phishing—A specific organization or individual is targeted.
- Whale phishing—Targets individuals such as CEOs.
- Vishing—Individuals are targeted over the phone.
- Smishing—Users of interest are targeted Over SMS.
- Pretexting—Presents oneself as someone else (e.g., IT).
- Tailgating (piggybacking)—Unauthorized person gains access to restricted area by following an authorized person.
- Water-holing—Hacker compromises a website in order to compromise a person who visits that website with malware.
- Dumpster diving—Collects information about a company and/or individual by going through the trash.
- Reverse social engineering—It's not a hacker who initiates a contact, but a potential victim (unknowingly) themselves.

- Baiting—Hacker drops malware-infected devices (e.g., a USB) hoping an employee will use it.
- Quid pro quo—Similar to baiting, quid pro quo involves a hacker requesting the exchange of critical data or login credentials in exchange for a service, prize or money.
- Scareware—To scare a person to install a malicious software to fix "a problem."
- Malvertizing—Spread malware through malicious online advertising on popular websites.

3.3.2.6 Network Attacks

In ethernet networked critical infrastructure industrial networks, the basic methodology of identify, enumerate, and penetrate still holds. To secure the such networks and industrial systems, the entry points, attack vectors, vulnerabilities, devices, and protocols must be understood. Security processes should ensure protection against the delivery of malicious payloads and protection of users, services, and hosts within the network from the payload.

Malware detection should also be present in any internal network communication, as well as in any software-based device, including users' devices. As a general remark, it should be pointed out that even in cases which a threat is contingent on a malicious software that could be possibly identified by a malware detection system (and, thus, such a system is in turn prerequisite to address these threats), the exploitability may become difficult but not impossible. There is no any malware detection technique/tool that suffices to identify in time any possible malware. In summary, considering the threat categories documented in Section 4.3.2, the analysis detailed in Table 3.1, regarding the effectiveness of antivirus/antimalware in limiting threat exploitability or technical impact is rated as follows:

3.3.2.6.1 Targeting a Traditional Enterprise Network

The attack process starts with the usual form of reconnaissance of a target in order to understand the critical infrastructure's security level. Information about company

TABLE 3.1
Overview of Antivirus/Antimalware Effectiveness for the Different Threat Categories

Threat category	Effectiveness in Limiting Exploitability	Effectiveness in Limiting Technical Impact
Network level threats	Low to Medium	Low
Cryptography related threats	Low	Low
Hardware/sensor level threats	Low to Medium	Low
Malware	High	Low
Threats for Smart grids	Low	Low
Technical/application development related threats	Low to Medium	Low
Threats necessitating actions by the victim user	Medium	Low
Generic/Miscellaneous	Medium	Low

employees, external and internal networks, internet presence, domains, and points of entry are within the typical information gathered. All information is gathered and evaluated since it could identify an entry point to the network, or it may be leveraged for social engineering purposes. Scanning is used to identify network devices, hosts, operating systems, services, etc. Enumeration is used to identify valid users, accounts, shared network resources, etc. Depending on the intentions of the attacker, the attack may take the form of disruption or infection and persistence. If the purpose of the attack is to disrupt an external process or a service, there is no need to penetrate the network since such service would be already exposed to the public network. If the service is not exposed externally then the attacker will need to penetrate the network in order to disrupt the service. If the attacker manages to access a system, the system can be infected through malware code which may include botnets, rootkits, or memory-resident malware in order for the attacker to open backdoors, escalate privileges, spread the infection, and establish command and control functionality. If the attacker's goal is to achieve persistence, then he will infect the network and would try to lie hidden, listen and wait in order to steal information and compromise the network, services, and infrastructure over time.

3.3.2.6.2 Targeting an Industrial Network of a Critical Infrastructure

Methods described in the previous section also apply to industrial networks found in critical infrastructures, since even critical infrastructures will utilize typical IT infrastructures. The difference with critical infrastructures is that these infrastructures will also utilize industrial control systems, constituting from specialized systems and protocols and thus a new attack surface is also presented to the attacker. When it comes to industrial systems there are reconnaissance, scanning and enumeration techniques which are specific to Supervisory Control and Data Acquisition (SCADA) and Distributed Control Systems (DAS).

When it comes to reconnaissance either for enterprise or industrial networks, a malicious user can carry out its own operations or can use online services like SHODAN. Subsequently, the malicious user can try to scan such exposed networks in order to find new or verify any information he has gathered. A scanning process can identify SCADA and DCS. Typical SCADA and DCS ports and services are presented in Table 3.2 [31].

SCADA systems would not be as progressed as they are today if large-scale networking and device (PLC, RTU, DCS) communication capabilities had not been developed.

- **Modbus:** Modbus can be considered as the grandfather of industrial networking protocols, but in today's connectivity and web services the simple request-response and unconnected message communication framework can be considered as bizarre and unreliable. Modbus in open standard allowing manufacturers to build it into their equipment making it the most preferable and widely used protocol in industrial automation and connection odd electronic devices in an industrial environment. Although it is a protocol that one can invest in, knowing that the overwhelming majority of manufacturers support it, it suffers from the limitations of the EIA-232 and EIA-422/485 serial

TABLE 3.2
Example of SCADA and DCS Ports and Services

Port	Service
102	ICCP
502	Modbus TCP
530	RPC
593	HTTP RPC
2222	Ethernet/IP
4840	OPC UA
4843	OPC UA over TLS/SSL
19999	DNP-Sec
20000	DNP3
34962–34964	Profinet
34980	EtherCAT
44818	EtherNet/IP

communications such as the limited speed of 0.115 Mbps unlike modern industrial networks of 16 Mbps and above and while interconnecting 20–30 devices is easy to do using EIA-422/485, connecting 500 devices or more can be a highly complex hierarchy of Master-Slave devices in Nested loops.

- **OPC**: OPC is essentially a standard that allows data exchanges between hardware and software data regardless of the manufacturing type. It is utilized in Industrial Environments and Automation. The OPC standard is based on the classic Client/Server architecture the OPC server collects data from the PLCs and transmits them to various on-demand applications (clients). Different types of PLCs can be connected to a server that provides data to surveillance and collection systems HMI, HISTORIAN, SCADA, etc). The implementation of Supervisory Control requires systems to have the OPC client in order to retrieve the data from OPC server. OPC servers NAPOPC and MX-AOPC UA represent Modbus OPC servers as they support Modbus RTU and Modbus TCP protocols. The most widely used OPCs ones are OPC DA (Data Access) servers, which have many features for real-time data exchange with PLCs and other devices, as well as modified OPC UA (Unified Architecture) servers that are cross-compatible.
- **EtherCAT:** EtherCAT is a fast, real-time master/slave network that is based on Ethernet. With EtherCAT, the Ethernet packet/frame is read, then translated and transmitted as data process at each node. EtherCAT slave devices read the data that is addressed to them while the "message" passes through them. Similarly, input data from the device is transmitted to the message. The frames are delayed less than 1µs at each node, so many nodes (typically the entire network) can be "queried" with each frame. The EtherCAT protocol is specifically designed for transporting process data and is transmitted directly within Ethernet IEEE 802.3 frame. Broadcast and multicast communication between slave devices is possible. The data exchange process of 1000 distributed digital I/O devices needs about 30µs. This protocol supports tree or star topologies and the maximum distance between two nodes is 100 m when 100BASE-TX is

used. It can connect up to 65536 devices making the size of the network almost unlimited. For integration with already existing fieldbus templates (e.g., DeviceNet, Profibus), EtherCAT incorporates special devices such as gateways. Also, other EtherNet based protocols can be used in combination with EtherCAT. Ethernet frames are "masked" through EtherCAT (a method known as tunneling), which is a common method for web applications (e.g., VPN, PPPoE (DSL), etc). The EtherCAT network is fully "transparent" to the Ethernet device, without damaging the real-time characteristics of the network, i.e., the response to real time. So all web applications can be used in the EtherCAT environment.

- **EtherNet/IP**: EthernetIP is an Ethernet-based protocol designed for industrial applications automation. At the application level EthernetIP is based on the CIP protocol used by DeviceNet. EthenetIP is a protocol that can be implemented on the industrial production sector and even on the network of business administration offices. Its design offers safety and performance features in real-time applications in accordance with standards CIPSafe and CIPsync. Ethernet IP has a growing installed base due to the simplicity of implementation using simple Ethernet, but also because of its compatibility with previous CIP standardizations. The Ethernet IP is mainly used in the United States.

- **Profinet**: Profinet is an industrial Ethernet standard that has an on-site fieldbus communications subnetwork as well as a subnetwork for office network communication. The profinet can simultaneously manage common ethernet packets and real-time packets within a maximum processing time of one millisecond. The communication of Profinet is scaled in three levels, the device level, which uses TCP/IP and achieves cyclic times of 100 ms. This level is usually selected in the communication between controllers. The second level is the real-time level, which operates at 10 ms cycle times and is mainly used to communicate I/O I/O units. The last level is the isochronous level, which achieves communication times of less than 1 ms. All three levels can coexist on a common channel together with a common Ethernet information network. PROFINET is lacking on installation compared to its counterparts, Modbus TCP/IP and Ethernet IP, given its shorter market life, but the nature of the network is expected to expand as more devices are available from different manufacturers.

Once an industrial system has been identified, one can use the inherent functions of the industrial network protocols, i.e., sniffing Ethernet/IP traffic to obtain Critical Infrastructure Protection (CIP) identifiers, sweep Distributed Network Protocol 3 (DNP3) requests to discover slave addresses, capture EtherCAT frames or SERCOS Master Data Telegrams, to obtain slave devices and time synchronization information, etc. When it comes to enumeration, many industrial systems are windows based and users can be enumerated through the standard techniques. Within control systems, useful authentication information includes HMI users, Inter-Control Center Communications Protocol (ICCP) server credentials, Master node addresses, and database authentication. HMI access can lead to process information theft, while ICCP server credentials would allow a malicious user to spoof the ICCP server,

Threats in Critical Infrastructures 119

leading to the theft or manipulation of information communicated between control centers. Figure 3.7 presents some of the possible entry points to a critical infrastructure's network. Possible entry points exist from public access (internet), WLAN access to the Business network, access from other facilities interconnecting to the supervisory network, access through links connecting field devices to the control system, etc. A malicious user will exploit either missing, misconfigured, or weak firewall rules.

3.3.2.6.3 Protecting an Industrial Network of a Critical Infrastructure

To increase security, industrial networks should be contained and isolated from public access. In case this is not feasible then secure zones should be established; thus, a malicious user will have to penetrate multiple security layers in order to reach the control system of the critical infrastructure and compromise its operation. Also, in order to avoid malicious enumeration, a combination of account verification methods should be employed. To harden the security of the critical infrastructure of Figure 3.7 several steps should be taken. These steps include properly controlling and monitoring inbound and outbound traffic, disabling all unnecessary ports and services to reduce the attack surface, enforce strong authentication and access control policies, minimize backdoor access, control the use of removable media, remote access and establish security awareness and policies. Firewall rules should be carefully

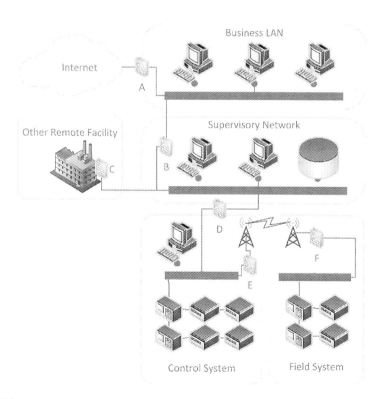

FIGURE 3.7 Entry points into critical infrastructures

configured to allow only legitimate communications. For example, many firewalls include only inbound traffic policies and allow any outbound traffic freely without any control, ignoring that a malicious user may be already sitting in the network either because he is a malicious insider or he has already infected the network through social engineering or other techniques.

3.3.2.7 COTS Drones

Drones can also be used to either target specific critical infrastructures or can be compromised during their usage by Critical Infrastructures. The availability of low-cost drones which can be equipped with a plethora of payloads, either electronic, explosive, or surveillance in nature, can turn the usage of such platforms in asymmetric threats for any organization or critical infrastructure.

3.3.2.7.1 Threats Arising from Malicious Drones

These types of threats arise by malicious drones operated on purpose by malicious operators. Drones may be equipped with cameras for malicious surveillance, malicious wireless systems for communications interception or remote wireless system hacking purposes, explosives with detonation devices, etc.

3.3.2.7.2 Threats Arising from Compromised Drones

These types of threats arise after a legitimate drone is compromised by a malicious individual. The legitimate drone may belong to an operator serving a critical infrastructure or other legitimate operator. By compromising a legitimate drone, a malicious user can access information, video footage or other information, which can be sensitive to a critical infrastructure.

3.3.2.7.3 Typical Components of a COTS Drone

Typical components of a drone include the propellers, the brushless motors, the landing gear, the electronic speed controllers (ESC), the flight controller, the receiver, the transmitter, the GNSS/GPS module, the battery, the antennas, the gimble, and the camera. From these components, the following components are of special interest since they can be either compromised, tampered with, or are related to a critical function of the drone operation that is of interest to malicious users or parties.

3.3.2.7.4 Electronic Speed Controller (ESC)

The Electronic Speed Controller is responsible for varying the speed of an electric motor; in this occasion, it controls the speed, and breaks, of the drone's motors, which represent the "engines" of the drone. Since the ESC controls multiple motors on a drone, it is also responsible for the direction of flight. Commercial-Of-The-Shelf (COTS) ESCs included with drones or sold as standalone devices are usually pre-programmed but can be customized and programmed on purpose through a vendor-specific Software Development Kit (SDK). This ability means that an ESC can be maliciously re-programmed if it is compromised by a malicious user.

3.3.2.7.5 The Flight Controller (FC)

A drone flight controller is the heart of a drone and controls and receives input from most onboard drone components. It acts like the motherboard of the drone. It receives input from the drone receiver, the GNSS/GPS module; it monitors the battery levels, any other onboard sensors, any inertial measurement unit (IMU), etc. It is responsible for regulating the motor speeds through the ESC, effectively steers the drone, and controls the autopilot mode. Similarly, if the drone is equipped with a camera and or a gimbal, is responsible for triggering and controlling the camera and any other related modules. Like the ESC, it is usually preprogrammed but can be customized and programmed on purpose through a vendor-specific Software Development Kit (SDK). This ability means that FC can be maliciously re-programmed by a malicious user.

3.3.2.7.6 The Receiver (Rx)

The receiver module is the module that is responsible for receiving drone commands that are sent through radio signals from the transmitter module. Tampering with the receiver by jamming its reception capabilities can compromise the operation and steering of a drone. Such an action can be carried out either by a malicious individual on the operation of a legitimate drone, or it can be carried out by a perimeter protection system in order to neutralize a malicious drone. Jamming principles, either for or friendly, will be further discussed in the drone neutralization section. Deepening on the COTS drone system used, receiver modules may support reception of 802.11 protocols and frequencies, or remote control (RC) signals at other ISM bands.

3.3.2.7.7 The Transmitter (Tx)

The transmitter module is the module that is responsible for sending the drone commands through radio signals to the receiver module installed on the drone. The transmitted signal should be encoded accordingly in order to differentiate itself from other transmitter modules operating on the same channel frequency. Deepening on the COTS drone system used, the transmitter modules may support 802.11 protocols and frequencies, or remote control (RC) signals at other ISM bands.

3.3.2.7.8 The GNSS Module

The Global Navigation Satellite System (GNSS) module is responsible for providing accurate location information to the drone system. The GNSS module could be GPS (US), Gallileo (EU), GLONASS (Russia), or Beidu (China) based. Typically, GPS is still used as the default provider. The GNSS module can be used to provide location but also time and return to home autopilot functionality. Depending on the system complexity of a drone system, tampering with the reception of GNSS signals can compromise its operation. GNSS/GPS Jamming principles will be further discussed in the drone neutralization section.

3.3.2.7.9 The Camera

Compact drone cameras are typically capable of real-time streaming and may have on-bard storage. Cameras can be integrated into the drone, along with a gimbal system, or could be externally attachable to a gimbal system. Drone cameras typically use 802.11 technologies for real-time video streaming and tend to prefer the 802.11 5 GHz band.

3.3.2.7.10 COTS Drone Detection

Detection of COTS drones can be attempted through several sensors. Such sensors include radar, acoustic, electro-optical, Radio Frequency (RF), and wireless protocol sensors. A brief explanation of these types of sensors is provided below.

- *Radar*: A radar sensor uses standard radar transmission techniques in order to detect the position and velocity of a drone. Typically, a Frequency Modulated Continuous Wave (FMCW) is used for this purpose. FMCW radars continuously emit periodic pulses whose frequency content varies with time. The range to the target is found by detecting the frequency difference between the received and emitted radar signals. Typical frequency bands used for this purpose include the X and K frequency bands.
- *Acoustic*: Acoustic sensors can be used to detect classify and locate drones by typically using spatially distributed arrays of sound pressure microphones. By using an array of spatially distributed microphones, one can calculate the direction of arrival of the sound wave based on the phase/time differences between the microphones at the different locations. Acoustic sensors are used to detect drones even if drones lack radio frequency links/transmissions, or their Radar Cross Section (RCS) is such that it's not detectable by radar sensors.
- *Electro-optical (EO)*: An electro-optical drone detection system is used to detect and identify drone threats based on the drone's visual signature. Typically, the EO system is supported with electro-optical components that can support day and night (EO/IR) observations for detection, identification, and classification of targets.
- *Thermal Imaging*: This type of sensor is used to detect the presence of drones based on their heat signature.
- *Radio Frequency (RF)*: This type of sensors detects the presence of drones, when drones transmit radio control signals or wireless video transmissions. These RF sensors continuously scan the frequency spectrum and may or may not use directional antenna arrays to detect the presence of drones.
- *Wireless Protocol*: This is a form of radiofrequency sensing with signal demodulation/decoding capabilities where the wireless protocol sensor can detect and decode a wireless transmission. An example of such wireless protocol sensor is an 802.11 sniffer where the wireless sniffer can detect and recognize an 802.11 RC or real-time video transmission from a drone. At the same time, such detection can be used to identify the MAC address of the transmission, if not spoofed, in an attempt to recognize the type of drone used.

3.3.2.7.11 Drone Neutralization

Multiple techniques exist to neutralize malicious drones. These include the usage of net guns, firearms, direct energy Weapons/high-energy lasers, signal disruptors, even trained predatory birds like eagles. The following sections will concentrate on how to

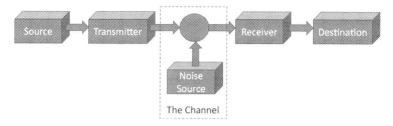

FIGURE 3.8 Basic wireless communication system

neutralize relatively low-cost COTS drones by disrupting the reception of their remote-control signals and any GNSS location information provided by their onboard GNSS receivers.

To understand how signal disruption through RF Jamming [32] works, one needs to understand the reception and demodulation processes in modern wireless communication systems. Figure 3.8 presents the basic components of a communications system.

The figure above presents the classic architecture of a generic communication system, originally described by Claude Shannon. A Source (computer data, voice, etc.) attempts to send information to a destination. Information is modulated and sent through a communication channel (wired or wireless) utilizing some form of a transmitter. Transmitted signal is modified by the channel (Noise source). The receiver should be able to overcome these modifications or impairments in order to translate the received signal back to the original data. In the wireless channel, noise sources can be subdivided into multiplicative, which refers to path loss, shadowing and fast fading, and additive, which refers to thermal, shot, atmospheric noise, and interference. The latter could be intentional or unintentional. To estimate the wireless range of a system, one must estimate the maximum distance at which the system's receiver, e.g., a drone receiver can successfully receive and demodulate the received signal. To achieve radio coverage over a certain range, one should transmit the appropriate power to overcome environment losses, use high sensitivity receivers and if the application permits it, use directional antennas. It must be noted that in the case of transceivers, both sides of the link should be able to successfully receive and decode the received signal. When trying to estimate the range of a wireless system, a wireless system designer estimates the path loss of a wireless link, which is the ratio of the transmitted to the received power, usually expressed in decibels (dB). The path loss includes all the possible elements of loss associated with interactions between the propagating wave and any objects between the transmit and receive antennas (reflected, refracted contributions, etc.) and refers to the power averaged over several signal fading cycles. To correctly define the path loss, the losses and gains in the communication system must be considered through a link budget.

Elements of a simple wireless link are presented in Figure 3.9.

The propagation (Path) loss L can then be expressed as

$$L = \frac{P_T G_T G_R}{P_R L_T L_R}$$

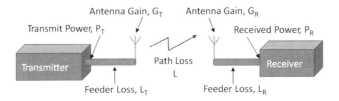

FIGURE 3.9 Elements of a basic wireless link

where
P_T: Transmit Power at the transmitter/power amplifier output,
P_R: Received Power at the receiver input,
G_T: Gain of the Transmitter antenna,
G_R: Gain of the Receiver antenna,
L_T: Feeder loss of the Transmitter, and
L_R: Feeder loss of the Receiver.

As it was already mentioned, wireless systems should be using sensitive receivers. To calculate the receiver sensitivity, one needs to calculate the impact of noise on a system This determines the system performance. Major noise contributions come from the receiver structure itself and/or external noise contributions are also possible.

The minimum detectable signal (MDS) at a receiver can be estimated by first considering the numerical value of the noise power N_{out}, which can be expressed as

$$N_{out} = F_{dB} - 204 + 10 log BW \ (dBW)$$

or

$$N_{out} = F_{dB} - 174 + 10 log BW \ (dBm)$$

where
F_{dB}: is the noise figure of the receiver in dB
BW: is the channel bandwidth in Hz

Practically, for successfully demodulating a received signal, the received signal level should be above a certain Signal-to-Noise to Interference Signal (SNIR) threshold, yielding that

$$\text{Successfully demodulated signal} = \text{MDS} + \text{SNIR}$$

Figure 3.10 defines receiver sensitivity levels of various 802.11 schemes, as these are defined in the relevant standards. Figures 3.11 and 3.12 present the constellation for a QAM16 signal under no RF jamming conditions and under RF Jamming conditions.

To be able to sustain a specific modulation scheme, as the one the received signal levels and SNIR requirements must be met. In simple words, RF jamming techniques

Threats in Critical Infrastructures 125

Modulation	Coding Rate (R)	Sensitivity (dBm) 802.11n, 20 MHz	Sensitivity (dBm) 802.11n, 40 MHz	Sensitivity (dBm) 802.11ac, 20MHz	Sensitivity (dBm) 802.11ac, 40MHz	Sensitivity (dBm) 802.11ac, 80MHz
BPSK	1/2	-82	-79	-82	-79	-76
BPSK	3/4	--	--	--	--	--
QPSK	1/2	-79	-76	-79	-76	-73
QPSK	3/4	-77	-74	-77	-74	-71
16-QAM	1/2	-74	-71	-74	-71	-68
16-QAM	3/4	-70	-67	-70	-67	-64
64-QAM	2/3	-66	-63	-66	-63	-60
64-QAM	3/4	-65	-62	-65	-62	-59
64-QAM	5/6	-64	-61	-64	-61	-58
256-QAM	3/4	--	--	-59	-56	-53
256-QAM	5/6	--	--	-57	-54	-51

FIGURE 3.10 Receiver sensitivity levels for 802.11n/ac schemes

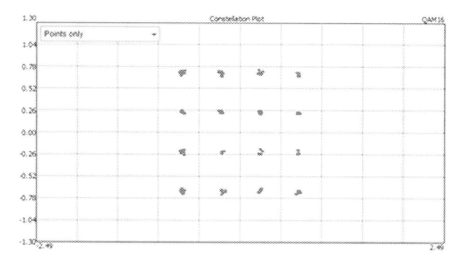

FIGURE 3.11 QAM16—No Jamming

maliciously interact with a receiver in order "break" SNIR thresholds, so the signal cannot be successfully demodulated at any given time. Figure 3.13 presents the 2.4-GHz RF Jamming signal that was injected to break the QAM16 constellation.

Having presented the basic elements of a wireless link and important parameters like transmit power, antenna gains, feeder losses, MDS, and SNIR, one needs to carefully consider how to accurately estimate the path loss parameter L. Correct estimation of L, between the jammer system and the target system to be jammed is critical. In the case of remotely controlled COTS drones, it is also useful to have an idea of the expected remote-control system transmit power from the malicious drone operator. A jammer system will need to overcome such transmissions.

To estimate correctly path loss L, one must consider the radio propagation environment where the jammer system is to be deployed. A critical infrastructure may be

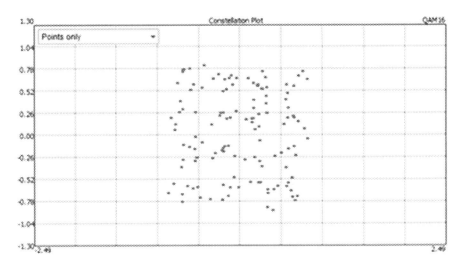

FIGURE 3.12 QAM16—Under RF

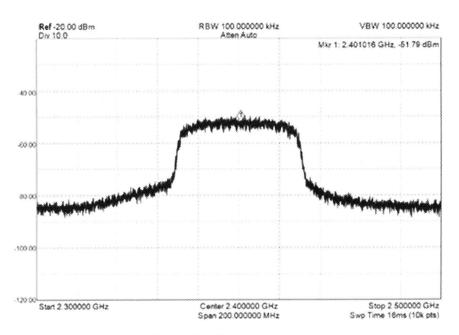

FIGURE 3.13 RF Jamming Signal at 2.4 GHz

situated in an area where there are no surrounding obstructions, building or other, in the close vicinity with the critical infrastructure. In this case, the path loss L between the jammer system and the malicious drone for any LOS distance can be easily estimated through the Free Space Loss (FSL) or the Plane Earth Loss radio propagation models. The Friis transmission formula defines the Free Space propagation Loss (FSL) between two antennas under certain conditions.

Threats in Critical Infrastructures

$$\frac{P_R}{P_T} = G_a G_b \left(\frac{\lambda}{4\pi r}\right)^2$$

where

P_T: Transmit Power,
P_R: Received Power,
G_a: Gain of antenna α,
G_b: Gain of antenna b,
λ: wavelength, and
r: distance.

In practical free space conditions, the received power may be less than what is predicted above, if the polarization states of the receive and transmit antennas do not match or if there is a mismatch between source or load impedance and antennas. In decibel form

$$L_{dB} = 10\log\left(\frac{P_T}{P_R}\right) = 32.5 + 20\log R_{km} + 20\log F_{MHz}$$

Another fundamental propagation situation is the two ray model or Plane Earth Loss model. In this case, the modeling process considers a direct unobstructed ray between a transmitting and receiving point and an unobstructed reflected ray between the transmitting and receiving point. These two rays are considered to sum up vectorially at the receiver point as demonstrated in Figure 3.14.

In this occasion and with reference to Figure 3.14, the Plane Earth Loss model can be estimated as

$$\frac{P_R}{P_T} = 2\left(\frac{\lambda}{4\pi r}\right)^2 \left[1 - \cos\left(k\frac{2h_1 h_2}{r}\right)\right]$$

Other radio propagation conditions of interest, w.r.t. jammer operation, include radio propagation over terrain. In such case one can use radio propagation models like the Longley Rice Model or parabolic equation model, or other appropriate models found in the literature, that predict radio propagation over terrain. Figure 3.15

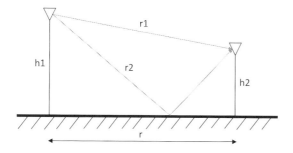

FIGURE 3.14 Plane Earth Loss Model geometry

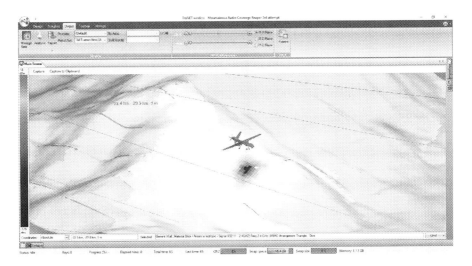

FIGURE 3.15 RF jamming from a UAV system

presents such radio propagation predictions over terrain, estimated by TruNET Wireless and refers to the RF jamming provided by a UAV system.

Another radio propagation environment of interest which is much more complex refers to scenarios where the critical infrastructure is situated in urban environments where the critical infrastructure will have to its surrounding multiple buildings which create short-, medium-, and long-range Non-Line of Sight (NLOS) conditions and zones and as such, simplified radio propagation prediction models cannot estimate path loss for such zones. This fact may create pathways within the urban environment where a malicious operator can operate a drone system to the vicinity of the critical infrastructure. In this occasion, ray tracing techniques and algorithms can be used to predict such danger zones and accordingly deploy multiple jamming systems at the right locations. Ray-tracing techniques are deterministic radio propagation techniques that take into consideration the environment clutter and its electrical parameters and can estimate with high accuracy the radio propagation conditions. Typically, practical ray tracing algorithms consider the reflected, refracted, and diffracted contributions which are generated in an urban environment and sum up such contributions arising at points of interest. Such points are typically the points where a wireless system, or on this occasion a drone system will exist at any single point in time.

Figure 3.16 presents such analysis where again TruNET wireless has been used to identify such weak spots, arising from the deployment of a single RF jamming system that is deployed to protect a critical infrastructure. This type of analysis can be used to answer the following questions: How many RF jamming systems should be deployed to cover the critical infrastructure grounds, where such RF jamming systems should be deployed around the critical infrastructure, how much power should be transmitted from the RF jamming systems, what kind of antennas should be used by the systems, etc. Proper analysis of the radio propagation conditions around the critical infrastructure can lead to the optimum deployment of an RF jamming system, thus efficiently protecting the critical infrastructure from COTS drones.

Threats in Critical Infrastructures   129

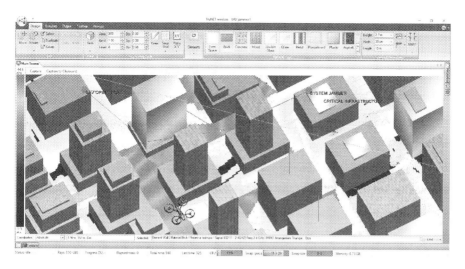

FIGURE 3.16 Analysis of RF Jamming System against Drone in an Urban Environment

3.3.3 Attacks on Smart Grid CIIs

3.3.3.1 Attacks on the Infrastructure Grid

Attackers seize control of energy infrastructure, which will provide them with the control necessary to execute attacks such as shutting down energy generation or exploiting network resources for botnets. The domains that are affected by such attacks are web applications, web services, IoT devices, and CIIs. The threat agents are Organized Crime and Criminals, Individuals, national and Intentional, Corporations, and Employees. The attack vector is using multiple methods with the attack exploitability being rated as difficult, and it is hard to detect. The prevalence is very common and the technical Impact of those type of attacks is disclosure of data to unauthorized parties, launching additional attacks, and denial of service. Due to the nature of these attacks, the technical impact is rated as severe, not easily observable attacks with system loss and data loss. These attacks can be mitigated by Firewalls, IDS, IPS, and fully secure communications stacks covering SCADA systems [33].

The ENISA taxonomy branches of attacks on the infrastructure grid are: Interception of information, Network Reconnaissance, Network traffic manipulation and information gathering, Identity Theft, Receiving unsolicited E-mail, Malicious code/software/activity, Manipulation of hardware and software, Unauthorized activities, Unauthorized installation of software, Compromising confidential information, Remote activity, Targeted attacks

The security dimensions that are affected are as follows: Spoofing of identity, Tampering with Data, Repudiation, Information Disclosure, Denial of Service, Elevation of Privilege.

According to Symantec, wave of attacks that started in December 2015 have hit energy companies in Turkey, USA, and Switzerland. The attacks started with hoax party invitations that were boobytrapped emails [27].

3.3.3.2 Attacks to Critical Infrastructure Components to Acquire Private Sensitive Information

Attackers exploit vulnerabilities in critical infrastructure components to steal data that affecting domains such as smart grid systems and IoT devices. As with the attack mentioned in section 4.3.3.1 the threat agents are Organized Crime and Criminals, Individuals, Humans—Intentional, Humans—Unintentional, Corporations, and Employees. The vectors of attacks are sniffers, hardware, and software vulnerabilities. The rating of the attack exploitability is average with a common prevalence and difficult detectability. The technical impact of these attacks is disclosure of information to unauthorized parties and is rated as severe.

The ENISA taxonomy branches of these attacks are, Interception of information and Compromising confidential information (data breaches)

These attacks can be mitigated with Device encryption, SSL/TLS protocols, user authentication, privilege escalation, re-authentication for sensitive devices, IDS [34, 35].

The attack can be observed by the reduction in the network performance

3.3.3.3 Compromise of Interactions of AMI Components with the Infrastructure

Interactions of AMI components with the environment could lead to unauthorized access to AMI communication information, modification of AMI data, denial of service to authorized users, and nonrepudiation. These attacks are considered uncommon in terms of prevalence, they are hard to detect and have a technical impact rated as severe. They can cause **data** corruption or loss, denial of service, complete system takeover, launching of additional attacks, disclosure of information to unauthorized parties

The ENISA taxonomy branches are Man in the middle, session hijacking, Denial of service, and Remote activity (execution). The Security dimensions affected by these types of attacks are Spoofing of identity, Tampering with Data, Repudiation, Information Disclosure, Denial of Service, and elevation of Privilege.

The mitigation involves tools that regulate the flow of channel, implementation of integrity checks on devices, limitation on network access to devices, and limitation on private loss [34, 35]. These attacks are observed from the abnormal network behavior, abnormal system behavior, and electricity consumption patterns

3.3.3.4 Hijacking Connections Between Meters and Demand Response Systems

Attackers conduct a man-in-the-middle attack to steal demand data, potentially influencing forecasting technologies. The attack vectors are malicious nodes and are considered a common prevalence attack that is difficult to detect. The technical impacts rating of such attacks is severe and it causes data corruption or loss and disclosure of information to unauthorized parties

Since this threat is based on Man-in-the-middle attack, the mitigation techniques include use firewalls, authentication of communication peers, and utilizing machine learning to establish baseline traffic patterns and detect anomalies [34, 35].

3.3.3.5 Altering of Meter Data When Transmitted Between Meter and Gateway, Gateway and Consumer or Gateway and External Entities

Attackers change these data to alter billing-relevant information or grid status information, the attack may be performed via any interface. This attack is mostly performed by Employees and the attack vector is interfering devices and software bugs. The technical impact is rated as moderate and it can cause data corruption or loss, but other impacts may exist depending on how data are processed. The Security dimensions affected are Spoofing of identity and Tampering with data. Mitigation consists of using network segregation and tamper-proof devices and cryptographic techniques [34, 36]. This type of attack can be observed when the System doesn't function normally, when known malicious payloads exists, and in case of an unusual system calls/network traffic patterns

3.3.3.6 Attacker Alters Meter Data, Gateway Configuration Data, Meter Configuration Data, CLS Configuration Data or a Firmware Update in the WAN

Attackers change these data to alter billing-relevant information or grid status information, disrupt system behavior or take over the system. The technical impact is rated as Severe and it can cause data corruption, or disclosure, loss, denial of access, complete system takes over. It is observed by unusual consumption patterns.

The mitigation is through the use network segregation, tamper-proof devices, and cryptographic techniques. Along with protecting the software update mechanism, user-friendly interfaces for device and services security management and relying on existing sources for security good practices in order to secure the infrastructure. Also providing secure backup and/or deletion of the data stored/processed by the device and associated cloud services [34, 36].

3.3.3.7 Compromise of an Existing Data Concentrator

Attacker compromises an existing data concentrator and causes sending wrong data to the central system. An attack that can affect any network-based application with the attack vector being connections of Gateway and meter data. The technical impact is data corruption or loss while other impacts may exist depending on how data is processed. The mitigation techniques include the use network segregation, use tamper-proof devices and cryptographic techniques, and limit the exposure of concentrators.

The Government of India-Ministry of Power (MoP) has announced National Smart Grid Mission (NSGM) in order to reduce AT&C losses by measuring and controlling each incoming and outgoing energy unit and each incoming and outgoing rupee.

[34, 36] These attacks are hard to observe, but the behavior includes anomalous integration of different smart meters to their data concentrators and Head End Systems (HES), especially in remote meter communication to HES.

3.3.3.8 False Data is Injected by an Attacker in the Smart Grid Traffic

False data are injected by an attacker in the smart grid traffic. The attacker injects false or malicious DR events in DRAS (Demand Response Automation Server), causing blackouts and instability of the grid. Due to the nature of this attack, the attack vectors are false packets.

The Technical impact is data corruption, or disclosure, loss, denial of service, and complete system take-over. Mitigation through the use identity management and false data detection techniques [34, 36]. It can be observed by abnormal traffic, unusual behavior of a network/IoT device, and when data do not observe physical laws

3.3.3.9 Injection of Realistic False Data

The adversary may inject false measurement reports to the disrupt the smart grid operation through the compromised meters and sensors. The Technical impacts are data corruption or loss, but other impacts may exist depending on how data is processed

The Mitigation is with Use of tamper-proof meters or reporting sensors; use false data detection techniques [34]. It can be observed by the measurements deviating from historical patterns

3.3.3.10 Load Redistribution Attack (Injection of Realistic False Data)

Load redistribution attack is an injection of realistic false data with limited access to specific measurement data. Attacker compromises an existing data concentrator and causes sending wrong data to the central system. It can affect any network-based application. The Technical impact of this attack is data corruption or loss while other impacts may exist, depending on how data are processed, and the attack vector is operator station. While it is hard to observe, be cautious of abnormal increase/reduce of load in buses. Mitigation includes the use network segregation, use tamper-proof devices, and cryptographic techniques [34, 36].

3.3.3.11 Monitoring Data of Other Customers

One of the most common attacks with easy exploitability is when Compromised data concentrators can be misused to monitor data of other customers. The attack vectors of this attack are information capturing devices and the technical impact is data disclosure to unauthorized parties. It can be mitigated with segregation and use cryptographic techniques [34, 37]. It can be observed through logging attempts and abnormal traffic

3.3.3.12 Time Modification of the Gateway

Attackers aim at changing the relation between date/time measured consumption or production values in the meter data records. The attack vectors are any attack changing the time with easy exploitability rating and a common prevalence. The technical impact of such an attack includes data corruption or loss. Mitigation includes segregation of Networks and the AMI, vulnerability surveys, protection of the software update mechanism, user-friendly interfaces for device and services security management, relying on existing sources for security good practices in order to secure infrastructures and provide secure backup and/or deletion of the data stored/processed by the device and associated cloud services

3.4 CONCLUSIONS

The significance of protecting critical infrastructures cannot be stressed enough, since their normal operation is vital to the welfare of a nation and its citizens. The complexity of today's cyber threats must be properly addressed in order to be able to defend them. Numerous real examples presented in this chapter for different domains, highlight that the danger is real, and it cannot be ignored. Cyber threats that must be addressed may range from your typical threats that any IT infrastructure must defend from, including social engineering attacks, which take multiple forms, malicious scanning activities, etc. With critical infrastructures, the attack surface does not stop there since specialized software and hardware like PLCs and SCADA are also in use, and these devices have their own attack vectors. Furthermore, depending on a critical infrastructure's mission, malicious drones can be a modern cyber threat, due to the capabilities they offer to a malicious individual, and they cannot be ignored. Finally, one cannot stress enough the importance of awareness and training activities that should be employed by critical infrastructure organizations in order to cultivate a security culture to all the personnel of their facilities and organizations. That firewall and all the specialized Cybersecurity systems and defenses may not mean much if they are not properly configured and a user decides to plug an infected USB device to a computer sitting in the organization's network.

REFERENCES

1. Department of Homeland Security, "Critical infrastructure sectors," 2019. [Online]. Available: https://www.dhs.gov/cisa/critical-infrastructure-sectors. [Accessed: Dec. 27, 2019].
2. European Commission, *EECSP Report: Cybersecurity in the Energy Sector: Recommendations for the European Commission on a European Strategic Framework and Potential Future Legislative Acts for the Energy Sector*, 1st ed. European Commission, 2017, pp. 5–7. [Online]. Available: https://ec.europa.eu/energy/sites/ener/files/documents/eecsp_report_final.pdf. [Accessed: Dec. 27, 2020].
3. N. Desai, "IoT in agriculture: Farming gets 'smart'," 2018. [Online]. Available: https://www.networkworld.com/article/3268971/internet-of-things/iot-in-agriculture-farming-gets-smart.html. [Accessed: Jul. 31, 2018].
4. Inductive Automation, "What is IIoT? (The industrial Internet of Things)," 2018. [Online]. Available: https://inductiveautomation.com/what-is-iiot. [Accessed: Jul. 29, 2018].
5. ENISA, *Methodologies for the Identification of Critical Information Infrastructure Assets and Services*, ENISA, 2016. [Online]. Available: https://www.enisa.europa.eu/publications/methodologies-for-the-identification-of-ciis. [Accessed: Dec. 27, 2020].
6. Deloitte, "Harnessing the power of Internet of Things to transform Industry in India," *Deloitte*, 2018. [Online]. Available: https://www2.deloitte.com/content/dam/Deloitte/in/Documents/manufacturing/in-mfg-harnessing-the-power-noexp.pdf. [Accessed: Aug. 3, 2018].

7. Sigfox, "IoT use cases for transport & logistics," 2018. [Online]. Available: https://vt-iot.com/wp-content/uploads/2018/02/SIGFOX-USE-CASE-OVERVIEW-Logistics.pdf. [Accessed: Aug. 3, 2018].
8. Microsoft, "IoT for transportation," *Microsoft*, 2018. [Online]. Available: https://www.microsoft.com/en-us/internet-of-things/transportation. [Accessed: Aug. 3, 2018].
9. Kaa, "IoT platform for smart supply chain solutions," [Online]. Available: https://www.kaaiot.io/solutions/logistics. [Accessed: Aug. 3, 2018].
10. L. Jukna, "The Internet of everything: IoT use cases," 2018. [Online]. Available: https://www.livingmap.com/technology/the-internet-of-everything-iot-use-cases/. [Accessed: Aug. 3, 2018].
11. R. Stevens and L. Zeltser, "IoT and security in the supply chain: Making smart choices," 2018. [Online]. Available: https://www.inboundlogistics.com/cms/article/IoT-and-security-in-the-supply-chain-making-smart-choices/. [Accessed: Aug. 3, 2018].
12. T. M. Fernández-Caramés, P. Fraga-Lamas and A. Suárez, "Reverse engineering and security evaluation of commercial tags for RFID-based IoT applications," *Sensors*, vol. 17, no. 1, 2017, pp. 1–12. [Online]. Available: https://www.mdpi.com/1424-8220/17/1/28. [Accessed: Aug. 3, 2020].
13. S. Kaplantzis and S. Y. Ahmet, *"Security and smart metering,"* in *18th European Wireless Conference (European Wireless)* Poznan, Polland, 2012.
14. M. Bellias, "3 ways IoT will change smart meters for utilities," *IBM*, [Online]. Available: https://www.ibm.com/blogs/internet-of-things/smart-meter-grid/. [Accessed: Aug. 3, 2018].
15. R. Anderson and S. Fuloria, *Smart Meter Security: A Survey*, Cambridge, United Kingdom: University of Cambridge Computer Laboratory, 2011. [Online]. Available: https://www.cl.cam.ac.uk/~rja14/Papers/JSAC-draft.pdf. [Accessed: Dec. 27, 2020].
16. European Commission, "Smart grids and meters-Energy," 2018. [Online]. Available: https://ec.europa.eu/energy/en/topics/markets-and-consumers/smart-grids-and-meters. [Accessed: Aug. 4, 2018].
17. O. Novo, N. Beijar, M. Ocak, and J. Kjallman, *"Capillary networks - bridging the cellular and IoT worlds,"* in *2nd World Forum on Internet of Things*, Milan, 2015.
18. CTTC, "Capillary networks," *CTTC*, 2017. [Online]. Available: http://technologies.cttc.es/m2m/technologies/capillary-networks/. [Accessed: Aug. 4, 2018].
19. Center for Advance Automative Technology, "Connected and automated vehicles," 2017. [Online]. Available: http://autocaat.org/Technologies/Automated_and_Connected_Vehicles/. [Accessed: Aug. 4, 2018].
20. SIEMENS AG., "Future of infrastructure—vehicle-to-X (V2X) communication technology," SIEMENS, 2015.
21. ERICSON, "Capillary networks—a smart way to get things connected," ERICSON, 2014.
22. I. Auge-Blum, K. Boussetta, H. Rivano, R. Stanica, and F. Valois, *"Capillary networks: A novel networking paradigm for urban environments,"* in *Proceedings of the First Workshop on Urban Networking*, New York, NY, USA, 2012.
23. J. Slay and M. Miller, *"Lessons learned from the Maroochy water breach,"* in *International Conference on Critical Infrastructure Protection (ICCIP 2007)*, Hanover, NH, USA, 2007, pp. 73–82.
24. McAfee,"Global energy cyberattacks:'NightDragon',"*McAfee*,2011.[Online].Available: https://www.mcafee.com/wp-content/uploads/2011/02/McAfee_NightDragon_wp_draft_to_customersv1-1.pdf. [Accessed: Dec. 20, 2019].

25. D. Kushner, "The real story of stuxnet," *IEEE Spectrum*, vol. 50, no. 3, pp. 48–53, 2013. Available: https://spectrum.ieee.org/telecom/security/the-real-story-of-stuxnet. [Accessed: Dec. 27, 2019].
26. R. M. Lee, M. J. Assante, and T. Conway, "Analysis of the cyber attack on the Ukrainian power grid," Mar. 2016, [Online] Available: http://www.nerc.com/pa/CI/ESISAC/Documents/EISAC_SANS_Ukraine_DUC_18Mar2016.pdf. [Accessed: Dec. 20 2019].
27. Symantec. "Internet security threat report: ISTR ransomware 2017," An ISTR special report. 2017.
28. ESET, "Machete just got sharper, Venezuelan institutions under attack," *ESET Research*, Jul. 2019. [Online]. Available: https://www.welivesecurity.com/wp-content/uploads/2019/08/ESET_Machete.pdf.
29. V. Rathod, "Smarter, safer, & sustainable fleet management" 2015. [Online]. Available: https://www.einfochips.com/blog/smarter-safer-sustainable-fleet-management/. [Accessed: Aug. 8, 2018].
30. X. Fan and G. Gong, "Security challenges in smart-grid metering and control systems," *Technology Innovation Management Review*, vol. 3, no. 7, 2013. pp. 43, [Online]. Available: http://doi.org/10.22215/timreview/702. [Accessed: Aug. 8, 2020].
31. E. D. Knapp, and J. T. Langill, "Industrial network security: Securing critical infrastructure networks for smart grid, SCADA, and other Industrial Control Systems," *Syngress*, 2014.
32. S. Stavrou and S. R. Saunders, "Apparatus for preventing the normal operation of a cellular mobile telephone," WO2000054538A1, 2000.
33. EUROPOL, "IOCTA 2016 Internet organised crime threat assessment," EUROPOL, 2016.
34. ENISA, "Smart grid threat landscape and good practice guide," ENISA, 2013.
35. California Energy Commission, *Smart Grid Information Assurance and Security Technology Assessment*, California Energy Commission, 2010.
36. ENISA, *"Reputation-based Systems: A security analysis,"* ENISA, 2007.
37. Microsoft, "Internet of Things security architecture," Microsoft, 2018.

4 Threats in Industrial IoT

N. Scheidt and M. Adda

CONTENTS

4.1 Introduction 138
    4.1.1 Definitions IIoT..140
    4.1.2 IIoT Application Domains 142
        4.1.2.1 Manufacturing 142
        4.1.2.2 Logistics and Transportation 142
        4.1.2.3 Banking 142
        4.1.2.4 Governmental Authorities 142
    4.1.3 IIoT Companies 143
        4.1.3.1 Siemens 143
        4.1.3.2 Tesla 143
        4.1.3.3 Airbus 143
        4.1.3.4 Caterpillar...143
4.2 Key Security Goals in IIoT 144
    4.2.1 Application Domain: Agriculture and Farming 144
        4.2.1.2 Key Security Goals..146
    4.2.2 Application Domain: Industrial Safety 146
        4.2.2.2 Key Security Goals..148
    4.2.3 Application Domain: Preventative Maintenance 149
        4.2.3.1 Risks within this Domain 149
        4.2.3.2 Key Security Goals..149
    4.2.4 IIoT Technologies 150
        4.2.4.1 Sensors 150
        4.2.4.2 Robots 151
        4.2.4.3 Wearables 151
        4.2.4.4 Drones 152
    4.2.5 IIoT Incidents...152
        4.2.5.1 Ransomware Attack Norsk Hydro 152
        4.2.5.2 Spyware Attack Bayer AG 153
        4.2.5.3 South Korea Cyberattack 153
        4.2.5.4 Honda Ransomware "SNAKE" Attack 153
        4.2.5.5 Climate Corp 153
        4.2.5.6 Sony Pictures...153
        4.2.5.7 SWIFT Banking Hack...154

4.3　Relevant Deployment Architecture in IIoT 154
　　4.3.1　Application Domain: Agriculture and Farming 154
　　　　　4.3.1.1　Relevant Deployment Architecture 155
　　4.3.2　Application Domain: Industrial Safety 157
　　　　　4.3.2.1　Relevant Deployment Architecture 157
　　　　　4.3.2.2　Aspects to Consider 159
　　4.3.3　Application Domain: Preventative Maintenance 159
　　　　　4.3.3.1　Relevant Deployment Architecture 160
　　4.3.4　Threats Within IIoT 161
　　　　　4.3.4.1　Man-in-the-Middle Attack 161
　　　　　4.3.4.2　Device Hijacking 162
　　　　　4.3.4.3　Inside Job 162
　　　　　4.3.4.4　Data Breaches 162
　　　　　4.3.4.5　Theft 162
　　　　　4.3.4.6　Outdated Systems 162
　　　　　4.3.4.7　Insufficient Training 163
4.4　Conclusion 163
References 164

4.1　INTRODUCTION

Worldwide, organizations, be it governmental, public, or corporate, are enhancing their businesses in a variety of areas by implementing and making use of the Internet of Things (IoT). In the IoT environment, an object interacts with systems as well as with the internet either directly or over another object. For example, a smart fridge can directly access the internet in this day and age, which enables to place orders for products that are running low and have them delivered directly to your home amongst other things. However, we are also able to access its systems and services via apps on our smart devices. Moreover, the systems and objects are able to be operated by one as well as multiple users. The IoT work environment can be put together of multiple levels [1]. These levels are, firstly the IoT overall and its user base, secondly threats towards this environment and appropriate solutions for a safer operation, thirdly detection methods of threats and traces of attacks which provide answers in terms of offender and attack aim and finally the level of the investigation process from receiving a case to the prosecution, as presented in Figure 4.1.

The IoT implementation into Industries (IIoT), therefore, increases the attack surfaces of these organizations especially to threats such as cyber-attacks, which alongside the technological development evolve greatly as well. Due to the amount of IoT devices increasing steadily, rapidly, and being easily accessible to everyone, these devices can also be used as a weapon in the cyber-world to implement threats such as launching distributed denial of service (DDoS) attacks by building botnets, which are multiple devices with the ability to connect to the internet to be utilized as a tool to attack and disrupt a specific object; these devices are then referred to as botnets whereas the attack itself is a DDoS, i.e., the Mirai botnet (malware) attacks on the Krebs website to get access to devices with low-level internet security protection [2].

Threats in Industrial IoT   139

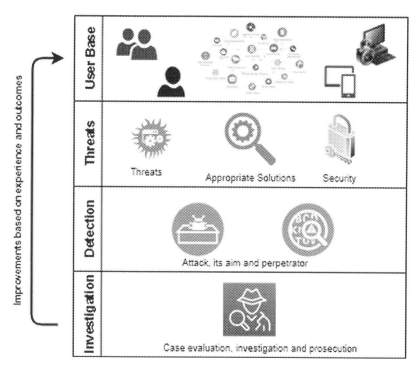

FIGURE 4.1   Levels of IoT Work Environment

Additionally, malware spreading and spamming are common threats which mostly are in the format of emails containing malicious content to access the device of the receiver [2]. Malware can come in different types such as viruses, spyware, and worms which describes the way a malware is distributed within the attacked systems.

Moreover, users interacting with devices are not only human, in today's Industry, but also can be intelligent machines. Therefore, utilizing the IIoT has a high number of advantages as well as disadvantages. The disadvantages are mainly linked to security issues and the vulnerabilities of the IIoT, which generally need to be tackled on three different levels, which are the Defense/Prevention, Detection, and Investigation/Prosecution. These three levels can and need to be interpreted as a continuous cycle as shown in Figure 4.2. The defense of the IIoT influences the detection of attacks on the different surfaces. This again needs to be investigated to not only prosecute the offender and solve the issue but also improve the defense methods for the future. This process also links to the concept of the IoT environment and its different levels of focus (Figure 4.2). Further investigation could be used to advise police forces on what to look out for or focus on more precisely, as previously stated the IoT devices are rapidly increasing a lot faster than the polices understanding of such devices. Therefore, this lack of understanding, availability, and education makes it difficult for them to pursue an investigation to the best of their abilities. A further issue is that

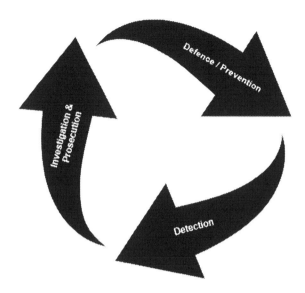

FIGURE 4.2 Levels to be tackled within IIoT

the law itself cannot keep up with the rapid change and rather than one precise act that addresses all of the offenses that can now be committed through the use of IoT devices; these are instead spread out across a multitude of acts causing difficulties for the prosecution.

4.1.1 Definitions IIoT

As slightly touched upon, the IIoT, also known as *Industry 4.0*, is the implementation of the IoT into the manufacturing industry such as governmental, public, or corporate organizations. Intelligent devices and objects build a network that is able to collect and share data over a central cloud-based service related to the necessary manufacturing and industry processes in the IIoT. According to [3] Industry 4.0 can be defined as

> the digitization of the manufacturing sector, with embedded sensors in virtually all product components and manufacturing equipment, ubiquitous cyber-physical systems, and analysis of all relevant data.
>
> [4]

Furthermore, IIoT is aiming for built-in cyber-physical systems into smart factories achieving better ecosystems in terms of [5]:

- self-configuration
  i.e., setup and updates

- self-monitoring
  i.e., recognizing issues and necessary areas for improvement
- self-healing
  i.e., fixing errors and allowing necessary system changes.

As these terms indicate, the systems are aimed to be designed to be independent and therefore improve and ease the workload of users. Structuring data of the processes is implemented and fitted according to a company's needs and will automatically be configured. Moreover, due to this, the systems need to be able to monitor themselves to structure but also heal the system independently if no bigger maintenance steps are needed.

These "autonomic self-properties," as defined by [5], will improve achievements and efficiencies in the industry and lead to new ways of technological development in production overall [5]. Therefore, the IIoT is expected to continue to change and improve the industry further by allowing access and gathering a larger data amount than before in a quicker and more efficient way [2]. This is supported by four additional aspects of technological implementation that has been suggested by [3]:

- Data, Computational Power as well as Connectivity
- Analytics and Intelligence
- Human and Machine Interaction
- Digital to Physical Transformation.

Finally, it can be said that the IIoT implements the leading developments of technologies while combining such with traditional aspects and methods of the industries to support the use of smart products which enhances processes of digital and physical kind [5].

Generally, due to the complexity of the IoT its threats are of variety as well. Characteristics of threats should be considered, to be able to somehow categorize the high diversity of threats and include the aspects of *target*, i.e., what is the aim of the attack and who is being attacked; *agents*, hence, who or what is responsible for the attack; the *technical impact* of the threat as well as its *severity* [2]. Therefore, according to [2], if threats are assessed by these four characteristics they can be categorized into eight groups as shown below, however, this list can also be expanded and updated with research developing within this area:

- Network-level threats
  i.e., overwhelming of network such as DDoS attacks
- Cryptography-related threats
  i.e., sensitive information such as bank details are not securely encrypted and allow easy access to third party
- Hardware/sensor-level threats
- Malware
- Threats for Smart grids
- Technical/application development-related threats

- Threats necessitating actions by the victim user
  i.e., clicking on infected links or visiting malicious websites
- Generic/Miscellaneous threats.

These groups are able to cover a vast variety of threats which help to understand the different levels and types of risks and vulnerabilities regarding the IIoT. Therefore, security is a highly important part of the IIoT which demands well thought out frameworks and deployment architectures [7].

4.1.2 IIoT Application Domains

The industry areas being implemented with IoT are also of a variety and provide a multitude of application domains within the financial as well as commercial sector to enhance interactions between organizations, enterprises, or other institutions [4].

4.1.2.1 Manufacturing

Inspecting if a machine is running and that without a problem as well as without being physically overseen or handled by a human has been improved since the implementation of IoT. Within this domain monitoring, maintenance as well as management of different aspect during manufacturing processes is greatly enhanced by IoT technologies and a key domain within the IIoT.

4.1.2.2 Logistics and Transportation

Management and monitoring are highly improved by the IoT implementation within this domain as well. Be it the tracking of products, location as well as inventory wise, management of warehouses, inventory thefts, self-driving vehicles within these processes, and the implementations of drones for delivery purposes.

4.1.2.3 Banking

Nowadays, everything is desired to be quickly and easily accessible in particular in terms of finances. Making payments online or dealing with own finances in general is just a few clicks away and possible from anywhere as long as an IoT device is available with an internet connection en par. Moreover, queries are able to be solved more precisely due to the easy connectivity of banking applications.

4.1.2.4 Governmental Authorities

All sorts of governmental authorities are utilizing the IoT for their benefits as well. Be it the police to support investigations by utilizing CCTV, bodycams, or devices for a more manageable and organized procedure. Information about weather, flooding's, and earthquakes can be detected as well as the management of public transportation. This domain covers a variety of areas that make efficient use of the IoT to improve the safety, maintenance, prevention as well as overall management of their areas overall.

4.1.3 IIoT Companies

Companies that are well known and have successfully implemented the IoT within their industry are continuously growing and aiming to adapt and improve to this technological day and age.

4.1.3.1 Siemens

Siemens, which is a German company, aimed to build and succeed in establishing a smart factory that is fully automated as well as internet based by developing the operating system "Mindsphere." The products Siemens manufactures are utilized within BMW and due to Mindsphere Siemens are able to easily collect and analyze all data regarding their production to improve procedures and end products within the facilities smoothly.

4.1.3.2 Tesla

As electric vehicles are getting more and more popular so is the success of this American company which is specializing in such. Vehicles can be easily charged, battery life got improved, and the control of vehicles as well as devices was made accessible and more convenient by Tesla utilizing the IoT efficiently during their production processes.

4.1.3.3 Airbus

This European aerospace company introduced "Factory of the Future" as an initiative of digital manufacturing. Tablets and additional smart wearables are in constant use by employees to ensure safety, increase productivity, and communicate with machinery or other employees as well as solving assigned tasks.

4.1.3.4 Caterpillar

Caterpillar uses augmented reality (AR) to handle their machines as being an American machinery and equipment company. This allows not only the employees but also users of their products to optimize processes. It is also suggested that they increased their production efficiency by around 45% due to the implementation of the IoT and its available technologies [6].

This chapter's main focus will be on the threats within the IIoT and specifically focus on three domains while exploring their key security goals as well as the relevant deployment architectures, which is the implementation of logical architecture into a physical environment, such as computing nodes, i.e., anything which can be allocated an IP address to, the Central Processing Unit (CPU) which is the main calculation area of a computer, its memory as well as storage, hardware and network devices [2]. The domains this chapter include and focus on are agriculture and farming, industrial safety as well as preventative maintenance.

4.2 KEY SECURITY GOALS IN IIOT

This section will focus on key security goals that are crucial within the IIoT with a special focus on the three application domains of agriculture and farming, industrial safety, and preventative maintenance. While exploring and defining these three different domains, this section showcases the importance of security if IoT is implemented and which need to be focused on specifically.

4.2.1 Application Domain: Agriculture and Farming

Implementation of IoT technologies into areas of agriculture and farming has been developed in a variety of research [7, 8]. This allowed to improve agriculture by analyzing issues, complications as well as challenges within that environment, such as water shortage, cost management, and productivity issues [7, 9]. Going by these implementation areas, generally it can be said that IoT platforms have the capacity to sense, process, and communicate with one another and data. When implemented well this can be used to improve performances within the agriculture and farming environment and therefore gather environmental data with high precision and efficiency. Therefore, with such sensors, farmers collect data from weather, soil, air quality as well as crop maturity. This enables individuals working within this domain to make smarter decisions with the help of technology and these techniques are being referred to as *precision farming* (PA) technology (see Figure 4.3) [10].

Additionally, sensors can be implemented in selected life stocks, such as cows or sheep. Therefore, this allows farmers to keep an overall picture of their animals' health to recognize possible health issues at an early stage. Recognizing and monitoring this in a quick and efficient manner without having to physically inspect the livestock straight away can be of benefit to the farmers. This process is also referred to as *pastoral farming* and supports farmers to keep their life stocks at the highest level of health at all times which lessens the financial losses occurring due to such issues.

Another possibility is the deployment of sensors into the production processes such as monitoring machine performances and the status of the content in tractors, trucks as well as tanks. A couple of the key aims cover the initiatives to increase efficiency and/or reduce the water usage [12]. Moreover, drones can be utilized to keep a live update on crops as well as livestock. These, in addition with global positioning systems (GPS) and image recognition, support farmers to monitor their crop's health periodically and therefore determine if and in what way it has been affected to ensure a timely reaction and take care of risks, such as insect infestation or water shortage, efficiently [13]. Overall, drones are commonly used in the following areas of agriculture and farming:

- Analysis of Soil and Field
- Planting of Seed
- Weed checking and if necessary, Crop and Spot Spraying
- Mapping as well as Surveying of Crop

Threats in Industrial IoT

- Managing and Monitoring of Crop and Livestock health in real-time
- Water Supply Monitoring and Management.

Generally, it can be said IoT is implemented into the agricultural environment not only to ease the monitoring of crops and livestock to ensure healthy and efficient farming but also to minimize costs occurring due to the fast recognition of issues which otherwise would end in losses farming wise and financially [14].

4.2.1.1.1 Risks within this Domain

The dependence on technology in all aspects of life cannot be denied; however, serious issues can arise if these technologies stop working or start malfunctioning. In terms of the agriculture and farming environment, these faults can have an impact on a variety of areas such as food and livestock safety, change of data relating to such, and easier access to control systems. The dependency on technology may lead to the decrease of traditional farming ways and the possible missing out on symptoms which affect the livestock, for example, cases such as swine flu outbreaks.

This sector does not necessarily focus on Cybersecurity and the implementation of such is not a high requirement for the design [13]. However, this compromises the safety of the whole system, especially because the systems most likely are implemented on unmonitored networks (such as a private network of the farmer instead of a huge companies network), which opens opportunities to a platform vulnerable to attacks and easily to access by hack attacks to implement malware, compromise data, and gain access to a third party to become botnets [2, 13].

One example showing that such risks are an issue within the agriculture environment is *PA* (Figure 4.3). According to [14], PA is a combination of technologies which gather growth, field and management data for sustainable farming, (which is the production of livestock and crops with a minimal effect on the environment and

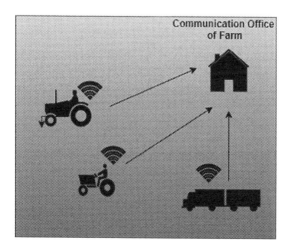

FIGURE 4.3 Precision Farming

include GPS), remote sensing technology (RS), and geographic information system (GIS) [15].

PA is very beneficial in improvement of the farming procedure; however, it is high in costs and requires an amount of training which can be time- as well as cost-consuming [15]. Moreover, the FBI also notes that the implementation of technologies into farming increases the risks as mentioned above due to their current priorities not being its Cybersecurity [16]. Such interference into the data can lead to over- or underwatering crops, misdiagnosing life stocks, or misinterpreting temperature data [10]. This then can lead to a loss of life stocks and decrease in crops and affects the industry greatly in a negative and dangerous matter. Putting this into a bigger picture such high-level security system issues could leave these vulnerable for larger scale attacks that aim to use illnesses carried by livestock to infect communities and in a worst-case scenario will be leading to a pandemic. This is not only endangering the life stock but also the farmers and their employee's livelihood.

4.2.1.2 Key Security Goals

Considering the risks stated and outlined in the previous Section 4.2.1.1, it can be said that the key security goals for agriculture and farming are mainly the following:

- Reliability of implementation
- Confidentiality in use of systems
- Integrity of systems
- Availability of the programs as well as the systems required.

Taking the aspects discussed in this section into account establishes the four key security goals for the agriculture and farming domain, which are highly crucial if the IoT is going to be securely and efficiently utilized.

4.2.2 APPLICATION DOMAIN: INDUSTRIAL SAFETY

Safety is a commonly used term, which is understood by the majority and generally defined as "the absence of unwanted outcomes such as incidents or accidents" [17]. Industrial safety, therefore, concentrates on the absence of such incidents within an industry domain, such as private Companies, governmental forces such as fire brigade, army forces, and police as well as institutions, i.e., education facilities, hospitals. Industrial environments always involve a number of risks and dangers, including but not limited to infections, slip, trip and falls, and moving objects (see Figure 4.4).

Plant managers are responsible to develop the most efficient solutions which minimize such casualties and accidents. Their key responsibilities focus on "reducing hazards, managing risks and preventing accidents" [2]. Moreover, legislation is enforced and provides regulations for safety in the industry, however, often these regulations are not as effective as necessary due to not utilizing and learning from past mistakes within these areas [18]. Furthermore, implementing new rules and regulations are also time consuming and in some cases come alongside money loss if the long term benefits are not considered because this would require constant updates

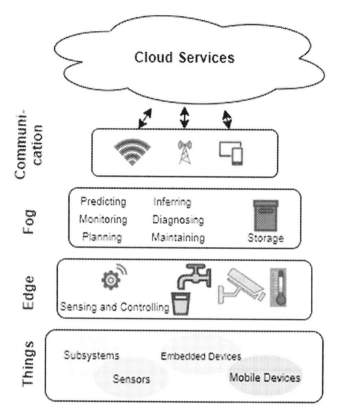

FIGURE 4.4 Hazard signs within the Industry

of regulations. Additionally, to some degree they are not able to put such measures in place due to the required funding which is relevant in order to keep up with the necessary updates adapting to the ever-changing landscape of IoT devices needed to maintain a high standard of safety [18]. The implementation of IoT sensors into the industrial safety domain allows employers and/or plant managers to deal with risks time efficiently due to having access to real-time data [2]. Therefore, utilizing the IoT to improve safety within the industry allows to recognize faults and ensure employees safety on a variety of terms by creating different scenarios and designing algorithms, which identify objects and velocities (speed) by utilizing processing powers, sensors, and systems like robots or drones [2, 18]. Risks which can be detected but are not limited to, by this machine-learning automated implementations are gas leaks, temperature changes, and humidity changes [2].

As outlined above, these changes can be detected by the implemented sensors which are available of different kinds. These work environmental changes and detection of such allow the work to be stopped immediately and/or inform the duty manager swiftly who then can initiate the security steps and protocol which should be put in place in every facility. Moreover, computer vision is an additional improvement and implementation area within the agricultural environment which enables to recognize situations more efficiently and minimizes the chance of accidents and improves prevention methods [19].

4.2.2.1.1 Risks Within this Domain

As slightly touched upon, the industrial environment enables risky situations such as infections, slip, trip and falls, and moving objects. Furthermore, working from heights can be of high risk if safety measures, such as safety nets, secure scaffolding, rails, and secure equipment's, are not sufficiently implemented. Additionally, noise and vibration due to heavy-duty tools and vehicles can affect workers negatively, being it health or productivity wise. Working with electricity, such as powerlines and improper wiring, can put staff at high risk which could be life threatening as well as possibly damaging to equipment or facilities. Employees hurting themselves due to wrongful manual handling are most likely the most common risk within industry. The lack of appropriate training or the rush to finish tasks are often the root cause of causalities from manual handling. Collapses of construction sites are a risk in itself; however, during collapses other risks are also involved such as falling and moving objects as well as slip and trips.

As demonstrated, majority of the risks involve the safety of workers; however, the implementation of the IoT can decrease such greatly through constant monitoring, managing, and efficient warning messages in cases of occurring risks and hazards.

4.2.2.2 Key Security Goals

As mentioned above, the utilization of the IoT for industry safety purposes can be beneficial leverage safety and furthermore minimize risks from unforeseen situations. Therefore, information must be accurate and timely regarding all processes and procedures. Furthermore, data that are being retrieved and transmitted for industry safety purposes are confidential due to containing device information about their status, whereabouts, and operational environment [2]. This would risk the data security of an environment within the industry, specifically regarding employees and material safety which can lead to a variety of issues as outlined in Sections 4.2.1.1.1 and 4.2.2.1. Therefore, the following goals and topics should be taken into consideration [2]:

- Safety
  i.e., operation of systems and sensors should perform without the chance to cause risk of material, service damage, or life loss
- Security
  i.e., protecting systems and sensors from accidental or "unauthorized access, change, or destruction" [2]
- Reliability
  i.e., sensors need to have the ability to supply precise and exact data to prevent wrong calculations which could risk/cost life or damage assets
- Accuracy
  i.e., prediction of risks and calculation of sensor data need to be accurate
- Confidentiality
  i.e., it is necessary to secure data from access which is not authorized.

If implemented correctly on top of the considerations of IoT implementation, the safety of Industry will be greatly improved while ensuring the efficiency of the domain.

4.2.3 Application Domain: Preventative Maintenance

To deal with IIoT and its security it is crucial to evaluate previous data and develop approaches to predict failure of equipment which is reliable and timely and therefore minimize security risks [20]. Furthermore, such procedures will also be highly beneficial to estimate the lifespan of equipment and when/how to tackle technical issues which may occur while extracting important data.

Therefore, IoT can greatly support the maintenance of prevention methods due to the gathering of real-life data from devices such as sensors within a network. A designed schedule for maintenance by the manager will demonstrate and improve data analysis and its efficiency, cost, and time saving to improve safety within the IIoT [21, 22]. These preventative methods and organized inspections are conducted while the equipment is still fully functioning to be able to predict areas which could fail in the future and to prevent an unexpected malfunction.

4.2.3.1 Risks within this Domain

Traditionally, regular maintenance is scheduled; however, it is possibly unnecessary due to the equipment still working efficiently or in other cases not sufficient due to the maintenance being scheduled too late. IoT implementation is aiming to counteract these issues. IoT technologies allow real-time monitoring, as outlined in the above sections, and therefore enable data to be predictive and analyzed to show if equipment are about to fail or need improvements, such as updates or repairs. However, even though these overall improvements solve the majority of issues it also can open the door for other issues. One of such is the high dependence on the IoT systems, which if a system failure would occur, i.e., power outage, hacking incident, could be fatal.

4.2.3.2 Key Security Goals

As touched upon, advancing preventative maintenance by utilizing the IoT can improve the safety for humans as well as the environment and furthermore; this implementation can also prevent the issues which can occur of equipment failure which then would mean they could not solve their tasks successfully. To meet these aims efficiently, the information must be precise as well as timely. As within the previous IIoT domain, information retrieved and transmitted is confidential because it contains data about status, whereabouts as well as operational environment [2]. The key goals of this domain which need to be taken into consideration are similar to the previous domain, industrial safety, and the additional key goal includes[2]:

- Resilience
  Hence, systems that avoid, absorb, and manage dynamic conditions while completing tasks assigned but are also able to restore systems and data.

Overall and considering the stated areas of importance, there are a high number of security goals which need to be paid attention to within the three chosen domains:

Agriculture and Farming, Industrial Safety and Preventative Maintenance, as well as Maintenance Methods. These domains should be designed and assessed accordingly [2]:

- Protection from harm
- Protection of the environment
- Resilience
- Operations reliability
- Continuity
- Maintenance of data integrity
- Confidentiality/privacy.

If the IIoT will continue to efficiently run and succeed toward threats in their domains, the key security goals need to be regularly evaluated to establish if methods need to be improved and to ensure the security of the domains.

4.2.4 IIoT Technologies

The following is a short breakdown of just a few of the technologies utilized to be implemented within the IIoT which will be more touched and clarified upon in Section 4.3 of this chapter.

4.2.4.1 Sensors

Sensors are the main contributors for the IIoT to be running smoothly. As the IoT supports improving the way of processes, such sensors are used to collect data for management or maintenance purposes, for example. There are different types of sensors that include level, electric current, humidity, pressure, temperature, heat, fluid velocity, flow, and infrared.

4.2.4.1.1 Level Sensor

These real-time sensors are implemented in areas such as waste supervision, feeding operations, or fuel and gas management.

4.2.4.1.2 Pressure Sensor

Specifically, in regard to monitoring pipelines and possible irregularities, them being stuffed or having leaks.

4.2.4.1.3 Temperature Sensor

Such sensors oversee equipment's to keep on track of overheating issues to ensure the safety of machinery, environment, and employees.

4.2.4.1.4 Infrared Sensor

These sensors are popular in all domains, within wearables these monitor health and within facilities these, just as the temperature and pressure sensors, provide information on possible heat or water leaks.

4.2.4.1.5 Risks of Sensors

Even though sensors have such a broad variety of advantages and are able to be utilized across domains, there are some risks that need to be considered when using such. Sensors can be affected and send faulty data if affected by environmental changes or possible contamination if the sensor was too greatly and too often exposed to different things, such as gas. Moreover, due to these reasons the life span of a sensor can be quite short lived and needs to be checked and replaced quite regularly. Furthermore, sensors can be relatively easy to be manipulated and data will be affected and change negatively and possibly put the domain at risk.

4.2.4.2 Robots

Mainly for the transportation of product purposes automated vehicles are implemented to be free roaming and ease the workload of employees. Maps and routes are programmed into these robots for them to be able to follow the layout precisely and carry the products from one location to another. These could also ensure the safety of employees in case of an area not being as easily accessible or not safe enough for them to enter first, i.e., for bomb detonation purposes.

4.2.4.2.1 Risks of Robots

The implementation can be of high financial burden at the start of an investment of a robot into the industry. These are more sophisticated tools than others and therefore need more focus on programming and updating. According to [13], there are also a number of hazards that are linked to the use of robots within the domain:

- Mechanical
  Hazards due to movements that are unintended or tools that release unexpectedly.
- Electrical
  Exposure to live parts or unsecure connections.
- Thermal
  Extreme Temperatures.

4.2.4.3 Wearables

Be it cameras or health sensors, wearables can and are mostly embedded in the protective wear and part of personal equipment of workers. This can be utilized in all kind of manufacturing facilities as well as the police to support them in terms of investigation purposes.

4.2.4.3.1 Risks of Wearables

Similarly, to other implementations, wearables can end up risking individuals or the environment using such greatly. Data need to be safely exchanged between wearables and systems and breaches during such can end in financial or health damages. Moreover, malfunctioning of wearables is not as uncommon as it should be and can therefore pose multiple threats as data protection issues or the loss of contact during crucial situations.

4.2.4.4 Drones
Drones can be implemented in a variety of domains, be it for product delivery, helping with task such as spraying fields or even planting as well as monitoring the work environment regarding facility standard check, well-being of livestock or crops as well as being an addition to hazard detection. As robots can be sent into dangerous areas to avert humans entering a challenging environment, drones can be used to cover and access a wider field of an environment, henceforth, provide a broader overview of a situation with the addition of cameras.

4.2.4.4.1 Risks of Drones
Initially, the purchase of drones is very high and updating such or purchasing new models if necessary is important to consider carefully financially. Furthermore, drones also need to be operated which needs to be carefully researched and educated on. Even though drones are beneficial in a number of areas especially with the agriculture and farming domain, it is very crucial to consider that weather circumstances will affect drones and are often very sensitive to such and therefore cannot be used constantly. Lastly drones need to be recharged and cannot cover an unlimited amount of ground for an unlimited time. If a third party is able to get access or take over management while it is in use, such could risk operation procedures enormously.

4.2.5 IIoT INCIDENTS
Establishing the key security goals within the three application domains, Agriculture and Farming, Industrial Safety and Preventative Maintenance of IIoT, following are some examples of global incidents and IIoT attacks which were on a bigger scale and are worth mentioning.

4.2.5.1 Ransomware Attack Norsk Hydro
In 2019, $40 million were lost by the Norwegian Norsk Hydro company, which produce aluminum, after a ransomware attack [21]. During this attack, the company's systems as well as its main site crashed and failed to work, and staff had to use personal devices to access work meetings or emails. Furthermore, the majority of tasks needed to be done analog until the situation was under control again.

This issue started by one individual, working for the company, opening an email that was infected even though coming from a trusted source. Opening this email enabled the hackers to plant the virus and therefore easily access the systems of the company, crash them, and demand bitcoins to repair the processes back to normal. Norsk Hydro decided on the following three steps to tackle this issue:

- They did not pay the ransom.
- With the help of Microsoft Security, they would restore the systems as well as improve security.
- Transparency of Attack.

From the start Norsk Hydro planned to be transparent and open about the whole process as well as challenges and losses during this attack. The company aims that these steps of openness will be beneficial for other industries and domains to be either a good framework in how to secure their systems before an attack even happens or guidelines how to deal during or right after such attacks to ensure the best as well as most efficient ways to protect employees, the business and rebuild if necessary.

4.2.5.2 Spyware Attack Bayer AG

In 2018, the German pharmaceutical company, Bayer AG, was hacked utilizing the malware Winnti to access and spy on the company's data. However, the security team was able to detect and monitor it timely to prevent any major data loss [24]. The whole system was thoroughly analyzed to detect affected areas and clean such to avoid any further attacks linked to this particular issue.

4.2.5.3 South Korea Cyberattack

A few South Korean television stations and multiple banks were affected by an apparent cyberattack in 2013. Systems shut down and sites were frozen and disrupted services which were suspected to be hacked into by North Korean hackers [25]. People were not able to withdraw money and during news broadcastings screens showed blank, but they were still able to broadcast. IP addresses used for this attack were suspected to be originated from China and, as mentioned above, possibly linked to North Korea. However, no evidence has been found or established in regard to these speculations.

4.2.5.4 Honda Ransomware "SNAKE" Attack

In June of 2020, a number of global Honda services were shut down due to a cyberattack using ransomware SNAKE which can collect data to then be used as blackmail or to be held hostage for a cryptocurrency exchange [26]. According to the Japanese company, no data in this case were able to be retrieved during the attack and the systems and production were restored.

4.2.5.5 Climate Corp

The farming industry is not one of the most primarily targeted domains for hackers. However, in 2014 a company has been hacked. Climate Corp. is a farm analytics business, the hack exposed information as well as details of credit cards of clients and employees. However, after additional and thorough investigation, Climate Corp. concluded that there was no misuse detected with the stolen data.

4.2.5.6 Sony Pictures

In 2014, film studio Sony Pictures got hacked by a group referring to themselves as the "Guardians of Peace" [14]. They accessed and leaked personal data which included the following:

- Information about employees and their families
- Work emails
- Company's salaries of executives
- Copies of unreleased movies

- Future movie ideas
- Scripts.

Additionally, they planned to erase the whole of the computer's infrastructure of studio Sony Pictures. Finally, the group demanded to withdraw the release of an upcoming comedy movie "The Interview" with the plot regarding Kim Jong Un's assassination. Responding to this, Sony canceled the release of the movie and changed it to digital release. After a thorough investigation, the United States established North Korean's Government to have funded this attack; however, as North Korea denied such these part of investigation outcomes are mainly alleged.

4.2.5.7 SWIFT Banking Hack

A number of cyberattacks were seen through in 2015 as well as 2016 using the SWIFT banking systems [17]. A group referred to as Apt 38 successfully stole millions of dollars and are suspected to be linked to the Sony Pictures attacker group. However, these were just suspicions. The attack itself worked by using vulnerabilities of the banking system itself by getting access to credentials and using such to communicate with a number of banks. Due to seeming legitimate, showing the credentials allowed the hacker groups to easily move funds around.

4.3 RELEVANT DEPLOYMENT ARCHITECTURE IN IIOT

Having outlined key goals within IIoT, in particular Agriculture and Farming, Industrial Safety and Preventative Maintenance, and its security it is also crucial to elaborate on relevant deployment architectures of these three domains, which aim and improve these goals as well as show how the IoT is implemented within the industry.

4.3.1 APPLICATION DOMAIN: AGRICULTURE AND FARMING

Research shows that there are a variety of relevant deployment architecture regarding agriculture and farming IoT. Most deployment architectures suggest different layers of communication and data acquisition especially in terms of environmental data [10, 28]. Technologies for smart harvesting are implemented while being cost-efficient and increasing production.

In Tokyo, January 2018, a prototype for a tomato picking robot was introduced by Panasonic Corporation. This robot includes cameras, sensors, and additional artificial intelligence features to identify and pick tomatoes that are ripe [29]. These implemented robots show the benefits of increased crop health, production value, and decrease cost in labor work as well as low staff numbers for vegetable and fruit harvesting. All these implemented aspects as well as the benefits of such are especially beneficial and increase efficiency specifically during the 2020 COVID-19 pandemic [30].

Research by [1] proposes a thorough model of the implementation of the IoT in the agriculture sector and shows a well-presented deployment architecture. This distributed model considers a variety of levels important for a reliable process within

Threats in Industrial IoT

agriculture and farming, such as the use of cloud computing. However, this can affect the performance of the system due to sensors and control levels requiring low latency and therefore [1] consider additional steps for an improved process [28]:

- *Fog Computing*
  On a network level, it pushes intelligence to the local area.
- *Edge Computing*
  Intelligence is directly pushed into devices such as programmable automation controllers.

Utilizing these will help to process data where it is closest to its origin storage and development such as information from sensors, pumps, or motors [2, 31].

4.3.1.1 Relevant Deployment Architecture

As slightly touched upon, an architecture of such kind and which considers the above ideas was presented in research by [1]. Their model is designed of a variety of layers to demonstrate the gathering of data and communication between layers presented just like a cake with different flavored layers as demonstrated in Figure 4.5. The bottom layer includes a variety of sensors and devices which provide raw data for further use. It is also referred to as the *Things Layer* which focuses on objects and content from the following environments [1]:

FIGURE 4.5 Deployment Architectures of Farming and Agriculture based upon [1]

- *Subsystems*
  These are hardware's of devices of systems.
- *Sensors*
  As more in-depth outlined in the introduction, a main task of sensors is to detect events.
- *Embedded devices*
  These are computing systems with a special purpose, such as cameras or printers.
- *Mobile devices*
  Mobile devices are portable devices such as smartphones, smartwatches as well as drones.

The next level, which is the Edge Layer and includes Subsystems control and interoperability, is an infrastructure adapted to the facility available and covers processes of:

- *Irrigation control*
  Examples for such are sprinklers or surface irrigation, hence water distribution without a mechanical pump.
- *Sensor data*
  i.e., accelerometer or photosensor which detects and responds to environmental information.
- *Image*
  Simply put, images are pictures or videos captured with CCTV or drones.
- *Climate control*
  Another way of air-conditioning.

The basic requirements for this level to work efficiently are interconnection and data access of all subsystems' data as well as configuration, operation, and modification processes [2]. This layer also includes the edge nodes, which filters, predicts climatic data calculations, and classifies services or detects environmental events [1].

The third layer has the fog nodes implemented and is therefore referred to as the Fog Layer which deals with the local analysis and expert rules management [1]. Tasks in this layer can include smart analysis as well as computing. Furthermore, it includes the implementation of "dynamic, real-time self-optimization data" and executing and adjusting policies such as [1, 2]:

- Prediction
  i.e., of water consumption which is necessary to plan finances for upcoming months or years
- Monitoring
- Planning
- Concluding that a possible system error has occurred
- Diagnosis of system error and specifying exact error
- Maintenance.

Data for these analysis methods and steps can be found in the local storage.

Following this is the Communication Layer, which is an interface in constant contact with cloud services and brings together the data gathering, analysis, and further maintenance for an improvement in the agriculture and farming environment and includes cell towers, satellite, and devices [1].

4.3.2 Application Domain: Industrial Safety

Focusing on the workplace security is highly necessary and utilizing the IoT to do so can improve such in this technological day and age. There are a variety of aspects in which the IoT can improve the safety of employees as well as the facilities overall. Fire prevention can be improved with IoT measures by using sensors to detect a fire but also be able to indicate which way the fire spreads to support firefighters in identifying the nature and also preventing further spreading. Moreover, employees with medical conditions would be able to be attended to more efficiently by wearing wristbands tracking their heart rate or sugar levels [32].

4.3.2.1 Relevant Deployment Architecture

Seeing the work environment security goals implemented into the IIoT architecture of industrial safety was demonstrated by Honeywell Industrial Safety, among others, which is applying sensors into protection equipment wear [2]. With the help of these utilized sensors live data from people and the environment can be collected to monitor this information and these will be forwarded to a cloud database. Then these data are being analyzed and the appropriate measure will be taken if and when the intelligent systems pick up any irregularities regarding the individual or the environment they are in while wearing the equipped protection wear or the sensors implemented in the surroundings. This proposed framework is a very beneficial model, has, however a variety of issues:

- *Sensor compatibility and being up to date:*
  Installation, connection, and calibration of systems as well as sensors need to be paid attention to thoroughly to make sure that the domains are protected efficiently. However, these also involve time and money which can be of a great issue especially in smaller businesses that just started to develop.
- *Data Processing*
  The data amount is of a high amount with such technological implementation. Therefore, systems have to be designed to be able to handle network traffic and/or need to be able to process data locally and only send the necessary information results so an analysis can be conducted efficiently.
- *Image vs reality*
  All implemented systems need to be programmed to be able to differentiate between situations of danger or safety. This can be challenging due to the algorithm and saved images possibly being different to an actual real-life situation, therefore, possibly resulting in false alarms or missing out on necessary signs and putting employees in danger.

Therefore, the model introduced from Honeywell Industrial Safety demonstrates three levels which are referred to as L1, L3, and L5. Therefore, to improve and ensure safety within the industry and in particular the protection wear employees are being equipped with [2].

L1

To highlight the nine crucial parts of protective equipment gear, which are also demonstrated in Figure 4.6, the following list is provided (similarly to [2]):

1. In Mask Heads Up Display
2. Communication enabled earmuff
3. QUIETPRO voice platform
4. Fit Check & End of life detection (Cartridge, cylinder)
5. Integrated Body Vitals Monitoring
6. Vibration Dosimetry in Gloves
7. Slip detection in footwear
8. Personal voltage detection (wearable)
9. Fall alert

L3

This *L3* level can be divided into two sections, which are the wearable gateway and the industrial gateway. These communication devices, i.e., two-way radio, are supporting a constant live update from the individual themselves with their manager and/or company to ensure an additional way of communication.

FIGURE 4.6 Personal Protective Equipment to improve safety and as suggested by Honeywell Industrial Safety [2]

L5

- Real-time health & exposure tracking
- Early hazard recognition and prevention
- End of Life replacement (sensors PPEs, i.e., Figure 4.6)
- Certification/compliance management
- Shift & ERT planning.

These devices and applications implemented within the protective equipment gear are in constant communication with a plant manager, emergency response, control room as well as an industrial hygienist. These collected and received information are shared via the cloud or device storage. This is to ensure a safer working environment for employees within the industry overall.

As mentioned above, the IoT can be implemented in various areas to improve workplace safety. Regarding Honeywell's three Levels, L1 would is a beneficial addition in regard to firefighters' equipment. Such helps them to not only be aware of own health and immediate surroundings but also be aware of their team members and their surroundings to either warn or support them in various situations. As L3 is in support of L5 tasks and can help to keep track on employees' conditions who have medical issues, recognizing safety hazards, time management as well as the product quality and/or availability.

4.3.2.2 Aspects to Consider

Exploring the deployment architectures in support of the industrial safety domain there are also additional aspects to consider, of which a few we have already outlined in our Key Security Goals Section. Therefore, as beneficial as the implementation of IoT in these areas may be, technology can also be prone to fail. Hence, it is necessary to evaluate additional solutions and backup methods in these areas in case of an IoT outage of the implemented security measures. Backing up data and having measures in place and on site which do not require the IoT are crucial to ensure the safety within industry even if an IoT failure is only of hypothetical measure.

Even though we can depend on technology in these domains greatly and technologies support and ease a variety of tasks, a full dependence on such is not recommended in any way. This is to ensure a safer working environment for employees in the industry and support meeting the key security goals as demonstrated in Section 4.2.2.

4.3.3 APPLICATION DOMAIN: PREVENTATIVE MAINTENANCE

Preventing threats to occur is just as crucial and part of keeping the working environment safe. Deploying the IoT in preventative methods can differ in each area of implementation. In the industrial environment, applied sensors audit the real-time performance of their systems to identify and repair issues.

Grade turbines are one area of IoT preventative maintenance within the industry where the sensors are distributed and utilized to collect and analyze real-time turbine data [2]. This enables employees responsible for the maintenance to have access and monitor health data from these turbines. Not only turbines can be monitored this way, but the engine status of vehicles can also be analyzed and reacted to by utilizing the IoT. However, there are a high number of relevant deployment architectures within the IoT environment for [33]:

- *Smart cities*
  i.e., waste is managed by container sensors and such will only be emptied if full, therefore, avoid unnecessary collection runs
- *Airports*
  i.e., tracking of lost baggage with RFID technology and the Hong Kong International Airport being the first to implement such
- *Highways*
  i.e., glow in the dark roads or electric lanes that are able to charge electric vehicles automatically
- *Healthcare*
  i.e., implantable Continuous Glucose Monitoring system which manages blood sugar; therefore, levels can be checked with a smart device.

As previously mentioned, there is a risk of failure of the implemented technologies; moreover, it is important to evaluate if the IoT technologies work well with one another especially if these are not manufactured or developed by the same company. Furthermore, due to dealing with high levels of private information as well as people's safety in these areas, there is the risk of cyber-enabled attacks such as hacking into the systems to attack such to fail or steal private information for purposes such as fraud or blackmail.

4.3.3.1 Relevant Deployment Architecture

The architecture model by [34] demonstrates a generalized demonstration of preventive maintenance architecture within the IoT and suggests that there are four levels as visualized in Figure 4.7. The following list also highlights these different aspects portrayed in Figure 4.7 and outlines each level and the processes crucial for this domain's deployment architecture to work efficiently:

- *Data sources-IIoT*
  All data relevant and usable for the specific infrastructure
- *Data Pipeline/Ingestion*
  Data transfer, transformation, and validation process
- *Data lake*
  Data repository which includes the steps of:
- *Time series*
  sequence of data
- *NoSQL*
  storing and retrieving data
- *Relational*
  data are categorized

Threats in Industrial IoT

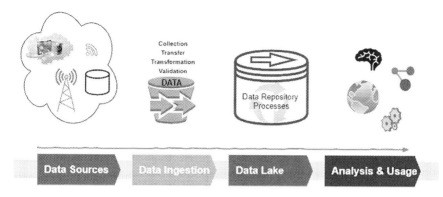

FIGURE 4.7 Preventative Maintenance within the IIoT

- *In Memory*
  reduces time to access data because it is compressed into the server's memory database which is quicker accessible
- *Files*
- *Search*
- *Processing*
- *Analysis and usage*
  Data analysis processes and techniques for collecting data and usable information.

Following this generic model gives an overview on how to efficiently collect data from a variety of sources which are then filtered and validated to evaluate the importance of data for specific aim of task. These data then will be processed on different levels to ease the last step of analysis and usage. Finally, data are analyzed further to establish a well-rounded maintenance and improve the security of equipment utilized in the industry by pre-evaluating possible future risks and dangers which can occur.

4.3.4 Threats Within IIoT

After elaborating the key security goals in the previous section and some of the relevant deployment architectures of IIoT within the three domains presented in this chapter's section, the following are some examples of possible threats that can occur within this environment and put employees and businesses at risk and outline why the above-demonstrated security goals and architectures are crucial within the industrial IoT.

4.3.4.1 Man-in-the-Middle Attack

Device and/or system control is overtaken by a third party to interrupt communication and breach information and potentially damage an industry such as farming and agriculture because crops and livestock can be negatively affected by such attacks. Information can also be changed regarding the products or the products can be controlled by the third party and therefore risk production and/or safety.

4.3.4.2 Device Hijacking

Hijacking is challenging to detect due to only one device being necessary to attack a system. Moreover, functions of the attacked devices stay the same as well as seem to be unaffected. Therefore, when such an attack is detected it may already be too late to analyze data correctly and livestock and crop may be lost during such attacks or employees may be put into dangerous situations due to such attacks.

4.3.4.3 Inside Job

A high number of attacks are conducted by ex-employees of a company who want to take revenge in some way or another and additionally who are still able to access the company's system for sensitive information or know valuable information to do so, hence whistleblowing. Therefore, inside jobs are considered one of the greater threats of IIoT due to the attacker being well knowledgeable of the environment as well as being aware of which area is the highest of risk when the system is attacked.

4.3.4.4 Data Breaches

Similarly, to the other threats an attacker accesses devices or systems to get different kinds of data for purposes such as ransom or fraud. Data which an attacker during such a breach may be looking for can include but are not limited to:

- *Client Data*
  Passwords, Clients of Companies' and their business partners, Clients systems, and security frameworks
- Personal Data
- Intellectual Property or Trade Secrets
  Vital frameworks and guidelines on how companies work
- Health Data
- Financial Data
  Banking and related login details as well as other related financial.

4.3.4.5 Theft

Most of the time, sensible data are not stored on devices' storage and only used for access purposes. However, portable devices which are used by the police for example, can store such sensitive data on the device itself. This is especially challenging and can result in information leaks where the content should have been kept private as long as possible in terms of investigation purposes.

4.3.4.6 Outdated Systems

Systems are being improved constantly and the need to update such is crucial to keep data management and processes efficiently as well as secure. This is time-consuming and likely a costly investment which not all industry domains are able to keep up to. Therefore, especially in terms of security, missing out on updates increases the vulnerability and can put these involved in risk.

4.3.4.6.1 *Exhaustion of Resources*

Adding to systems being outdates are resources being exhausted. Such can likely link to systems not being updated as well as not being managed efficiently. Furthermore, if the services or systems are not precisely programmed from the beginning the usage of IoT technologies within a domain are inefficient.

4.3.4.7 Insufficient Training

The majority of threats and/or attacks on domains that are utilizing the IoT can be linked to mistakes made by people handling the systems. In-depth training can be time-consuming as well as costly and often is considered not as of a high necessity when having to switch to a new system as quickly as it is seen appropriate for a business. Moreover, educating employees and people, handling the forever improving IoT systems, on risks and threats is still not as common it should be. This adds to more vulnerabilities of the system and threatens the domains on a level, which may be less considered or thought about than other possible threats.

4.4 CONCLUSION

Threats within the IIoT can take different forms. Therefore, focusing on the three different areas of agriculture and farming, industry safety and preventative maintenance demonstrated key security goals which are necessary for a better and safer environment. However, the goals also indicate and show the areas an improvement in regard to Cybersecurity is of high necessity within these domains.

Considering the different factors over the three different areas presented in this chapter, it has been demonstrated that system safety and reliability are among the common and highest key security goals which need implementation and focus on. Furthermore, the deployment architectures which are relevant and implement the key security goals are mostly different in all industry areas. However, generally the deployment architectures work on different levels which focus on a variety of important aspects in how to improve the security of the specific industry domain. Elaborating on a number of incidents as well as threats it can be concluded that these deployment architectures presented are of highest efficiency. Nevertheless, it is also necessary to continuously consider the failing of the IoT implementation within a domain and therefore have backup guidelines and frameworks for the industry domains which are not in need of functioning by utilizing mainly the IoT. This is to ease processes and efficiency within the environment and simultaneously decrease the dependency on IoT.

Therefore, this chapter aims to build a foundation to evaluate the threats within the IIoT environment. To allow this, we focused on the three following domains of agriculture and farming, industry safety, and preventative maintenance. To better demonstrate the threats in these specific areas this chapter focused on two crucial parts to evaluate these which were the Key Security Goals and Relevant Deployment Architecture of chosen domains while utilizing past IIoT incidents and threats for a more efficient elaboration.

This approach demonstrated that issues such as crop, livestock, product, and employee safety can be highly affected even with the implementation of the IoT. The presentation and elaboration of cases was able to show that the security of the IoT within these areas is not of the highest importance as of yet and therefore, allows attackers to easily access and manipulate systems and data. However, this is often linked to the lack of finances or time within the domains.

Overall, it can be said that the following security goals are being considered key and as of high importance within the three IIoT domains discussed in this chapter, which were Agriculture and Farming, Industrial Safety and Preventative Maintenance:

- Reliability
- Confidentiality
- Integrity
- Availability
- Safety
- Security
- Resilience
- Accuracy.

Improvement and implementation of the key security goals are linked to the deployment architectures which are considering on a variety of levels within the industry to utilize the IoT efficiently; however, it also demonstrates that a full dependency on the IoT is not recommended.

REFERENCES

1. F. J. Ferrández-Pastor, J. M. García-Chamizo, M. Nieto-Hidalgo, and J. Mora-Martínez, "Precision agriculture design method using a distributed computing architecture on internet of things context," *Sensors*, vol. 18, no. 6, pp. 7–10, 2018.
2. EuropeanUnion, "Cyber trust. Advanced cyber-threat intelligence, detection, and mitigation platform for a trusted Internet of Things grant agreement: 786698. D2.1 threat landscape: Trends and methods," Co-funded by the Horizon 2020 Framework Programme of the European Union, 2018.
3. D. Wee, R. Kelly, J. Cattel, and M. Breunig, "Industry 4.0-how to navigate digitization of the manufacturing sector," *McKinsey & Company*, vol. 58, pp. 7–11, 2015.
4. G. Lampropoulos, K. Siakas, and T. Anastasiadis, "Internet of things in the context of industry 4.0: An overview," *International Journal of Entrepreneurial Knowledge*, vol. 7, no. 1, pp. 4–19, 2019.
5. R. Schmidt, M. Möhring, R. C. Härting, C. Reichstein, P. Neumaier, and P. Jozinović, "*Industry 4.0-potentials for creating smart products: empirical research results*," in *Conference on Business Information Systems*, Poznan, Poland, 2015.
6. K. Taylor, "Examples of Industrial Internet of Things(IIoT)," *HiTechNectar*, 2020. [Online]. Available: https://www.hitechnectar.com/blogs/examples-industrial-internet-of-things/#Siemens. [Accessed: Oct. 10, 2020].
7. M. Stočes, J. Vaněk, J. Masner, and J. Pavlík, "Internet of things (iot) in agriculture-selected aspects," *Agris on-line Papers in Economics and Informatics*, vol. 8, pp. 83–88, 2016.

8. M. S. Farooq, S. Riaz, A. Abid, K. Abid, and M. A. Naeem, "A survey on the role of IoT in agriculture for the implementation of smart farming," *IEEE Access*, vol. 7, no. 156237–156271, 2019.
9. P. P. Ray, "Internet of Things for smart agriculture: Technologies, practices and future direction," *Journal of Ambient Intelligence and Smart Environments*, vol. 9, no. 4, pp. 395–420, 2017.
10. L. Junka, "The Internet of everything: IoT use cases," 2018. [Online]. Available: https://www.livingmap.com/technology/the-internet-of-everything-iot-use-cases/. [Accessed: Feb. 21, 2020].
11. C. Y. Lin, S. J. Chang, M. H. Lai, and H. Y. Lu, "*Overview of precision agriculture with focus on rice farming*," in *International Workshop on ICTS for Precision Agriculture*, 2019.
12. T. Grau, A. Vilcinskas, and G. Joop, "Sustainable farming of the mealworm Tenebrio molitor for the production of food and feed," *Zeitschrift für Naturforschung*, vol. 72, no. 9–10, pp. 337–349, 2017.
13. V. Murashov, F. Hearl, and J. Howard, "Working safely with robot workers: Recommendations for the new workplace," *Journal of Occupational and Environmental Hygiene*, vol. 13, no. 3, pp. 61–71, 2016.
14. K. Zetter, "Sony got hacked hard: What we know and don't know so far," 2014. [Online]. Available: Examples of Industrial Internet of Things(IIoT). [Accessed: 10, 2020].
15. T. Kitten, "Inside look at SWIFT-related bank attacks," 2016. [Online]. Available: https://www.bankinfosecurity.com/interviews/inside-look-at-swift-related-bank-attacks-i-3285. [Accessed: 10 2020].
16. L. Thames and D. Schaefer, "Industry 4.0: An overview of key benefits, technologies, and challenges," in L. Thames and D. Schaefer (Eds.) *Cybersecurity for Industrie 4.0*, New York: Springer, 2017, pp. 1–34.
17. IndustrialInternetConsortium, "Industrial Internet of Things volume G4: Security framework," 2016.
18. M. Guerra, "3 Ways the IoT revolutionizes farming," 2017. [Online]. Available: https://www.electronicdesign.com/analog/3-ways-iot-revolutionizes-farming. [Accessed: Feb. 1, 2020].
19. N. Desai, "IoT in agriculture: Farming gets 'smart'," 2018. [Online]. Available: https://www.networkworld.com/article/3268971/internet-of-things/iot-in-agriculture-farming-gets-smart.html. [Accessed: Feb. 15, 2020].
20. P. Jayaraman, A. Yavari, D. Georgakopoulos, A. Morshed, and A. Zaslavsky, "Internet of things platform for smart farming: Experiences and lessons learnt," *Sensors*, vol. 16, no. 11, pp. 15–17, 2016.
21. R. Roberts, "FBI warns from smart farm risk," 2016. [Online]. Available: https://securityledger.com/2016/04/fbi-warns-of-smart-farm-risk/. [Accessed: Jan. 18, 2020].
22. E. Hollnagel, *Safety-I and Safety-II: The Past and Future of Safety Management*, CRC Press, 2018.
23. Safeopedia, "Industrial safety," [Online]. Available: https://www.safeopedia.com/definition/1052/industrial-safety. [Accessed: Feb. 20, 2020].
24. Indatalabs, "Computer vision," [Online]. Available: https://indatalabs.com/services/computer-vision. [Accessed: Feb. 20, 2020].
25. J. Maktoubian and K. Ansari, "An IoT architecture for preventive maintenance of medical devices in healthcare organizations," *Health Technology*, vol. 9, pp. 233–243, 2019.
26. BetterBuys, "3 top IoT applications in the maintenance industry," 2017. [Online]. Available: https://www.betterbuys.com/cmms/iot-applications-in-maintenance/. [Accessed: Jan. 15, 2020].

27. I. Lee and K. Lee, "The Internet of Things (IoT): Applications, investments, and challenges for enterprises," *Business Horizons*, vol. 58, no. 4, pp. 431–440, 2015.
28. T. Jowitt, "Ransomware attack costs norsk hydro $40m," 2019. [Online]. Available: https://www.silicon.co.uk/security/cyberwar/ransomware-norsk-hydro-40m-242775. [Accessed: Aug. 10, 2020].
29. T. Jowitt, "Pharma giant bayer 'contains' cyber attack," 2019. [Online]. Available: https://www.silicon.co.uk/security/cyberwar/bayer-contains-cyber-attack-243573. [Accessed: Aug. 11, 2020].
30. T. Branigan, "South Korea on alert for cyber-attacks after major network goes down," 2013. [Online]. Available: https://www.theguardian.com/world/2013/mar/20/south-korea-under-cyber-attack. [Accessed: Aug. 12, 2020].
31. J. Whittaker, "Honda global operations halted by ransomware attack," 2020. [Online]. Available: v. [Accessed: Aug. 11, 2020].
32. T. Bell, R. Chamberlain, M. Chambers, B. Rieck, and T. Steinbrueck, "*Security on the farm: Safely communicating with legacy agricultural instrumentation*," in *2019 15th International Conference on Distributed Computing in Sensor Systems (DCOSS)*, pp. 192–194, 2019.
33. H. Claver, "Smart harvest tech becoming more cost efficient," 2019. [Online]. Available: https://www.futurefarming.com/Machinery/Articles/2019/4/Smart-harvest-tech-becoming-more-cost-efficient-411514E/. [Accessed: Jul. 27, 2020].
34. K. Hodge, "Coronavirus accelerates the rise of the robot harvester," 2020. [Online]. Available: https://www.ft.com/content/eaaf12e8-907a-11ea-bc44-dbf6756c871a. [Accessed: Jul. 29, 2020].
35. D. Linthicum, "Edge computing vs. fog computing: Definitions and enterprise uses," 2018. [Online]. Available: https://www.cisco.com/c/en/us/solutions/enterprise-networks/edge-computing.html. [Accessed: Feb. 21, 2020].
36. Digiteum, "How to improve workplace safety with IoT," 2019. [Online]. Available: https://www.digiteum.com/iot-workplace-safety#2. [Accessed: Jul. 28, 2020].
37. ClouderaInc., "Top 5 IoT use cases," 2017. [Online]. Available: https://www.slideshare.net/cloudera/top-5-iot-use-cases. [Accessed: Feb. 15, 2020].
38. P. Duhem, "Predictive maintenance by analysing acoustic data in an industrial environment," 2016. [Online]. Available: https://www.slideshare.net/capgemini/predictive-maintenance-by-analysing-acoustic-data-in-an-industrial-environment. [Accessed: Jan. 18, 2020].

5 Threats in IoT Supply Chain

S. A. Kumar, G. Mahesh, and Chikkade K. Marigowda

CONTENTS

5.1 Introduction	168
5.2 IoT Application Domain in Logistics, Tracking, Fleet Management	169
5.2.1 Background	169
5.2.2 The Key Security Goals	169
5.2.3 Relevant Deployment Architectures	171
5.2.3.1 IoT in Secure Logistics Management	171
5.2.3.2 IoT in Secure Tracking Management	171
5.2.3.3 IoT Secure Fleet Management	178
5.3 IoT Application Domain #8: Asset Tracking	183
5.3.1 Background	183
5.3.1.1 Asset	184
5.3.1.2 Types of Asset	184
5.3.1.3 Asset Management and Tracking	184
5.3.1.4 Where can Asset Tracking be Used	184
5.3.1.5 Goals of Asset Management	185
5.3.1.6 IoT-Based Asset Management and Tracking	185
5.3.1.7 Advantage/Benefits of IoT-based Asset Tracking	186
5.3.2 Key Security Goals	187
5.3.2.1 Attack Objectives	187
5.3.2.2 Vulnerabilities on IoT-Based Asset Tracking Systems	187
5.3.2.3 Attacks and Counter Measures	187
5.3.3 Relevant Deployment Architectures	188
5.3.3.1 Asset Tracking System Architecture	188
5.3.3.2 Case Study 1: Health Center	189
5.3.3.3 Case Study 2: Laptop	189
5.3.3.4 Case Study 3	190
5.3.3.5 Case Study 4: Steel Inventory Tracking	190
5.3.3.6 Case Study 5: Mobile X-Ray Machine	191
5.4 IoT Application Domain: Manufacturing	191
5.4.1 Background	191
5.4.1.1 Adoption of IIoT in Manufacturing	193
5.4.1.2 Smart Manufacturing a Use Cases	194

5.4.2 Key Security Goals 195
5.4.3 Possible Security attacks and its Countermeasures 196
5.5 Conclusion.. 197
References.. 197

5.1 INTRODUCTION

Internet of Things (IoT) is the cutting-edge technology used in various applications, to reduce our work, and work intelligently. There are various applications that use IoT; some of them are smart security and surveillance, smart agriculture, smart home, smart city, smart healthcare, smart grid, and smart water supply. These applications are common and will be used in day-to-day life by everyone. This chapter discusses about application of IoT for supply chain and its associated threats. The various subdomains of supply chain considered in this chapter include logistics management, fleet management, asset tracking, and manufacturing.

IoT plays a major role in logistic management and is used for locating, route management, inventory tracking, and warehousing. In this chapter, logistic management based on radio-frequency identification (RFID) and secure object tracking for smart logistics is discussed. The next application discussed in this chapter is IoT-based tracking systems, which finds its applicability in mobile tracking, animal tracking, and consignment tracking. Even in some sports application also IoT-based tracking system is used. In this chapter, tracking systems based on handled devices, multiservice ambulance tracking, and secure object tracking are discussed. Section 5.2 discusses about the application of IoT for logistics, tracking, and fleet management along with the security goals and relevant deployment architecture.

IoT for supply chain is all set to revolutionize the operational efficiencies and revenue opportunities of an organization. It not only helps an organization to keep track of its assets but also provides a way out to gain an edge over its competitors. Asset management and tracking brings in accountability of every important item in the organization and thereby streamlines the operational structure of the company. In this chapter, asset management of a health center, laptops, humans, steel inventory, and mobile X-ray machines is discussed. Section 5.3 discusses about the application of IoT for asset tracking along with the security goals and relevant deployment architecture.

The use of IoT sensors in the manufacturing industry enables condition-based maintenance observant. By ensuring the prescribed working environment for machinery, manufacturers can conserve energy, reduce costs, eliminate machine downtime, and increase operational efficiency. Potentially, manufacturing companies can use IoT data to improve productivity, predict equipment failure, spot trends, improve employee safety, and more, but only if they have the right data integration tools to make sense out of the raw data they gather. Large manufactures will utilize analytics information tracking using connected IoT sensor devices to analyze and optimize the manufacturing process. Industries have a considerable prospect at hand where they can not only supervise but also automate numerous complex processes concerned in manufacturing. However, there have been systems that can track development in the plant but the IIoT provides far more complex details to the supervisor to manage the activities effectively. Section 5.4 discusses about the application of IoT for Manufacturing along with the security goals and relevant deployment architecture.

5.2 IOT APPLICATION DOMAIN IN LOGISTICS, TRACKING, FLEET MANAGEMENT

5.2.1 Background

The sudden development and progress of the IoT has led to various cutting-edge IoT applications such as Intelligent Transport Systems, smart home systems, and Smart shopping systems, smart buildings. IoT provides us a platform for connecting small actuators, smart devices, and people anytime and anywhere to the Internet. This can be done with the help of some of the supporting technologies like WSN, RFID, Bluetooth, GSM, GPRS, and Wi-Fi. All the communications are Distributed in storage and other functionalities through Cloud Computing and Big data. We know the other major and vast applications are on smart logistics management, Object tracking, and Smart Fleet management. For all this application, first involves the automatic identification or smart identification of an object and tracking the location of the object. So, Figure 5.1 contains the classification of various technologies involves in smart identification of an object [2–4].

All these applications are identified and tracked the location of an object through an IoT technology. All the interconnecting devices are accessed by anyone in this world; it leads several vulnerabilities and treats to the devices as well threads and attacks to the information which is transferring inside the networks. So, security is one of the major challenges in the above-mentioned applications. This section is going to address various security challenges and how these challenges overcome with certain solutions for the smooth communication in smart environment. Section 5.2.2 is key security goals, Section 5.2.3.1 is going to discuss about the secure smart logistics management, Section 5.2.3.2 is discussing about smart secure tracking management, and Section 5.2.3.3 is discussion about smart secure fleet management.

5.2.2 The Key Security Goals

The challenges in IoT communication are vulnerability, that is, authentication, data protection, and recovery issues, compatibility issues in application layers, secure communication in network layers, fake nodes, node capture, and DoS attacks in physical and datalink layers. The key solution for the smart secure applications is fuzzy-based approach for wormhole attacks and hello flood attacks. Cluster-based scheme is used for secure communication. Trust management is playing an important role in secure communication used for secure storing, blowfish cryptographic-based algorithm for encrypting the data to avoid from DoS attacks, Block chain-based technique for authentication and secure storage [5]. Figure 5.2 describes the various technologies in secure transmission in IoT.

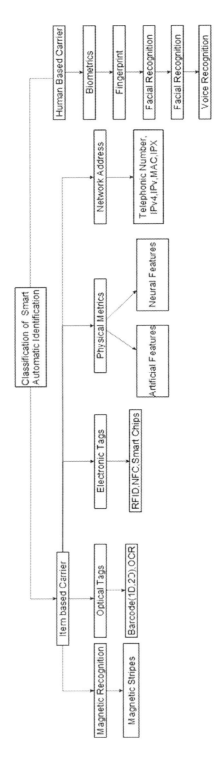

FIGURE 5.1 Classification of Smart object Identification [1]

5.2.3 Relevant Deployment Architectures

5.2.3.1 IoT in Secure Logistics Management

IoT plays a major role in the field of smart communication, smart buildings, monitoring, tracking and transportation, and logistics that technology is going to integrate into service-oriented process.

5.2.3.1.1 RFID in Logistics Management

In general, supply chain management (SCM) deals with material purchase, production, transport, and consumption. In this process, the accuracy of the data and timeliness place an important role in Logistics SCM. The delay in sharing of data leads to a huge loss to the organization. The solution for this is RFID, by using TAG it collects the data locally and store the information but the given information is static. But in an logistics organization due to changing environment dynamic information is required to the supply chain. So, the RFID and wireless Sensor Network is deployed in all the sites in different locations, which forms an IoT environment and gives the dynamic information about who, what, when, and where to SCM [6].

5.2.3.1.2 IoT Enabled RFID Authentication and Secure Object Tracking System for Smart Logistics

The basic issues in SCM is manual counting, identifying, and tracking the location of the objects, data managements, and securing the logistics information. Technologies like Wireless Sensor Networks, IoT, and RFID are providing the solutions; with this technology, we can make thesmart logistics management, meaning that we can automate the counting, and the location of the objects will be tracked easily using GPS and other supporting technologies. Initially finding the characteristics of the objects in logistics was difficult, IoT and RFID technology made it easy. This method requires a high computational resource. To avoid the computational resources, secret key is shared within the users, for authentication. Figure 5.3 shows the Application of Smart Logistics in various fields [7].

5.2.3.2 IoT in Secure Tracking Management

5.2.3.2.1 Tracking Context of Smart Handheld Devices

One of the applications in wireless sensor network is tracing the location of elements, which can be used in sports applications, animal tracking and consignment tracking.

We have certain android applications to track the lost or stolen mobile like GPS Tracking Pro, LocateMyDroid, SeekDroidLite, AntiDroidTheft, where is my Droid, Plan B, etc

All the application software is used to identify the GPS location, integrated with Google Map interface. But the issues are all these applications are high power consumption. Based on the existing system, a context-based tracking of the handheld devices is developed. The main function is we can track the location and state of the stolen devices from other smart or handled devices. The basic structure is client-server model; the server module is installed in lost smart devices, which is protected using system and SMS password. Client software is the any devices used to identify.

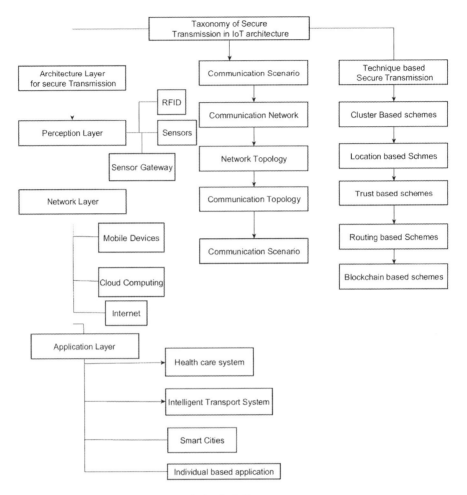

FIGURE 5.2 Various secure transmission in IoT

Through the client software, we can identify the location of devices with the help of in-built accelerometer sensor. The client will be received two messages from the server/lost device: the new mobile number of the lost device and current location of the lost phone [8].

5.2.3.2.2 Secure Real-Time Vehicle Management Framework

Through this research, we aim to secure management of smart vehicles and vehicle fleets through Real-time Vehicle Management Framework (RtVMF), by monitoring various operational parameters using security, privacy, and dependability metrics. A proof-of-concept implementation was deployed on real vehicles; performance overhead results validate the framework's feasibility.

IoT is one of the biggest emergence & developments in the technological world, which has led to several devices & technologies being developed and updated over the years, such as the Bluetooth, GPS (global positioning system), Smart Home, and

Threats in IoT Supply Chain

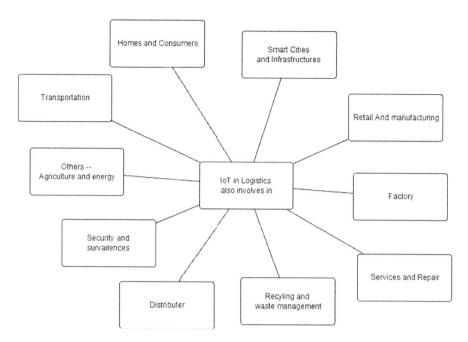

FIGURE 5.3 IoT in Logistics

many more. Here, Smart Vehicles, the ease of developing them & other parameters are discussed. Modern vehicles are present today with various latest/modern features, that of safety or luxury, which will ensure the safety and comfort of the driver as well as that of the passenger; also, the features help in achieving energy-efficient operation of the vehicle, and maximizing vehicle lifetime. The main approaches of RtVMF contain the speed of the operation and geo-fencing, meaning that location of the vehicle will be intimatedto the owner (Figure 5.4). Second intimates the vehicle health and engine conditions [9].

5.2.3.2.3 Secure Object-Tracking Protocol for the Internet of Things

This object-tracking protocol provides the security as visibility, traceability along with the travel path to support the secure IoT. The security features are attained with the help of RFID [10].

5.2.3.2.4 IoT-enabled Secure Multi-service Ambulance Tracking System

In this paper, IOT technology has been implemented on the ambulance van which carries patient from one place to another place merely hospitals. The tracking system has been introduced over van to keep a constant watch to get time-to-time updates of the patients carrying over. The ambulance is embedded with the multiple services to help doctors diagnosing while carrying the patient. These services include the medical equipment's which measures the vital parameters such as ECG machine to get ECG report, defibrillator to regulate cardiovascular activities, and ventilator and spirometer to measure respiratory volumes. The tracking system includes the cloud

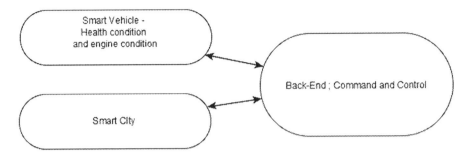

FIGURE 5.4 RtVMF Architecture

server, IOT module, and Android Application. The protocol incorporated is AES CCM which in turn has algorithm of message authentication code. It enables the encryption and decryption of vital data over GPRS system while tracking. Cloud server retains the database of the patient which can be easily accessed through mobile device application. IOT module (Arduino MEGA) interfaces the server with application. Application fetches the data through cryptographic method and routes the information via SMS or HTTP [11].

Advantages:
- Fast and accurate treatment of the patient.
- Easily accessible of the past patient record through cloud server which would help in diagnosing.
- Tracking system would help to have continuous watch over patient.
- SMS Alert on the mobile phone.

Disadvantages:
- Limited facilities and medication in ambulance for patient.
- Server loss could happen.
- Data breach due to cryptographic key changes to public by default.
- Data decryption would take more time and also to activate the generated key.

5.2.3.2.5 IoT-based Vehicle Tracking system in Bangladesh

Initially, Bangladesh used Finder vehicle tracking system, which is commercially available in the market contains GPS sensors, GPRS/GSM modules are used to track the vehicles, Later a little intelligent system called as LinkIt on board devices which is an IoT device containing GSM/GPRS, Bluetooth, Wi-Fi and other supporting devices was used. It's a prototype-based device, which is used to track the vehicle and along with that vehicle Devices RFID-based Driver authentication system is used for Security and proximity card is used for additional functionality of fuel line status.(Figure 5.5). With this system, the mobile application can easily track the vehicles, track the authenticated drivers, and fuel status [12].

5.2.3.2.6 Smart Secure Human Detection and Tracking

Traditional surveillance system is not scalable; at the same time, computation is very complex. So, the new technology is emerged to manage the security and tracking the human is IoT-enabled edge computing, where the surveillance cameras are embedded with smart communication process. Hence the computation is done using CNN-based human classification feature selection, at the edge devices(Figure 5.5).

As shown in Figure 5.6 the outcomes of IoT enbled edge computing is Communicated to Fog domain which is further communicated to cloud for historical data analysis [13].

5.2.3.2.7 IoT Health Care Monitoring and Tracking

This paper describes the various ways in which IoT devices can be used to ensure proper functioning of health care institutions and monitoring and treatment of patients using RFID technology. It can also be used in: [14]

- Control of medication and medical equipment: RFID tags can be used to monitor production, delivery, and tracing of medical equipment and medications.
- Information management in hospitals: IoT in hospitals can enable storage of data related to patients and help for future references and follow-up of treatment. RFID technology can be used in managing blood bag statistics ensuring less contamination and increasing efficiency of disposal of blood bags during emergencies. RFID tags can also be used for hospital waste disposal.

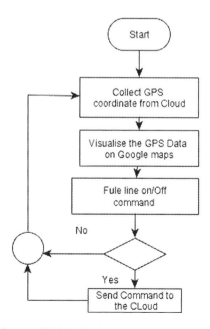

FIGURE 5.5 IoT-Based secure Vehicle Tracking System in Bangladesh

FIGURE 5.6 Edge Computing-based Secure Human Detection and Tracking

- Telemedicine: Helps in reaching to a diagnosis early, reduces the health care cost, and enables continuous patient monitoring thereby increasing the quality of life of patients.

The currently used E-health forms are:

- EHR (Electronic Health Record)—Acquires detailed information about the patient's diagnosis and treatment for further evaluation.
- PHR (Personal health Record)—Private, secure, and confidential space to store patient's information.

The devices currently available in market as follows:

- HEARTFAID—Detecting and managing heart diseases in elderly.
- ALARM-NET—Combines information from wearable devices and devices placed inside living spaces.
- CAALYX—Combined Ambient Assisted Living Experiment detects whether elderly person has fallen and if so their location.
- TeleCARE—Supervision and assistance for the elderly.

Thus, collection of data through IoT can be used for the betterment of human life.

Advantages:
- Large amount of data storage possible.
- Data stored will be precise → due to less manual error.
- Easy to access the data from anywhere.

Disadvantages:
- Leakage of confidential data.
- Mismatch of patients reports
- Network traffic and server get hang.

Cell phones have an imperative impact in everyday exercises of individuals. Without that individuals feel hard to endure. Despite the fact that it is a lot of valuable, it has influenced the conventional correspondence among individuals. It permits getting to of web for some reason, for example, getting to of messages, online installment, web-based shopping, person-to-person communication.

It assists with interfacing with all different remote gadgets, for example, Bluetooth, Wi-Fi. Since it helps in putting away all data identified with portable client, making sure about of these data is an important thing. Despite the fact that advanced cells are made sure about utilizing passwords, it tends to be burglary or lost. Looking of lost or stolen cell phones is the troublesome errand. Here a framework is given, which helps looking for lost or stolen mobiles utilizing area identification. The area of advanced mobile phones is sent as an SMS to the enrolled portable number [15].

5.2.3.2.8 IoT Edge Computing-Enabled Collaborative Tracking System for Manufacturing Resources

Manufacturing is an important process involved in the marketing world; it is the process of making goods by hand or by machine which upon completion, they are sold to the consumers by the business which made the products for use and sale purposes. It usually happens on a large-scale production line of machinery and skilled labor. This research discussed about the problems in manufacturing and how IoT and other-related devices helped make the process easier. In manufacturing, the availability of machines, the amount of labor, the workload distribution, the delivery timeline, and other factors are important in manufacturing, but sometimes there might be inconsistencies in the process, as in if any one of the factors become inconsistent, then the process becomes tedious & long, and it may sometimes halt in between, thereby affecting the deadline and targets of production and sales. There is one key manufacturing resource called material trolley since it conducts the major transportation tasks among different parties throughout the industrial park. However, the recycle of the vacant material trolley becomes a serious challenge for the industrial park [16].

To automate location tracking on manufacturing resources, there have been recent attempts in implementing this using IoT, encompassing devices & technologies like GPS, RFID, Bluetooth, and combination of various technologies. But despite the technologies & their applicability tests in identifying the manufacturing resources location, there are several concerns/drawbacks, such as inefficiency of a tracking device, mostly a GPS, to not work properly indoors as compared to outdoors, due to signal penetration & block issues, the requirement of large amounts of signal data collection & calibration work for the traditional indoor tracking devices and many more. The location can be automatically tracked and traced based on Bluetooth signals and edge gateways [17].

Arrhythmia and cardiac failure are the leading cause of death worldwide. Artificial cardiac pacemakers when used along with controllers help in normalizing the heart rate. This paper aims at using controllers to generate electrical impulses to maintain the heart rate by minimizing fluctuations.

The existing controllers in the market are FO-PID controller, F-PID controller, and PID controllers. The maximum error percentage is 0.72% for FOF-PID controller, while the same is 3.14%,6.18%, and 23.6% for FO-PID, F-PID, and PID, respectively, in a male heart. The corresponding error percentages in a female heart for FOF-PID, FO-PID, F-PID, and PID controllers are 1.55%,2.13%,11.25%, and 24.38%, respectively.

Clearly, this FOF-PID controller along with the pacemaker makes a big difference in reducing the error overshoot percentages in the heart. Real-time tracking of heart rate through IOT network enables easy communication between patient and healthcare unit possibly preventing a mishap.

The FOF-PID controller uses the fractional-order fuzzy logic principle to minimize the error overshoots in the heart. The paper has the mathematical modeling of the heart using a pacemaker, closed-form expressions for fractional orders, fuzzy logic working, data security in healthcare unit, and Simulink model. Anonymization technique is used for secured communication of cardiac data to the dedicated healthcare unit. A prototype using hardware is also being developed.

5.2.3.2.9 Smart School Bus Monitoring and Notification System

The safety of the children and women have been a question as the ratio of violence towards them have increased for the past few decades. The technological solution to this problem is discussed in this paper. The main objective is the safety of people at any cost. Designing an IoT device which is Arduino and this Arduino will be connected to GPS module and Wi-Fi module and a Bluetooth module so this Arduino will be connected to a database and it gives the location info to location tracker this unfix device also connected to embedded 1D 2D Barcode Scanner module. CCD Barcode Scanner Engine module with USD 2.0 Interface and as php barcode generator will generate barcode from either the webpage or the android app barcode will be pasted to bus and it scans and records the timestamp in bus table in database and sends the latitude and longitude location to the parents and school management.

As today increase of crime rate in every country the safety for children is getting vulnerable to everyone who is misusing and attacking innocent children and over 90% of kidnaps and child killings have been increased since last 5 years so there is a necessity that technology comes in a place and help the children in communication and tracking their locations so it would be helpful for both the parents and their caretakers [18].

Advantages:
- School children will be tracked, and we can safeguard them easily
- This tracking system will help parents to overcome the criminal cases happening against those children
- The main problem faced by parents and school management is due to increase in crime rate of mafia and child kidnaps parents and management are worried about their child safety

5.2.3.3 IoT Secure Fleet Management

5.2.3.3.1 Car e-Talk: An IoT-enabled Cloud-Assisted Smart Fleet Maintenance System

The paper presents a system for fleet management using technologies such as IoT and Cloud computing. The proposed system monitors the vehicle's health with the help of various dedicated sensors such as MQ-3, MQ-7, and MQ-135 gas sensor,

DHT-11 temperature and humidity sensor, tire pressure sensor, fuel level sensor, water level sensor, F031-06 voltage sensor, etc., along with HC-05 Bluetooth module and RF 433 MHz receiver module. Arduino controller is used to acquire the data from the sensors, which is then processed before storing the information on cloud server, deployed on Microsoft Azure platform, and displaying it on the driver's phone. The data are processed, analyzed, and classified using the Naïve Bayes classifier. The cloud-end and the driver-end of the system communicate using Simple Object Access Protocol (SOAP) and Hyper Text Transfer Protocol (HTTP). Any anomaly, if found, is also reported to the driver after analysis based on previous data using data virtualization techniques. The number of times the given vehicle had critical state, the frequency of occurrence of critical states, all possible health states, etc., are taken into consideration in anomaly detection. Air pollution has also been considered in the system by monitoring the CO and CO_2 concentrations in vehicle exhaust. Experimental results have shown that this system would be very useful for vehicle owners and fleet managers to get an insight about their respective vehicle health [19].

5.2.3.3.2 Improving the Fleet Monitoring Management, through a Software Platform with IoT

The fleet management system proposed in the article aims to effectively monitor of every vehicle in the fleet, by incorporating technologies such as IoT, Docker container, different microservices, and On-Board Diagnostics (OBD) Amazon Web Services (AWS) was used for cloud deployment of data as it is supports Docker container platform, along with a web and mobile application for monitoring. The fleet management in the software platform is intended to obtain real-time location of vehicles, alert the users through notifications in case of anomalies in vehicle health or whereabouts, interact with drivers and configure related statistics so as to obtain a comprehensive platform to monitor vehicles, and evaluate drivers' behavior simultaneously. The data collector modules used are XML, MQTT, LCDP, etc., along with technologies such as Cassandra and Scala for web application, IONIC2 for mobile application, Google Maps API for REST protocol, and D3 as data visualization library. The usability of the system has been analyzed based on road tests [20].

5.2.3.3.3 Smart Fleet Management System Using IoT, Computer Vision, Cloud Computing, and Machine-Learning Technologies

The features of the fleet management system presented in the paper are vehicle maintenance, multiple shipments tracking, vehicle telematics and speed management, driver management, health and safety management, and fuel management. The technologies employed for the proposed system include IoT (Watson IoT), Heroku for Cloud computing, OpenCV for computer vision, and other Machine Learning and Deep Learning (CNN) techniques. Sensors such as DHT-11 temperature and humidity sensor, reed switch sensor, LDR sensor, etc., were used with Raspberry Pi 3 model and MQTT protocol, various APIs, etc. The challenges that are addressed by the system proposed are driver behavior and safety, efficiency and fuel costs of vehicles, tracking of fleet to prevent theft, and detection of fake insurance claims by customers. The driver behavior and safety are ensured by the system by authenticating the driver and keeping track of his driving patterns, while the trucks are continuously monitored to prevent theft or anomalies in vehicle health. The challenges with respect to efficiency and fuel costs are solved by considering factors such as speed, idle wait time, and distance covered with cruise control in comparison with fuel consumption and driving and truck and driver recommendation modules were developed [21].

5.2.3.3.4 IOT-based Predictive Maintenance for Fleet Management

Technological advancement in IoT and data analytics made predictive maintenance feasible. Predictive maintenance reduces the cost of vehicle maintenance, ensure long life of equipment by avoiding unscheduled services, and fault management. The article highlights an existing fleet management system for the public transport buses. It is named as Consensus self-organized models approach abbreviated as COSMO. This model identifies faulty buses from the rest of the fleet. Predictive maintenance or smart transportation is not an easy task as labeled datasets of equipment faults are not readily available. And creating a dataset of vehicles deliberately making faults is not economical. The article points out another fleet management system, MineFleet, which gathers information from all the buses of a fleet using a gateway that connects sensor devices on each bus to the Internet. With the understanding obtained from these two systems, the authors have proposed a novel predictive maintenance fleet management system focused on the maintenance of buses. IoT can be seen as a layer's structure with five layers namely, sensors, network, storage, learning, and application. The sensors and actuators in the sensor layer acquire data that are stored to cloud by the wireless networks of the network layer. These stored data are fetched and processed using learning algorithms at the learning layer to give desired output to the applications on remote devices. When it comes to deviation detection, most labeled datasets only determine if an equipment is abnormal or normal. It is not possible to find the extent of deviation from one abnormal vehicle to another.

Prognostics is a term used for processing health status of an equipment. Prognostics is done in three ways: model based, expert system, and data driven. Out of these three, data driven is most used because it is both technically and economically feasible. Data-driven approach is generalized and can be applied to all types of equipment, whereas the other two approaches have to be configured separately for each type of equipment. COSMO used data-driven approach. However, it had few drawbacks. It resulted in repairs although there were no deviations or faults, which proved to be added to the economic burden. Thus, the authors proposed a system which has three-layered architecture, namely perception layer, the middleware layer, and the application layer. The perception layer holds the vehicle nodes (VN), gateway, and J1939 network connecting all VN nodes. The gateway acquires all sensor data and MQTT protocol is used for communication with the fleet system. The middleware layer has a server leader node (SLN) and each SLN has an MQTT broker and its own distributed database (DDB) fragment. The application layer has a root node (RN) accountable for entire system management. It has an MQTT broker to control all data access requests. The algorithm they proposed is Improvised Consensus self-organized models (ICOSMO) with semi-supervised learning techniques for selecting the apt sensors and features. The features should be apt in order to overcome the drawback of false fault predictions. The model does not promise to remove such false predictions but only aims to reduce them. A black box Document Retrieval algorithm (BBDRA) is used to dynamically adjust COSMO sensors over time.

A minimal viable product (MVP) is developed using raspberry pi, a UPS Pico, Laptop as RN and GSM communications, and Wi-Fi network. A lot of data have been acquired by this system when applied to 19 buses. The information includes network routing data, proprietary data and vehicle velocity, angle, deviations, and acceleration data. This is just a prototype and future work of this article includes the full implementation and experimentation [22].

5.2.3.3.5 Open Interfaces for Connecting Automated Guided Vehicles to a Fleet Management System

Advancement in digitalization meant acceptance of more IoT-based cyber-physical systems (CPS). Most control architectures are based on the standard ISA 95 and IEC 62224, but problem arises with the interfacing of different hardware. This paper highlighted the significance of vertical and horizontal integration and proposed a model for interfacing between different nodes at different levels of hierarchy. "Industry 4.0" is a highly discussed and researched topic, which demands a change in the manufacture of equipment in order for them to be able to communicate and enhance flexibility and productivity. This paradigm also includes the management of the data flowing. Many manufacturing companies are unable to accept the automated solutions due to lack of standardization and high cost of equipment. Most commercially Automated Guided vehicles (AVG) are not flexible and compatible with Industry 4.0. Thus, through this paper the authors proposed open interfaces to connect AVGs to Fleet Management systems.

Prior to Industry 4.0 paradigm, automation of production systems followed a pyramid-shaped hierarchical model. This pyramid model had five layers namely: FIELD (layer 0, consisting of all hardware involved in the production), PLC (Layer 1, the programmable Logic control), SCADA (Layer 2), MES (layer 3, Manufacturing Execution System), and ERP (Layer 4, Enterprise Resource Planner layer). Each layer needs to communicate with the layer below and above it. Each layer also has to communicate with the zeroth layer, that is with all the hardware nodes. Thus, the hierarchical structure is changed to accommodate the required communication patterns. These communications are achieved using open interfaces thereby dealing with the scalability issues as well. Task Planners (TP) are a core for any Fleet Management system. These TPs are configured to automate the processes of equipment. The authors propose this open interface with many objectives. Some of the objectives include creating a context/message oriented open middleware, development of an AGV interface, message exchanges between middleware and AGV interface, and message exchanges between AGV interfaces and AGVs. For the development of these interfaces, many cutting-edge technical advancements like Logistic Task Language Domain (LoTLan), orion context Broker with connection to global database via a DDS, and ROS (Robot Operating System) platforms are used. The LotLan Task identified tasks actions and motions using Unique UserId (UUID) given to each of the nodes. For message exchange between AGV interface and middleware, five channels are built namely, action_channel, motion_channel, movement_channel, status_channel and description_channel. And for message exchange between AGV interface and AGV, three channels were developed namely, cmd_vel, status_channel_AGV, and description_channel_AGV.

The proposed architecture was tested on different hardware environments and the test included two tasks one with and other without obstruction in the path of a node. The results showed the benefits of the open interfaces ensuring vertical integration. Future development involves more work on horizontal integration among node of same level and of different systems in the same hierarchical levels [23].

5.2.3.3.6 IoT-based Interoperability Framework for Asset and Fleet Management

The paper outlines a technical framework for a fleet management system. There is a need for cutting-edge innovative solutions to service and maintenance solutions. Frameworks for heterogeneous systems have been developed but not many were based on IoT. Moreover, when they did, interoperability remained a problem. The framework proposed in this paper takes all basic functionalities of a fleet management system into consideration. Both frontend and backend systems functionalities are pointed out. Some of the backend functionalities include data storage, analytics, optimization algorithms, etc. FMS operations are mostly confidential, but a few were pointed out namely data mining, distributed data storage, communication, data visualization, access control, and remote monitoring. The proposed framework consists of data management models, protocols and standards, user characteristics, security parameters, and essentially interoperability meters. These interoperable frameworks highly depend on the specific requirements of the companies. The requirements themselves are used as frameworks and help in choosing the fleet management system suitable for the requirements. The usability and completeness of the proposed requirements framework is yet to be determined in future work [24].

5.2.3.3.7 Empirical Path Loss Model for Vehicle-to-Vehicle IoT Device Communication in Fleet Management

Vehicle fleet management systems manage vehicle information to maintain and service the fleet. While using IoT as a base it is essential that the vehicles communicate by themselves at essential times. This requires modeling and configuration of signal propagation for in-vehicle-to-in-vehicle communication. In this paper, one such model is proposed. In the experimental setup, IoT nodes were scattered. Each of these nodes has receiver sensitivity, and polarized omnidirectional antennas. The modulation technique used by them is a direct sequence spread spectrum. This was used to shield noise and other hindrance. Zigbee or IEE 802.15.4 is used as implementation protocol. One of the nodes is configured to be the sink node and the rest were routers. Experimentation was carried on to determine the path loss linear regression model. The metrics determining the model were far-field distance, reference distance (predetermined), and distance between sender and receiver nodes. Some other statistical models like MAPE (mean absolute percentage error) are used. The results were obtained for sample vehicles like Sedan and Suv. Their research showed

Threats in IoT Supply Chain

that the proposed model deviates from theoretical model by 6–23%. In future, the authors aim to continue the study for vehicles at different speeds [25].

5.2.3.3.8 Smart Fleet Monitoring System using Internet of Things (IoT)

The paper highlights the need for resource management in terms of saving fuel by avoiding economical loss. The author points out that having good control over maintenance of buses and other vehicle fleets can help save resources. Thus, a smart fleet management system is proposed. The objectives of the model were to monitor the fuel levels of each vehicle through fuel sensors, check fuel consumption levels per distance traveled, minimize fuel wastage, flexibility and feasibility for entire system monitoring, and increase reliability at low cost. The model developed used IoT with simple monitoring sensors, distributed databases, and reliable interfaces. Control flows are represented using flow charts and pseudo-codes are elaborated. The hardware components used for implementation are a microcontroller, odometer, fuel sensor, Arduino UNO, and GSM SIM900 Modem. This FMS measures fuel consumption and wastage details providing control measures to the drivers. In future, an RFID sensor can be added to link the vehicle information with the driver [26].

5.2.3.3.9 Clone-Resistant Joint-Identity Technique for Securing Fleet Management Systems

The paper presents a fleet management system with all its sensitive entities made clone resistant. The need for this secure FMS is evident in the present scenario where monitoring vehicle activities has become mandatory. Central fleet management has become an essential for almost all manufacturing units. However, fake devices are a hurdle in the proper functioning of these FMSs. Thus, the authors proposed a secure method to prevent faking. The paper points out few threat scenarios for the fleet management system. It is then pointing out the updates required. Some of the updates are individual device authentication, jointly clone resistant entities, location certainty, unclonable time scale, nonrepudiation, confidentiality, availability, data integrity, and authenticity. The authors proposed a sample solution in which all above measures have been applied. Digital clone-resistant units were used, Secret Unknown Ciphers were deployed, and Random stream ciphers were put into action. The proposed framework had ELD devices, device gateway, end-user services, platform services, and several security components. The ELDs and vehicles are made clone resistant. The proposed system is flexible and scalable and is extendable to any attack scenario not just the one tested [27].

5.3 IOT APPLICATION DOMAIN #8: ASSET TRACKING

5.3.1 Background

From the operational point of any business, security of assets in the organization is important. This calls for classification, management, and tracking of assets in the organization. Asset management and tracking brings in accountability of every important item in the organization and thereby streamlines the operational structure of the company. In this section, a brief introduction to asset management and tracking is discussed.

5.3.1.1 Asset

From the business point of view, any item that has a value and supports the business operations can be viewed as an asset. Asset can be a device, data, people, or any other item that exists in the business structure.

5.3.1.2 Types of Asset

- **Physical Asset:** This includes all the physical items that exist in the business structure which supports the business operations and adds value to the organization. Traditionally a tag is affixed to each of the physical asset and a document like stock register or ledger is maintained to keep track of such assets. Examples of physical assets are IT devices, vehicles, machinery, equipment, tools, etc.
- **Data Asset:** This includes all the pieces of information that exist in the business structure which supports the business operations and adds value to the organization. Typically, this includes details of employees, clients, salary statements, design documents, intellectual properties, etc.
- **Human Asset:** This includes all the people in the business structure who support the business operation and add value to the organization. Typically, this includes all the employees of the organization.

5.3.1.3 Asset Management and Tracking

Asset management and tracking is the process of maintaining and monitoring the assets of the organization which includes physical assets, data assets, and human assets. Traditionally asset management is done using ledger documents or inventory management systems, and asset tracking is done predominantly by human intervention. Human involvement in tracking/estimation of assets around a business facility is often time-consuming and generally adds more cost and time to the organization. Asset tracking systems typically avoid human intervention on physically counting the assets and solve this problem of tracking/estimation of assets with low cost and time.

5.3.1.4 Where can Asset Tracking be Used

Asset tracking can be used in various domains. Few of the domains are listed below

- **Construction**—To keep track of different machinery and tools used in multiple worksites and warehouses.
- **Hospitals**—To keep track of medical equipment and other medical tools across hospital networks or a single large hospital.
- **Retail**—To keep track of different items in huge go down or warehouses, manage inventory, and to keep track of shopping carts and baskets.
- **Travels**—To keep track of vehicles in the parking lot and on the move, generate invoice based on actual vehicle usage.

- **Leasing equipment**—To keep track of the equipment and generate invoice to clients based on actual usage of equipment.

5.3.1.5 Goals of Asset Management

The ultimate goal in asset management is to ensure seamless business operation while safeguarding the valuable assets of the organization. Asset management helps in planning and maximizing returns on investments made, as well as minimizing losses.

The goals of asset tracking include:

- **Optimal asset utilization**—Asset management should help in optimal utilization of assets in an organization. A good asset management system must give a clear picture of the total owned assets of an organization at any given point of time. It should also be able to give an insight about the assets that are utilized to the maximum and assets that are unused. This would help the organization to plan and reconfigure their working strategies by reallocating portions of workload from max utilized assets to unused or idle sitting assets.
- **Track asset usage**—Asset management should help an organization in tracking the asset usage across the organization. A good asset management system must give a clear picture of who is using what, where, and when about all the assets of the organization.
- **Prevent losses**—Asset management should prevent or minimize losses. A good asset management system must be able to identify idle time of assets as well as inefficient workflows in the systems, thereby helping the organization to plan better for minimizing the losses. Also, a good asset management system must see to that the incidents of thefts are lowered and there are no items lost in the business facility.
- **Facilitate data collection and decision-making**—Asset management should help in data collection and decision-making. A good asset management system must be able to collect and store huge volumes of data generated in the business process. The collected data should be helpful in decision-making for example, when purchasing a new asset, previous data collected must help the organization in making an appropriate decision about which vendor to choose for purchasing the asset. An asset purchased from a specific vendor might have performed better over a period of time or might have a lifespan which is longer than other assets purchased from a different vendor.
- **Maximize return on investment**—Asset management should be able to maximize the return on investment of an organization. A good asset management system must be able to help the organization with decision-making, reduce losses, and optimally utilize the available assets reducing the need for purchasing additional assets unnecessarily.

5.3.1.6 IoT-Based Asset Management and Tracking

One of the characteristics of IoT is providing a unique identity to the physical objects. This characteristic makes use of IoT in the development of a consumer asset tracking system [28]. Currently mobile phones have unique identity and GPS tracking system to trace/locate the lost phones; IoT is enabling this feature to assets. Organizations with the help of IoT can now find the location of their products/assets that are stationary or in transit.

IoT-based asset management and tracking systems connect all the assets (physical, data, and people) into a single system. It does all the job of traditional asset management software's like tracking the location of an asset, giving the current condition of the asset, inventory status of the asset and complete life cycle of assets. In addition to this, IoT-based systems add intelligence to the life cycle of assets and gives real-time alerts. The purpose of using IoT for asset management and tracking is to improve the overall operational efficiency and productivity.

The IoT-based asset tracking systems typically utilize sensors to track the location of assets without human intervention. The sensors are attached to the assets along with any other traditionally used tags like RFID, Barcode, or QR code. In these systems, the information about the real-time location of an asset is obtained from the sensors and the same is sent to the cloud through a router over a secure communication channel.

5.3.1.7 Advantage/Benefits of IoT-based Asset Tracking

Like traditional asset management systems, IoT-based systems also improve staff efficiency and reduce the waste of assets in an organization. IoT-based systems have a number of other advantages as compared to traditional systems that include:

- **Lesser track time**—Typically finding assets in a large business facility using traditional methods takes more time. Asset tracking systems using IoT can instantly give the location of the asset in no time.
- **Real-time access**—Traditional systems only documents or maintain data about assets in an organization; however, IoT-based systems give real-time access to the assets.
- **Accurate asset location**—Traditional methods only give the details of the asset location but there is no assurance of the asset being physically present at that location. Since IoT-based systems give location data on real-time basis, asset location can be 100% assured.
- **Data generation**—Traditional methods only maintain the data in the system and do not generate any new data. IoT-based systems periodically generate data that can be stored and used for further analysis and optimization.
- **Prevents theft and loss of assets**—Traditional methods do not have any mechanisms to alert the owner in case of theft of an asset. IoT-based systems add intelligence to the system and can alert the owner or concerned person in case of a theft, thereby preventing thefts and loss of assets.
- **Live inventory management**—Traditional methods only give inventory details based on the relevance of database updating. This can go wrong sometimes when the concerned person misses to update a transaction. IoT-based inventory management is live and the inventory status is obtained in real time. This in turn will result in a better asset life cycle.

5.3.2 Key Security Goals

5.3.2.1 Attack Objectives

Typically, an attacker may plan to attack an IoT-based asset tracking system with one of the following objectives:

- **Cause physical damage to the equipment**—This can happen in scenarios wherein IoT is used to control some of the physical activities in an organization. The attacker may take control of the equipment and create physical damage to the equipment.
- **Snoop, corrupt, or destroy data**—This can happen in scenarios wherein IoT is used to capture data about the various activities of an organization. The attacker may snoop on sensitive data of the organization without causing any damage to the data or he may corrupt the data making the organization take wrong decisions or he may destroy the data to cover up some malicious activity in the organization.

5.3.2.2 Vulnerabilities on IoT-Based Asset Tracking Systems

As IoT devices are connected to Internet 24 hours a day, it is easy to break its vulnerability by the cyber criminals [29]. The following are some of the vulnerabilities on IoT-based systems [30]:

- **Deficient physical security**—Since majority of the IoT devices operate on unattended environments, an attacker may easily take control of the device and cause physical damage to it.
- **Insufficient energy harvesting**—Since IoT devices have a limited amount of energy, an attacker may flood messages to it and drain the energy of the device.
- **Inadequate authentication**—Because of the constraints on energy and computational power of IoT devices using complex authentication schemes is challenging. Hence, it would be easy for an attacker to break through the system.
- **Improper encryption**—Encryption is an important mechanism for data protection, however, because of the resource limitations of IoT devices, using robust encryption algorithms is challenging. Hence, the attacker would be able to circumvent the encryption algorithm to access sensitive data.
- **Unnecessary open ports**—Typically IoT devices have a number of open ports, enabling an attacker to barge into the system and exploit vulnerabilities.

5.3.2.3 Attacks and Counter Measures

The following are some of the attacks and countermeasures for IoT-based systems [31]:

- **Denial of service attack**—In this type of attack, the attacker floods messages to the system thereby attacking on the network bandwidth and services. This prevents legitimate users from accessing the system. Typically, this can be controlled using firewalls or intrusion detection systems.

- **Man-in-Middle attack**—In this type of attack, the attacker places himself in between the origin and destination of the message over the communication channel, intercepting and modifying the message. Typically, this can be controlled using one-time passwords.
- **SQL Injection**—In this type of attack, the attacker injects malicious input to the database to retrieve data from the database or destroy the database. Typically, this can be controlled using parameterized query or stored procedures.
- **Remote code execution**—In this type of attack, the attacker exploits the vulnerability of the system and places a malware that can remotely control the system. Typically, this can be controlled using malware prevention techniques or checking for bounds of the data buffer.
- **Brute force attacks**—In this type of attack, the attacker tries to gain access to the system using different sets of input until a valid one is found. Typically, this can be controlled by means of lockout feature after certain number of invalid inputs.
- **Data sniffing**—In this type of attack, the attacker sniffs the data passing over the network to gain access to confidential information of the organization. Typically, this can be controlled by means of encrypting sensitive data before transmission over the network.

5.3.3 Relevant Deployment Architectures

5.3.3.1 Asset Tracking System Architecture

Figure 5.7 is the architecture of IoT-based asset tracking systems. Typically, it includes the following components.

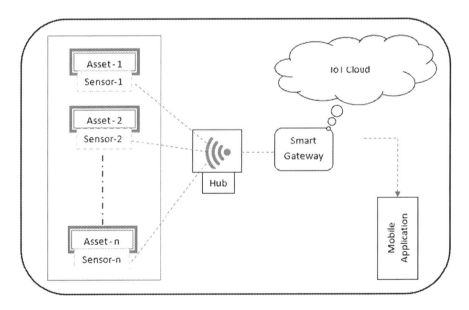

FIGURE 5.7 Architecture of IoT-based asset tracking systems

- **Assets**: This includes the physical components of the organization like devices, machinery, packages, etc. Each asset is uniquely identified and tracked with the help of sensors attached to it.
- **Sensors**: The sensors used can be of varied type based on the specific purpose of asset tracking. Location sensors can be used in order to track the location of the specific asset. These types of sensors typically send the location of the asset to the IoT cloud through Hub and Gateway. Other type of sensors also exists to measure humidity, proximity, temperature, pressure, acceleration, etc. Based on the specific needs of the organization, appropriate sensors need to be selected for the system.
- **Hub**: This receives the data from sensors and forwards the same to the gateway for storing into the IoT cloud.
- **Smart Gateway**: This receives the data from Hub and stores the data into IoT cloud. It can also perform some of the local computation before storing into cloud.
- **IoT Cloud**: This is the place where all the data is stored and details about any asset of the organization can be retrieved on a real-time basis. The data here can be further analyzed using machine learning algorithms for appropriate decision-making in an organization.
- **Mobile Application**: This is the application interface available to the user of the asset tracking system for data display, surveillance, notification, and alert during critical times.

5.3.3.2 Case Study 1: Health Center

The authors of [32] have proposed an IoT-based asset tracking systems for a health center named LoCATE—Localization of Health Center Assets Through an IoT Environment. LoCATE is a near real-time tracking system based on IoT for tracking patients, medical staff, and medical devices in a health center using 802.11 WIFI infrastructure. The system continuously logs the real-time location data of patients, medical staff, and all the devices into the cloud using which the administrators can find the location of patients, doctors, or devices with help of a mobile or web application. LoCATE also analyzes the collected data and exposes the inefficiencies in the workflow of the health center. The hardware used in LoCATE is Raspberry Pi equipped with a USB wireless adaptor capable of monitoring 802.11 wireless packets. The software used in LoCATE includes EC2 cloud instance running on AWS utilizing the Ubuntu server operating system. To build the user interface of web application, the application framework used is LAMP. To sniff the network traffic TShark utility is used in LoCATE. In summary, LoCATE is a low-cost asset tracking system designed using cloud, IoT, and 802.11 technologies which include the benefits of flexibility, scalability, and improved health center work operations when compared to traditional systems.

5.3.3.3 Case Study 2: Laptop

The authors of [33, 34] have proposed an IoT-based asset tracking systems for laptops. They have proposed a generic middleware architecture for securing laptops in an office environment. The middleware for this is implemented in the Windows operating system using a technique called middleware in-lining, which demands writing the middleware code in modules that are injected directly into the application. The authors suggest the use of a GPS and GPRS integrated hardware in the laptop with a nondetachable SIM card. The serial number of the laptop is not displayed; instead, it is given as a separate printout to the laptop owner along with the SIM details. Now, in case of theft the owner simply sends the serial number of the laptop to the contact number given in the printout. The GPRS module requests the GPS module to find the location data and it then sends this information to the owner of the laptop. In this way, the theft of laptops can be tracked by the owner. In summary, the middleware supports a number of services like laptop monitoring, tracking the location of laptop, bidirectional SMS communication, and generated data management using SQL commands. This solution can be extended and incorporated into any other valuable assets like televisions, machinery, and other costly devices. The solution is also a cost-effective one as it involves already mature and existing technologies like the GPS and GPRS, taking advantage of network roaming, the assets can be tracked on a real-time basis anywhere by anyone across the globe.

5.3.3.4 Case Study 3

The authors of [35] have proposed an IoT-based asset tracking system for securing women, children, and people with mental disorders. The hardware used includes a nano microcontroller, vibration sensor, sound sensor, LCD display, RF module, GSM module with inbuilt GPS, and WIFI module. The software includes the development of a widget and website to determine the location of assets using ThingSpeak—an open-source IoT platform. The system operates in two phases tracking phase and monitoring phase. In the tracking phase, the transmitter kit is placed with the asset that needs to be tracked. The data from the sound and vibration sensors are continuously collected, as soon as the sensed value exceeds the predetermined threshold value, a call and a predefined alert message is sent to the preregistered mobile numbers through GSM indicating that the asset is in danger. Also, the transmitter and receiver modules continuously communicate with each other using the radio signals. When these two moves away from each other the signal strength decreases and even this case a call and alert message is sent the predefined mobile number indicating that the asset is getting lost. The monitoring phase is done with the help of the receiver kit. Typically, this will be the administrator monitoring the asset. The LCD screen on the receiver kit displays the alert message when the asset is in danger or when it is lost. The buzzer attached to the receiver kit emits sound indicating an alert situation to the administrator. In summary, this is a low-cost asset tracking system typically proposed to secure human beings with slight modifications the same can be used to secure any other valuable assets of an organization.

5.3.3.5 Case Study 4: Steel Inventory Tracking

The authors in [36] have presented an intelligent steel inventory tracking system using IoT. This is based on the automation of one of the warehouses of Steel Mill company by extending the RFID-based solution to IoT domain. The width of the warehouse is 40 meters, length is 400 meters and height is 15 meters. An overhead crane is used to stack up the products (steel beams) of different length and weight in the warehouse. The steel beams are uniquely identified using a label and RFID tag. The products are rolled in and rolled out of the warehouse from a fixed point with the help of the overhead crane. The traditional approach followed before this system was, on receiving an order from the ERP, the crane operator manually had to pick the corresponding products from their respective locations and load into the transport vehicle. Product location was purely based on the skill and memory of crane operator as well as a spreadsheet document. In the IoT-based system, each product is uniquely identified using an EPC (Extended Product Code) embedded in the RFID tag. The overhead bridge is used to carry several products simultaneously and place them in the transport vehicle. Location of the products in the warehouse is obtained using the location engine in the bridge. The products can also be tracked after they are dispatched from the warehouse to the client with the help of IoT sensors attached to the product. In summary, the intelligent steel inventory tracking system which started in one of the warehouses of Steel Mill was approved by Steel Mill to be expanded to the corporation.

5.3.3.6 Case Study 5: Mobile X-Ray Machine

The authors in [37] have presented a mobile X-ray tracking and monitoring system using IoT. The system was developed to track and monitor mobile x-ray machines with higher efficiency and accurate services. The system is capable of searching and displaying the location of the machine in real time and also capture the usage details of the machine. The hardware part of the system includes MCU nodes with Wi-Fi sets, sensor modules, and wireless communication modules. The software part consists of a program for sending/receiving data and an application interface for configuration and reporting results. The efficiency in correctness of location identification is tested at Songkhla Hospital and is found to be 88.13 and 90.63%. In summary, this system can be used to locate and track mobile medical device services in the building more efficiently. It can also evaluate statistical data thereby helping the administrators in making decisions for various aspects of medical devices like repair, maintenance, and efficient new purchases.

5.4 IOT APPLICATION DOMAIN: MANUFACTURING

5.4.1 BACKGROUND

To appreciate the benefit of Industrial Internet of Things (IIoT) in manufacturing it's critical to initially understand how precisely manufacturing has advanced since first industrial revolution till today's industry 4.0.

The initial three revolutions of industry are depicted as being driven by mechanical assembly and fabrication relying upon on steam and water concentrating on manual performed by labor and heavily supported bythe use of animals, utilization of large-scale manufacturing and electrical energy empowered industries to expand improved efficiency and helps machinery more portableand due to which the mass production concepts like assembly line were introduced to increase the productivity, and the utilization of semiconductor, computer technology, computerized production, respectively [38]. The fourth technological industrial revolution was proposed in the year 2011 [39]. This revolution is portrayed by its dependence on the adoption of CPS capable interacting with each other without human intervention and of building autonomous, de-centralized decisions, through which theincrease in industrial productivity, efficiency, transparency, safety and as IoT is the core component of industry 4.0 refers to the developmental procedure in the management of manufacturing and production chain.

The characterization of the IIoT incorporates two key segments: The association of mechanical machine sensors and for local processing using actuators to the Internet. The forward association with other significant industrial networks that can generate a value independently.

Back in the 2000s, IIoT appears with the RFID [40] tags for routing function, reviewing and loss prevention [41]. The higher approximate value of IIoT applications concerns manufacturing plants IIoT, integrate Physical assets and Cyber world, prompting the formation of CPS, which enables among others, Real-Time production control, activity optimization, and predictive maintenance [42]. Major influences of IIoT are the hardware affordability, reasonable computing power, consistency, architectures, and the Tactile internet. Then again, IIoT challenges are technological competence, nonscalable systems, security issues, blended criticality in industrial systems and latency.

Major difficulties for the apprehension of IIoT are the Interoperability because of the Heterogeneous information and Variation in information translation, the Scalability in terms of merging the existing sustainable networks with the new network and most important security issues in terms of Confidentiality and Integrity. So as to deal with challenges associated with the implementation of IIoT. The reference architectures of IIoT will be discussed in the next section.

There are different IoT definitions, those of significance to manufacturing application make explicit the sorts of smart components that get attached to ordinary objects with the goal that those ordinary objects can consider as IoT devices, and structure constituents of CPS.

IoT technologies used in the manufacturing are called as The IIoT; to characterize IIoT by adopting two basic features: (1) the types of advances that are adopted in an IIoT setting and (2) the particular aims and intention to which those advanced technologies are put. An advantage of the basic idea is that because it makes it clear that the related technologies are adopted for the industry, it fulfills the essential model of permitting us to differentiate IoT devices from IIoT devices. For instance, devices like smart car locks and smart oven are not valuable from the perspective of the industry as such, the basic conception correctly categorize those things as non-IIoT devices. In spite of this benefit, the definition stays uninformative all things considered.

The vision of IIoT world is to connect assets with smart to function as part of the sophisticated system or group of systems that enable the manufacturing as smart enterprise. Since smart manufacturing is very much essential for an industry to exemplifies the characteristics of IIoT, this definition is circulated uninformatively.

The IIoT is comprised of a large number of smart devices associated by communications software. The subsequent systems, and even the individual devices that contains it, can gather, exchange, monitor, examine, and immediately act on data to intelligently change their behavior or their condition without a human intervention.

By considering all the above points IIoT is defined as a system including network of smart objects, assets on CPS, related basic information technologies and discretionary cloud platforms, which allow intelligent, real-time and the autonomous access, gathering, analysis, communications, and exchange the information of service and/or process, within the background of the industry, in order to optimize overall production. This may include delivery service, improving product, reducing labor costs, enhance productivity, reducing the build to order cycle and energy consumption.

5.4.1.1 Adoption of IIoT in Manufacturing

The adoption of Industrial IoT provides the following benefits [43]:

- **Cost reduction.** By adopting IIoT, it drastically optimizes asset and catalog management. It helps in lowering stock carrying expenses, cut down search times, reduced machine downtime, increase agile operations, and effective use of energy, thus organization diminish operational expenses and make new sources of income. For example, smart, associated products permit to move from selling items to selling experience item utilization and post-deal services.
- **Shorter time to market.** Quicker and progressive manufacturing and flexibly chain operations permit in reducing the cycle time of production. As per the study, Harley Davidson adopted IIoT to reconfigure its manufacturing division at York, PA, due to which it managed to produce a bike in 6 hours instead of 21 days by reducing the production time.
- **Mass customization.** The mass customization process requires a remarkable enhancement in the assortment of produced SKUs, which makes inventory to escalate and turn out to be more assorted. Manufacturing tasks get increasingly mind boggling; also the creation of 20 items of SKU X can be immediately followed by the manufacturing of 10 items of SKU Y. Following the inventory and the assembling activities gets difficult and, at times, not feasible. IIoT encourages mass production with the help of real-time information required for attentive anticipating, shop floor planning, and steering.
- **Improved safety.** IIoT assists with guaranteeing a more secure workplace. Matched with wearable gadgets, it also allows monitors the health state of workers and dangerous exercises that can prompt to injuries. Alongside guaranteeing employees' safety, IIoT addresses security issues in potentially unsafe situations. For example, in industries of oil and gas, IIoT is deployed to screen gas spillages as it goes through the funnel.

Various IoT devices are jumping up these days, the measure of IoT units in various companies will raise to 29.7 million divisions by the year 2025 as per the survey. The investment in various domains like utilities, Transportation, Consumer electronics and cars, and Manufacturing are $61, $71, $108, and

$197 billion, respectively, as per the survey conducted by IDC Corporate, USA [44]

Several worldwide industries have effectively deployed IoT devices into their infrastructures. It permits them to enhance the process of production, decrease in release time, and reduce the investment. This makes the overall IoT market share grow exponentially throughout the next couple of years. As per the report of IDC, IoT impression is relied to grow up to 11.1 trillion U.S. dollars by 2025.

- **Asset tracking:** Increasingly the manufacturing industries tend to adopt asset practices. IoT technology joined with the advancement of local web and mobile applications for Android or iOS makes it conceivable to acquire real-time asset management and make realistic decisions [45].

The key job of tracking lies in finding and regulating such critical resources as the parts of the supply chain like containers, raw materials, and products. Similar applications can definitely improve logistics, to keep stocks of the progress of the work, and unveil violations and thefts.

IoT-based resource tracking enables the manufacturer to figure out the use of portable components and initiate measures to reduce idle period and improve utilization.

5.4.1.2 Smart Manufacturing a Use Cases

Most ideal approaches to comprehend the idea of smart manufacturing better are to consider how it could be applied to the business. This section includes three use cases to understand the benefit of adopting IIoT in manufacturing operation [46]:

- **Supply chain optimization and management.** Industry 4.0 solution provides industries more prominent insight, control, and information visibility over their whole supply chain. Besides utilizing SCM, the board abilities, industries can deliver items and services to market quicker, less expensive, and with improved quality to amplify over less productive competitors.
- **Predictive analytics/maintenance.** Industry 4.0 solutions enable industries to anticipate when potential issues will emerge before they actually occur. Without IIoT set up in the industry, preventive maintenance happens dependent on routine or time. At the end of the day, it is a manual task. With IIoT frameworks set up, preventive maintenance is significantly increasingly automated and streamlined. Systems can detect when issues are emerging, or machinery should be fixed and can allow to tackle potential problems before they turn out to be serious issues. Predictive investigation permit industries to not simply pose conversation questions like, "what has occurred?," or "for what reason didit occur?," yet additionally proactive inquiries like, "what will occur," and, "what would we be able to do to keep it from occurring?" These sorts of analytics can empower the manufacturers to turn from preventive upkeep to predictive maintenance with the IIoT framework.
- **Asset optimization and tracking.** Industry 4.0 setup assist industries with getting progressively productive with asset at every phase of supply chain, permitting them to maintain an enhanced strike on stock, quality, and streamlining

opportunities associated with the logistics. With IIoT setup at industries, workers can show signs of improvement in terms of distinctness into their product around the world. Standard resource management errands, for example, transfer of the asset, disposals, rearrangements, and modifications can be updated and overseen centrally and in real-time.

The purpose of discussing these use cases assist the manufactures to understand and to think how IIoT will enable the smart production

5.4.2 Key Security Goals

IIoT applications associated with sensors attached to equipment in the manufacturing industries. As discussed in Section 5.3.1.1, adoption of IoT will reduce the operational cost and will increase the productivity and achieves more profits in the manufacturing industry. However, it will also increase security threats by adopting IoT in manufacturing. Recent attacks on Cyber systems affect a vast assets loss and might also prompt hazardous circumstances.

This section includes the different security threats and the possible countermeasures to prevent the attacks [47].

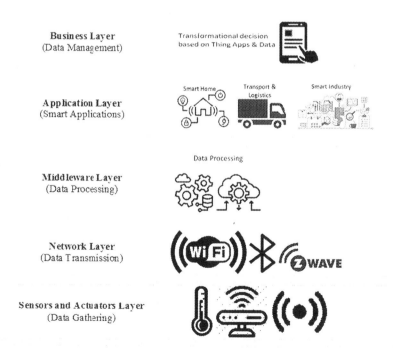

FIGURE 5.8 Five-layer Architecture of IIoT

- **Perception layer:** The bottom-most layer of IoT architecture is the Perception layer also called as physical layer where it includes physical components for example, RFID tag, Sensors, Actuators which are used to gather the desired information from the surrounding
- **Transmission layer:** Second layer is the Transmission layer and also called as network layer; it will forward the sensed information securely to the next higher layer. It includes the communication infrastructure like Home Area Networks (HAN), Wide Area Networks based on the dimension of the area
- **Middleware layer:** Third layer is the Middleware layer which mediated between the network and application layer, which facilitates the database and the services of cloud computing. It will process the gathered information in terms of decision-making.
- **Application layer:** Fourth layer is the Application layer gives different high-quality service and various smart applications like smart home, grid, health, etc., for the user. This layer additionally can give the interface for clients to collaborate with a physical device to get to the assigned information
- **Business layer:** Fifth and top layer of the architecture is Business layer is fit for dealing with the general framework. This layer is noticed that the achievement of an innovation relies upon both the advances in innovation and the sound plan of action. It facilitates the protection and security, which are the difficulty components in the improvement of IoT. For instance, this layer just gives the entry to the approved limitative users.

5.4.3 Possible Security attacks and its Countermeasures

- **Denial of Service (DoS) attack:** This assault denies the accessibility for the legitimate users from access the resources like a system, a framework, devices, memory, files, and process. The few most recent prominent IoT botnets like Mirai, Bricker, Reaper exhibiting the danger of setup and overlook way to deal with IoT devices.
- This attack can be restricted by Intrusion Detection and Prevention system as it will monitor the behavior of the activities and it distinguishes the behavior of the attacker and legitimate users [48]
- **Eavesdropping:** This attack checking the network, an attacker gains sensitive data about the conduct of the network. This attack will monitor and analyze the network traffic even the data in the cipher text can uncover the data privacy. This attack can be countered with the help of establishing secure connection like web-based client server uses the protocol HTTPS instead of HTTP
- **Man-in-the-Middle attack:** The attack will capture the sensitive data which is exchange between the device by intercepting the network communication channel. This attack can be addressed by allowing the user to connect once after the authentication like Mutual Authentication. Transport Layer Security (TLS) is the best way to avoid this attack [47]
- **False node injection attack**: Attacker will inject the malicious node to the existing network this is with an intension of injecting the false information to the network. This can be avoided by checking the unique ID of the device at the time of establishing the connection with the nodes for communication.

- **Time delay attack:** Attacker infuses additional time delays into estimations and control estimations of the network which can upset dependability of the network and cause hardware to crash. This will be addressed by computing the estimated time, if it exceeds the threshold the request will be automatically dropped instead of waiting for the infinite amount of time
- **Side Channel attacks:** To carry out a side channel the attacker will induct the Virtual machine on cloud to gain the access of the system. Generally, in IIoT cloud service provider will enable the user with the security functionality to avoid this attack [48]
- **Cloud Malware Injection:** Attacker infuses a malicious worm on virtual machine of the cloud. As worm has the ability to propagate by itself on the infected machine without human intervention this will contaminate the cloud service. checking the integrity to restrict this attack

5.5 CONCLUSION

IoT for supply chain is all set to revolutionize the operational efficiencies and revenue opportunities of an organization. It not only helps an organization to keep track of its assets but provides a way out to gain an edge over its competitors. When it comes to operation efficiency in supply chain, IoT offers logistics management, fleet management, asset tracking, and manufacturing process management. When it comes to revenue opportunities, IoT offers to understand the customers and vendors better. It provides greater insights for the administrators of an organization to make decisions. In this chapter, a brief introduction to logistics management, fleet management, asset tracking, and manufacturing process management is provided. Building a secure architecture is one of the important objectives when using IoT, as data storage and processing can make the system vulnerable to outside attacks and data leaks bringing down the organization's reputation. The security goals in each of these domains of supply chain along with the relevant deployment architecture for various sectors in the domain is discussed. In summary, IoT for supply chain will improve the operational efficiency and revenue of an organization, however, the system would be vulnerable for attacks. A number of secure architectures have been proposed for various domains of supply chain in literature. Choosing the right one would depend on the business needs of an organization.

REFERENCES

1. M. Liukkonen and T. -N. Tsai, "Toward decentralized intelligence in manufacturing: recent trends in automatic identification of things," *The International Journal of Advanced Manufacturing Technology*, vol. 87, no. 9–12, pp. 2509–2531, 2016.
2. R. Xu, L. Yang, and S.-H. Yang, "*Architecture design of internet of things in logistics management for emergency response*," in *2013 IEEE International Conference on Green Computing and Communications and IEEE Internet of Things and IEEE Cyber, Physical and Social Computing*, Beijing pp. 395–402, 2013.
3. A. Bassi, M. Bauer, M. Fiedler, T. Kramp, R. van Kranenburg, S. Lange, and S. Meissner "IoT in practice: examples: IoT in logistics and health," in *Enabling Things to Talk*, Springer, Berlin, Heidelberg, pp. 27–36, 2013.

4. P. Tadejko, "Application of Internet of Things in logistics--current challenges," *Ekonomia i Zarządzanie*, vol. 7, pp. 54–64, 2015.
5. S. N. Mahapatra, B. K. Singh, and V. Kumar, "A survey on secure transmission in Internet of Things: Taxonomy, recent techniques, research requirements, and challenges," *Arabian Journal for Science and Engineering*, vol. 45, pp. 6211–6240 Apr. 2020.
6. W. Youssef, A. O. Zaid, M. Sami, and M. H. Kammoun, "*RFID-based system for secure logistic management of implantable medical devices in tunisian health centres*," in *2019 IEEE International Smart Cities Conference (ISC2) Casablanca, Morocco*, pp. 83–86, 2019.
7. S. Anandhi, R. Anitha, and V. Sureshkumar, "IoT enabled RFID authentication and secure object tracking system for smart logistics," *Wireless Personal Communications*, vol. 104, no. 2, pp. 543–560, 2019.
8. S. K. Mazumder, C. Chowdhury, and S. Neogy, "*Tracking context of smart handheld devices*," in *2015 Applications and Innovations in Mobile Computing (AIMoC)*, pp. 176–181, 2015.
9. K. Fysarakis, G. Hatzivasilis, C. Manifavas, and I. Papaefstathiou, "RtVMF: A secure real-time vehicle management framework," *IEEE Pervasive Computing*, vol. 15, no. 1, pp. 22–30, 2016.
10. B. R. Ray, M. U. Chowdhury, J. H. Abawajy, and S. Member, "Secure object tracking protocol for the Internet of Things," *IEEE Internet of things Journal*, vol. 3, no. 4, pp. 544–553, 2016.
11. J. Mathew, "*Design and evaluation of an IoT enabled secure multi-service ambulance tracking system*," in *2016 IEEE Region 10 Conference*, Singapore pp. 2209–2214, 2016.
12. M. S. Uddin, J. B. Alam, and M. Islam, "*Smartanti-theft vehicle tracking system for Bangladesh based on Internet of Things*," , 4th International Conference on Advances in Electrical Engineering (ICAEE). IEEE, 2017, Dhaka, Bangladesh, pp. 28–30, 2017.
13. S. Y. Nikouei, Y. Chen, T. R. Faughnan, and C. Engineering, "*Poster: Smart surveillance as an edge service for real-time human detection and tracking*," *2018 IEEE/ACM Symposium on Edge Computing*, Bellevue, United States, pp. 336–337, 2018.
14. S. S. Mishra, "IoT health care monitoring and tracking: A survey," in *2019 3rd International Conferenceon Trendsin Electronics Informatics*, no. Icoei, Tirunelveli, Tamil Nadu, India pp. 1052–1057, 2019.
15. G. Nalinipriya, J. Isravel, and N. L. Kumar, "A dynamic tracking system for smart phones-A secure approach," in *2019 International Conference on Smart Structures and Systems (ICSSS)*, Chennai, Tamil Nadu, India, pp. 1–4, 2019.
16. Z. Zhao, P. Lin, L. Shen, M. Zhang, and G. Q. Huang, "Advanced engineering informatics IoT edge computing-enabled collaborative tracking system for manufacturing resources in industrial park," *Advanced Engineering Informatics*, vol. 43, no. October 2019, p. 101044, 2020.
17. P. Khan, Y. Khan, and S. Kumar, "*Tracking and stabilization of heart-rate using pacemaker with FOF-PID controller in secured medical cyber-physical system*," in *2020 International ConferenceonCommunication Systems & Networks*, Bengaluru, India, pp. 658–661, 2020.
18. J. T. Raj, and J. Sankar. "*IoT based smart school bus monitoring and notification system*," in *2017 IEEE Region 10 Humanitarian Technology Conference (R10-HTC)*, Bangladesh, Dhaka, pp. 89–92. IEEE, 2017.
19. S. Hussain, U. Mahmud, and S. Yang, "Car e-talk: An IoT-enabled cloud-assisted smart fleet maintenance system," *IEEE Internet of Things Journal*, vol. 4662, no. c, pp. 1–11, 2020.
20. M. Falco, F. De Ingeniería, I. Núñez, and F. Tanzi, "*Improving the fleet monitoring management, through a software platform with IoT*," *2019 IEEE International Conference on Internet of Things and Intelligence System (IoTaIS). IEEE, 2019*, Bali, Indonesia, pp. 238–243, 2019.

21. P. Singh, M. S. Suryawanshi, and D. Tak, "*Smart fleet management system using IoT, computer vision, cloud computing and machine learning technologies*," in *2019 IEEE 5th International Conference for Convergence in Technology*, Pune, India, pp. 1–8, 2019.
22. P. Killeen, B. Ding, I. Kiringa, T. Yeap, and I. Edi, "ScienceDirect IoT-based IoT-based predictive predictive maintenance maintenance for for fleet fleet management management," *Procedia Computational Science*, vol. 151, no. 2018, pp. 607–613, 2019.
23. W. Quadrini, E. Negri, and L. Fumagalli, "Sciencedirect sciencedirect open interfaces for connecting automated guided vehicles to a fleet management system," *Procedia Manuf*, vol. 42, no. 2019, pp. 406–413, 2020.
24. J. Backman, J. Väre, K. Främling, and A. F. S. Project, "*IoT-based Interoperability framework for asset and fleet management*," *IEEE 21st International Conference*, Berlin, Germany, pp. 0–3, 2016.
25. I. Oraibi, C. E. Otero, and T. O. Olasupo, "*Empirical path loss model for vehicle-to-vehicle IoT device communication in fleet management*," in *2017 16th Annual Mediterranean Ad Hoc Networking Workshop (Med-Hoc-Net)*, pp. 1–4, 2017.
26. M. Penna, B. Arjun, K. R. Goutham, L. N. Madhaw, K. G. Sanjay, and others, "*Smart fleet monitoring system using Internet of Things (IoT)*," in *2017 2nd IEEE International Conference on Recent Trends in Electronics, Information & Communication Technology (RTEICT)*, Bengaluru, India, pp. 1232–1236, 2017.
27. E. Hamadaqa, S. Mulhem, A. Mars, and W. Adi, "*Clone-resistant joint-identity technique for securing fleet management systems*," in *2018 NASA/ESAConferenceon AdaptiveHardwareandSystems*, The University of Edinburgh, United Kingdom, pp. 327–332, 2018.
28. S. R. J. Ramson, S. Vishnu, and M. Shanmugam, "*Applications of Internet of Things (IoT)—An Overview*," in *5th International Conference on Devices, Circuits and Systems (ICDCS)*, Coimbatore, India, pp. 92–95, 2020.
29. R. Gurunath, M. Agarwal, A. Nandi and D. Samanta, "*An overview: Security issue in IoT network*," in *Second International Conference on I-SMAC (IoT in Social, Mobile, Analytics and Cloud)*, Palladam, India, pp. 104–107, 2018.
30. N. Neshenko, E. Bou-Harb, J. Crichigno, G. Kaddoum, and N. Ghani, "Demystifying IoT security: An exhaustive survey on IoT vulnerabilities and a first empirical look on internet-scale IoT exploitations,"*IEEE Communications Surveys & Tutorials*, vol. 21, no. 3, 2019, pp. 2702–2733.
31. A. C. Panchal, V. M. Khadse, and P. N. Mahalle, "*Security issues in IIoT: A comprehensive survey of attacks on IIoT and its countermeasures*," in *IEEE Global Conference on Wireless Computing and Networking (GCWCN)*, Lonavala, India, pp. 124–130, 2018.
32. T. D. McAllister, S. El-Tawab, and M. H. Heydari, "*Localization of health center assets through an IoT environment (LoCATE)*," in *Systems and Information Engineering Design Symposium (SIEDS)*, Charlottesville, Virginia, USA. April, 2017.
33. A. Mhlaba, and M. Masinde, "*A hardware based model for an asset monitoring and tracking system: Case of laptops*," in *International Conference on Emerging Trends in Networks and Computer Communications (ETNCC)*, Windhoek, pp. 155–161, 2015.
34. A. Mhlaba and M. Masinde, "*Implementation of middleware for Internet of Things in asset tracking applications: In-lining approach*," in *IEEE 13th International Conference on Industrial Informatics (INDIN)*, Cambridge, pp. 460–469, 2015.
35. R. Indira, G. Bhavya, S. Dheva Dharshiniand R. Devaraj, "*IOT asset tracking system*," in *SSGR—International Journal of Computer Science and Engineering*, Vol 7, pp. 45–50, Feb. 2019.
36. F. J. Valente and A. C. Neto, "*Intelligent steel inventory tracking with IoT/RFID*," in *IEEE International Conference on RFID Technology & Application (RFID-TA)*, Warsaw, pp. 158–163, 2017.

37. N. Sangpraserand K. Inthavisas, "Mobile X-Ray tracking and monitoring system using the Internet of Things (IoT) technology," *KasemBundit Engineering Journal*, vol. 10, no. 1, pp. 149–163, Jan.–Apr. 2020.
38. D. Lukac, "*The fourth ICT-based industrial revolution*," in *23rd Telecommunications Forum Telfor, IEEE*, Belgrade, Serbia, pp. 835–838, 2015.
39. Industrie 4.0, Available: https://www.bmbf.de/de/zukunftsprojektindustrie-4-0-848.html.
40. K. Alexopoulosa, S. Koukasa, N. Bolia, and D. Mourtzisa, "*Architecture and development of an industrial Internet of Things framework for realizing services in Industrial Product Service Systems*," in *51st CIRP Conference on Manufacturing Systems, Procedia CIRP*, Stockholm, Sweden 72, pp. 880–885, 2018.
41. P. Suresh, J. V. Daniel, V. Parthasarathy, and R.H. Aswathy. "*A state of the art review on the Internet of Things (IoT) history, technology and fields of deployment*," in *2014 International Conferenceon Science Engineeringand Management Research*, Chennai, India pp. 1–8, 2014.
42. J. Manyika, M. Chui, P. Bisson, J. Woetzel, R. Dobbs, J. Bughin, et al. *The Internet of Things: Mapping the value Beyond the Hype*, McKinsey Glob Inst, Jun. 2015, p. 144.
43. https://www.scnsoft.com/blog/iot-in-manufacturing.
44. https://www.byteant.com/blog/5-best-use-cases-of-iot-in-manufacturing/.
45. https://www.byteant.com/blog/5-best-use-cases-of-iot-in-manufacturing/.
46. https://www.epicor.com/en-in/resource-center/articles/what-is-industry-4-0/.
47. A. C. Panchal, V. M. Khadse, and P. N. Mahalle, "*Security issues in IIoT: A comprehensive survey of attacks on IIoT and its countermeasures*," in *2018 IEEE Global Conference on Wireless Computing and Networking (GCWCN) Lonavala*, India.
48. N. Tuptuk and S. Hailes, "Security of smart manufacturing systems," *Journal of Manufacturing Systems*, vol. 47, pp. 93–106, 2018.

6 Threats in IoT Smart Well-Being

E. Darra, V. Mantzana, V. Giovana Bilali, and D. Kavallieros

CONTENTS

- 6.1 Introduction 202
- 6.2 IoT Application Domain #11: Smart Cities 203
  - 6.2.1 Background and Driving Toward Smart Cities 204
  - 6.2.2 The Concept of a Smart City 204
  - 6.2.3 Smart Cities Components, Attributes, and Characteristics 204
  - 6.2.4 Threats in Smart Cities 206
  - 6.2.5 IoT-Based Smart Cities 207
    - 6.2.5.1 IoT Architecture 207
    - 6.2.5.2 IoT Technologies for Smart Cities 208
    - 6.2.5.3 IoT Applications for Smart Cities 209
  - 6.2.6 Smart City and IoT: Challenges and Opportunities 210
- 6.3 IoT Application Domain #12: Smart Homes 211
  - 6.3.1 Key Achievements of Smart Homes (Automation, Security, Sustainability) 212
  - 6.3.2 Smart Home Infrastructure and Technologies 213
    - 6.3.2.1 Smart-Home Security Ecosystem 214
    - 6.3.2.2 Smart-Home Sustainability Ecosystem 215
  - 6.3.3 Smart Home Devices 215
  - 6.3.4 SHE Architecture 217
    - 6.3.4.1 Centralized Architecture 217
    - 6.3.4.2 Distributed Architecture 217
    - 6.3.4.3 Open Architecture Blueprint 217
  - 6.3.5 Connectivity and Protocols 218
  - 6.3.6 Application Areas 218
  - 6.3.7 Key Security Concerns and Challenges 219
    - 6.3.7.1 Security Concerns and Challenges 219
    - 6.3.7.2 Threat Landscape in SHE 219
    - 6.3.7.3 Threats and Vulnerabilities 220
  - 6.3.8 Security Solutions 222
    - 6.3.8.1 Design of SHE Architecture and Convergence Services 222
    - 6.3.8.2 Device Security Measures 223
    - 6.3.8.3 Network and Communications Security Measures 223

6.3.9　Social and Economic Factors..223
6.4　IoT Application Domain #13: Healthcare...224
　　　6.4.1　Background and Driving Forces..224
　　　6.4.2　IoT in the Healthcare Sector..225
　　　6.4.3　Benefits of IoT in the Healthcare Sector...227
　　　6.4.4　Barriers...229
　　　6.4.5　Security Issues of IoT in the Healthcare Sector..................................229
　　　6.4.6　Security Measures of IoT in the Healthcare Sector............................232
6.5　Conclusions..234
6.6　Acknowledgment..234
Notes..234
References...234

6.1　INTRODUCTION

IoT devices in the well-being umbrella provide great benefits, enhanced functionalities and automations. However, they have introduced security concerns, by providing a new attack surface that is characterized by insecure design, limited computational power, heterogeneity of protocols and connectivity between multiple networks etc.. Furthermore, within the well-being framework, the data which are being transmitted are sensitive and personal in most cases, exposing the daily life of individuals to attackers.

Three domains of IoT application will be presented and thoroughly explained:

- Smart cities and the adoption of IoT devices;
- Smart homes and IoT devices;
- Healthcare and IoT devices.

To gain a better understanding the concept of a Smart City will be presented mainly focusing at first on the smart city environment and secondly on the main components, attributes, and characteristic of the smart city. In addition, the threats in smart cities will be presented, with an emphasis on the threat categories taking into consideration the ICT architecture of smart cities. Moreover, IoT technologies for smart cities are presented as well as the IoT applications which have been identified in a smart city with a main focus on the Smart Homes and Smart Health care.

The Smart Home domain falls under the smart city, providing various automations (e.g., turn the air conditioner on remotely, the refrigerators can track which goods you are using and automatically create a shopping list, submit, and order, etc.) focused in making daily life easier. The adaption of smart appliances is gaining more ground each year and unfortunately malicious attackers are taking advantage of this in order to compromise and gain access to smart homes. It is significant to enhance the security of Smart Homes and thus, security practitioners and experts must understand the overall infrastructure and the respective main components, their characteristics, security threats possible measures.

The third domain is focusing in the healthcare sector, which has adopted a plethora of new technologies that will result in enhanced services provision and will save and improve human lives. These technologies range from IoT, wearable external and implanted medical devices, order entry and administrative Information Systems (IS) to laboratory and operation theatre IS, internet-based telemedicine, personal health records, picture archiving communication systems, etc. These play an increasingly crucial role in the healthcare sector advancement, by providing an infrastructure to integrate people, processes, data, and technologies. The secure adoption and use of IoT devices in the healthcare sector is crucial. Security practitioners must understand how these devices are interconnected in order to provide the appropriate services. In doing this, we will initially present the way that different IoT connected medical technologies support health and well-being services, as well as some published models/frameworks. In addition, the benefits and barriers of IoT in the healthcare sector will be presented, with a focus on security issues that is a top challenge for healthcare organizations planning, implementation, and adoption of IoT solutions. Finally, organizational and technical security measures that healthcare organizations should take to prevent or at least reduce unauthorized access, use, disruption, deletion, corruption, etc.; to respond effectively, timely, and efficiently and; minimize the impact of attacks to their network, information technology, and systems are presented.

This chapter is structured as follows: In Section 6.1 the Smart city ecosystem is presented. The Smart home domain is presented in Section 6.2 followed by the Smart healthcare domain in Section 6.3. Finally, Section 6.4 concludes this chapter.

6.2 IOT APPLICATION DOMAIN #11: SMART CITIES

In the past few years, there has been a growing development of the ICT (Information and Communication Technology) due to the advancement of digital technologies. The use of ICT in cities in different forms has led to the increased efficacy of city operations. These cities have been named using the term "smart cities." Additionally, as the world population has increased significantly it has been reported that around 70% of the world population will live in urban areas by the year 2050 [1]. It is also mentioned that cities currently consume 75% of the world's resources and energy. In that way, there is a scientific research stating and proving that this will have a severe damage in the environment. Here comes the creation and necessity of a smart city. The smart city has proven to be a mitigation plan for rapid urbanization growth. The only concern that is associated with a smart city is the high cost of its implementation. Based on its establishment it can reduce energy consumption, water consumption, carbon emissions, transportation requirements, and city waste [1].

The ICT technology is defined as the core part of a smart city and based on that the key component of the smart city includes the Internet of things (IoT). The IoT is distinguished by a network of physical devices that consists of several sensors, software, and electronic devices which can intercommunicate between each other. The evolution of the IoT is extremely rapid due to the excessive development of the Internet and digital technologies passing through the last decades.

6.2.1 Background and Driving Toward Smart Cities

The smart city concept started appearing as a means to define urban technological evolution. In order to investigate on the adoption behind this technology, there is a need to delve into some of the drivers. At first, the infrastructure of many cities was degrading starting from the roads, the railroads, and the bridges. Smart technology embeds sensors in order to determine a broad range of things, including the extent of degradation, daily traffic flow increases, temperature extremes that may exacerbate damage, safety issues that could lead to mass injuries or loss of life, and many more. In that way, smart technology is utilized and can monitor the conditions of the infrastructure leading to improvements in safety. In this aspect, the need for improved security using facial recognition systems, biometric systems, and many more is defined as critical for the IoT evolution. The smart technology offered a reduction of energy consumption as cities were using much of nonrenewable sources particularly wasted lighting/heating/cooling. The improvement of communication in a city is one of the key aspects in the adoption of smart technology as the installation of smart networks allowed the interaction between the infrastructure and the human beings. Another driver behind the adoption of the IoT technology is the emergency preparedness and environmental awareness. The installment of smart sensors can help the prediction of environmental changes. Another important driver of cities adopting smart technology has to do with the ability to manage traffic flow within urban centers and on highways [2].

6.2.2 The Concept of a Smart City

The concept of a smart city is distinguished on the fact that it is dependent on embedded systems, smart technologies, and the IoT. In general terms, a smart city relies on information technology and the embedded infrastructure to facilitate it for a better living standard.

A smart city is equipped with several components ranging from applications to citizens and smart technology to wide areas of smart governance appliances and smart infrastructure. In addition, this can spread the usage of individual mobile devices. Therefore, by considering the heterogeneous environment, different terms, such as features of objects, contributors, motivations, and security rules should be investigated. Some of the main aspects of a smart city are introduced in Figure 6.1 [3].

6.2.3 Smart Cities Components, Attributes, and Characteristics

A smart city is comprised of quite diverse characteristics, attributes, and components which are summarized in Figure 6.2. Some of these components include the smart infrastructure, smart buildings, smart transportation, smart energy, smart health care, smart home, smart governance, etc. On the other hand, the smart cities attributes include the sustainability, quality of life (QoL), urbanization, and smartness [1]:

- Sustainability of a smart city is mainly related to the infrastructure of a city as well as the governance, energy and climate change, social issues, economics, and health.

Threats in IoT Smart Well-Being

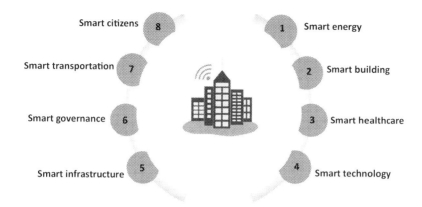

FIGURE 6.1 The main components of a smart city

Components	Attributes	Characteristics
• Infrastructure • Smart buildings • Transportation • Energy • Health care • Smart home • Smart governance	• Sustainability • Quality of life • Urbanisation • Smartness	• Infrastructure (Physical, ICT, Services) • Themes (Society, Economy, Environment, Governance)

FIGURE 6.2 The main components, attributes, and characteristics of a smart city

- The quality of life includes the emotional and financial well-being of the citizens.
- The urbanization includes different aspects of a smart city, such as technology, infrastructure, governance, and economics.
- The smartness is the ambition to improve economic, social, and environmental standards of the city and it includes smart economy, smart governance, smart mobility, and smart living.

Three characteristics are mainly indicated as regard to the infrastructure: physical, ICT, and services. The physical infrastructure consists of the physical entities of a smart city, including buildings, transportation, and mainly all the components of a smart city. On the other hand, the ICT infrastructure is the smart city component that brings together all the other components of the smart city. The service infrastructure is based on physical infrastructure and may be included by some ICT components.

There are four main core topics for a smart city: society, economy, environment, and governance. The society of a smart city is related to the citizens while the economy signifies that the city is able to thrive with continuous job growth and economic growth. On the other side, the environment of a smart city indicates that the smart city will remain as operational as possible. Finally, the governance of a smart city suggests that the city is robust in its ability to administer policies and combine the other elements.

6.2.4 Threats in Smart Cities

As described in ENISA's report [4], there is a combination of the threat categories taken into account with a simplified view of the ICT architecture of smart cities that was developed after analysis of smart city characteristics:

- Availability threats include DoS and DDoS attacks
- Integrity threats include the unauthorized access to restricted information (e.g., through masquerade attacks or malware) as well as loss, manipulation, and corruption of information.
- Authenticity threats Authenticity is a major challenge in ITS as usually all system stations have the ability to send, receive, and replay most types of messages.
- Confidentiality threats include the illicit collection of data through eavesdropping or the analysis of message traffic
- Nonrepudiation/accountability threats: it is important to ensure that nobody can deny that particular messages were sent or received, or that specific services or data were modified.

Specific threats that are relevant to smart cities are provided as an overview of the threat landscape in the following list. It is worth mentioning that all these threats are applied to all smart infrastructures of a smart city:

- Threats from intentional attacks
  o Eavesdropping/wiretapping is an act of capturing network traffic and listening to communications between two or more parties without authorization or consent. It may affect availability, integrity, and confidentiality of data and information systems, respectively.
  o Theft refers to the unauthorized appropriation of information/data or technology. Theft may affect availability and confidentiality. Theft can be associated with the theft of cryptographic keys, theft of mobile devices of operator employees, theft of credentials, or other sensitive information, theft of information/data.
  o Tampering/alteration aims at altering information/data, applications, or technology with direct and potentially significant effect on availability and integrity. It is also relevant from the perspective of nonrepudiation/accountability.
  o Unauthorized use/access can be at the source of other threats. Apart from eavesdropping/wiretapping, theft, and tampering/alteration, it may also be that information/data, applications, or technology are used/accessed in an unauthorized way. This includes unauthorized connection to a network, data leaks, browsing files, acquiring private data, controlling field components, and using resources for personal use. Moreover, unauthorized use/access

Threats in IoT Smart Well-Being 207

may affect integrity, confidentiality, authenticity, and nonrepudiation/accountability as attackers might have obtained comprehensive possibilities.
- o Distributed Denial of Service (DDoS) consists of the usage of several sources connecting simultaneously to one destination, with the objective of overflowing the connection. A DDoS usually deprives a target from Internet connectivity; it can also be preliminary to other attacks.
- Threats from accidents
  - o Hardware failure/malfunctioning can occur due to, for instance, old age, lack of maintenance, and overheating.
  - o Software errors are comparable with hardware failure/malfunctioning. Any extraneous or erroneous code in the operating system or an application that result in processing errors, data output errors, or processing delays is considered a software error. Software errors mostly affect availability but may also affect integrity.
  - o Operator/user error refers to "an improper or otherwise ill-chosen act by an employee that results in processing delays, equipment damage, or lost or modified data." Operator errors often occur during maintenance when for instance hardware and software are modified and/or updated but also during operation.
  - o End of support/obsolescence may lead to serious vulnerabilities. Often, manufactures, solutions providers, and vendors stop supporting applications and technology as they become obsolete.
  - o Electrical and frequency disturbance/interruption may affect availability.
  - o Acts of Nature (including bad weather) is due to unexpected events that impact the service. Such acts of nature include extreme drought or flood, snow, strong wind. Acts of nature usually impact systems that cease to operate.
  - o Environmental incidents, such as major electrical failure and liquid leakage (e.g., burst or leaking pipes, discharge of sprinklers), are similar to acts of nature and can cause destruction of field components, vehicles, and infrastructure [4].

6.2.5 IOT-BASED SMART CITIES

6.2.5.1 IoT Architecture

The IoT architecture is distinguished in six layers as indicated below [5]:

- Coding Layer: It is related to the basic layer of the IoT architecture where the object of interest is provided with a code for the sake of unique identification.
- Perception Layer: This layer is also known as device layer or recognition layer. The devices are usually RFID sensors, IR sensors, and sensors that are responsible for the temperature, pressure, moisture, speed, location, etc. The data sensor gathers the information, converts it to digital signal, and transmits it to Network layer.
- Network Layer: It is responsible for secure data transmission between Perception and Middleware layer. This layer receives the information from the

Perception layer in digital form and then sends it to the Middleware layer for further processing.
- Middleware Layer: This layer uses ubiquitous computing, cloud computing, fog computing, edge computing, etc., to access the database directly and store the required information in it.
- Application Layer: This layer provides the personalized service on the basis of user needs, using the result of the processed data.
- Business Layer: The Business layer is the higher level of the IoT architecture, where various business models are generated for the effective business strategies (Figure 6.3).

6.2.5.2 IoT Technologies for Smart Cities

The development of IoT network utilizes various communication protocols and consists of several objects that can be measured, inferred, understood, and can change the entire environmental conditions. Based on that, the IoT ecosystem consists of smart devices and other relevant technologies that can be described as follow:

- Radio-Frequency Identification (RFID): The RFID systems consist of readers and tags which are playing a key role in the IoT. This technology can be applied to any IoT object carrying out the automatic identification and assign a unique digital identity to each object. The reason behind that is to be incorporated in the network and related to the digital information and service [6]
- Wireless sensor network (WSN): The advantage of a WSN is that it can be used in many cases such as healthcare, government, and environmental services. Furthermore, the WSN can be integrated with RFID system to obtain information regarding the position, movement, temperature, etc [6].
- Addressing: The interconnection between people and objects, in order to establish smart environments is crucial for favorable outcomes of the IoT. This is because uniquely addressing the large-scale combination of objects is vital for controlling them via the Internet [6]

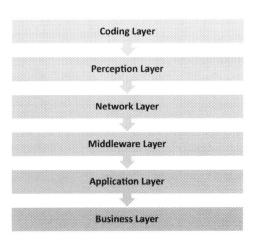

FIGURE 6.3 IoT Architecture Layers

- Middleware: The middleware plays a key role in the interconnection of the objects to the application layer. The key objective of the middleware is to concisely integrate the functionalities and communication capabilities of all involved devices [6].
- Cloud Computing: The cloud computing technology is one of the core parts of the IoT as it stores, processes and presents the analysis of all aggregated data from various IoT devices. It can converge many servers on to one cloud platform with an objective to share the resources and access them from anywhere and anytime. As the IoT devices increase rapidly and fully depend on the cloud, there is a need for more development of this technology to unleash its true power [5].
- Data Storage and Analytics: The critical factors that affect the growth of data are its storage, data ownership, and data expiry. Algorithms that make use of intelligent data and are either evolutionary, centralized, or distributed should be designed for effective decision making. The systems thus created must possess characteristics such as interoperability, integration, and adaptive communications. The system is based on modular architecture for hardware and software development [5].
- Visualization: Visualization allows the user interaction with the virtual environment. The visualization of the IoT application can be easily achieved through the advances in the touch screen and speech recognition technologies. The conversion of data into information to knowledge will lead to faster decision-making process [5].

6.2.5.3 IoT Applications for Smart Cities

The IoT utilizes the Internet to incorporate heterogeneous devices with each other. In this regard and in order to facilitate the accessibility, all available devices should be connected to the Internet. In order to achieve this target, sensors can be developed at different locations for collecting and analyzing data to improve the usage.

Smart Homes: As indicated in the ENISA's report [7], smart home environments have the ability to integrate multiple IoT devices and services that collect, process, and exchange data. They provide users with the possibility to control and adapt the status of their home, either manually or automatically. There exist interactions that take place between the internal and external actors of smart home devices and services. Smart Home Environments being an emerging domain and because the liabilities are not well defined, it becomes important for all actors to develop adapted security measures to prevent cyber-threats. For that purpose, there is a need to secure Smart Home Environments and effectively reduce the threats.

Smart Healthcare: Smart healthcare is an intelligent infrastructure that uses intelligent infrastructure that uses sensors to perceive information, transmits information through the IoT, and processes the information using supercomputers and cloud computing. Smart healthcare is defined as a health service system that uses technology such as wearable devices, IoT, and mobile internet to dynamically access information, connect people, materials, and institutions related to healthcare. Smart healthcare can also promote interaction between all parties in the healthcare field, ensure that participants get the services they need, help the parties make informed decisions, and facilitate the rational allocation of resources. In a nutshell, smart healthcare is a higher stage of information construction in the medical field [8]. The various components of smart healthcare include emerging on body sensors, smart hospitals, and smart emergency response. In smart hospitals, various mechanisms are used, including ICTs, cloud computing, smartphone apps, and advanced data analysis techniques [1].

6.2.6 Smart City and IoT: Challenges and Opportunities

Smart cities challenges are categorized into city traffic, citizen behavior, and city planning. Sensors for GPS, GIS attached with vehicles can analyze the traffic in real time. Big data research challenges of smart cities are categorized into two. Business challenges (Planning, Sustainability, Cost, Integration with Cloud computing) and Technological challenges (Privacy, Data analytics, Data formats, and QoS). Smart city challenges originate from its design to operation such as design & implementation cost, technology identification, heterogeneity of devices, volume of data, Cybersecurity issues, dynamic future adoption, and connectivity speed. With the available network technologies and mechanisms, smart city is getting matured and more realistic [9].

- Design and Implementation Cost: Smart IoT requires procurement and installation of new devices and software for stepping into cloud & big data paradigm. Most of the existing city administrative procedures have to be changed accordingly. Placing the sensors in the appropriate locations without affecting the people's convenience and privacy is still a concern [9].
- Heterogeneity: The devices, technologies, software, and platform are different for each IoT system. All these data should be integrated and processed in the cloud. Interoperability issues impact the IoT integration and device communication. Though standards are existing, it is not evolved completely to manage the different IoT platforms [9].
- Volume of data & devices: Considering multiple thousands of sensors installed in the city and monitored almost of the day in real time, volume is so huge to handle. Data generation, communication and processing require separate tools and strategy for each IoT system. Sensor and IoT devices deployment at large scale in a city poses significant challenges in managing, processing, and interpreting the big data they generate [9].
- Cybersecurity: Prevention of intrusion is important in cyber-physical systems. As the internet is the backbone of IoT, cyber-attacks are very common and hackers try to steal the citizen data. Hence, it is necessary to design and implement IoT system against these types of attacks [9].
- Dynamic adoption: IoT technologies are getting matured every day and adapting to the newer with the existing IoT system is quite challenging with minimal changes. IoT sustainability is required for the long run of smart cities by identifying the system which accepts the new technical viability [9].
- Connectivity speed: Most of the IoT systems generates real-time data and should be processed and available immediately. Hence, high-speed networks such as 4G/5G is essential in the IoT communication gateway [9].

- Security Issues: Smart IoT should be secure since it involves lot of citizens network is the major challenge in the smart cities project. Though there were many technological solutions available, intrusion may happen, which is inevitable. Both proactive and reactive measurements have to be placed in the IoT design [9].

6.3 IOT APPLICATION DOMAIN #12: SMART HOMES

The idea of smart home (or smart house), an automated smart solution which provides to house holders a variety of comforts, started its first steps almost at the end of the 20th century. Appliances and applications related with smart homes reached the market after many technological evolutions that took place during the last decades. The major achievement was the creation of internet in the early 2000s. This was the primary step in order smart home vision to become reality. Then other sectors of technology followed, in brief some of them were a) the evolution of information and communication toward to computer networks technologies, b) the establishment of embedded systems concept as well as the components development, c) the enhancement of automation through artificial intelligence methods [10]. All the above-mentioned innovations enhance the evolution of the IoT technology, a widespread approach of "smart" devices interconnected with the internet. Nowadays, the vision of smart home is firmly identified with the more settled region of Home Automation, otherwise called domotics, just as to later territories of Ambient Intelligence and Smart Environments. More information regarding the Smart Home Environment (SHE) ecosystems, will be described in the respective part of Section 6.2.2.

Generally, Smart Home Market (SHM) influences and adjusts the economical and operational framework of the overall Business Market (BM). In fact, an increased buying trend in SHM is rigidly linked to economic growth and then to market demand for smart home appliances and technologies. More specifically, the increase purchasing demand in a specific smart home domain brings growth in:

- the demand of smart-home appliances and technologies which are going to be constructed by industries.
    o the amount of job positions which are related with the product construction and the delivery of the product to the marker (e.g., transfer and supply services, IoT devices manufacturers, etc.)
    o the number of employees in each domain
- the funding that are invested to each domain.
    o the research programs which are conducted under smart home development research frameworks (e.g., IoT research programs, etc.)
    o the occupational sectors related with IoT and smart home development (e.g., Research and Development (R&D), Digital Marketing, Cybersecurity, Information Security, etc.)
    o the amount of job positions which are related with the abovementioned sectors (e.g., Information Security Analysts, Research Associate, Testers, etc.)
    o the number of employees in each domain

Reaching today, SHM domain has being directly affected due to smart home devices' increasingly demanding power. Based on a variety of surveys [11–13], in the forthcoming decade it is estimated to be one of the major "players" in the business market. For instance, based on Smart Home Market Analysis 2018–2026 in 2018 the amount of money was spent in SHM globally reached 79.90 billion dollars and is expected to be increased in 779.21% by 2026. Thus, the total amount of money will reach the amount of 622.59 billion dollars. As a result, within the next six (6) years the public and online market will overflow by smart home devices and applications.

6.3.1 KEY ACHIEVEMENTS OF SMART HOMES (AUTOMATION, SECURITY, SUSTAINABILITY)

Smart home's popularity emerges from the need of householders to create a reliable environment and develop it tailor made to their demands and expectations. Smart home technologies have achieved to provide numerous comforts and utilities to home dwellers. These achievements are related to smart home provisions and are separated to three (3) categories.

High-level smart home automation provisions:

- an efficient and automated "smart system"
- improved householders' comfort,
- fitness activities and healthcare assistance monitoring,
- recommended information and options based on householders needs,
- automated and rapid execution of householders' tasks

High-level smart home security provisions:

- security of householders' personal data,
- security of householders' physical system (including devices),
- safety of householders,
- security of SHE assets

High-level energy sustainability provisions:

- reduction in energy consumption and operating costs,
- improved life cycle of utilities,
- eco-friendly environment providing good impact to the environment

Nevertheless, the positive impact on the smart-home, provisions creating also negative aspects such as intrusion of SHE, data breaches, device, or system corruption, etc. More information regarding the security challenges will be analyzed in Section 6.2.7 (Figure 6.4).

Threats in IoT Smart Well-Being   213

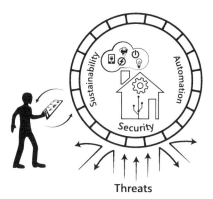

FIGURE 6.4 SHE is capable to provide sustainability, automation, and security to householders. They monitor and interact with SHE assets through a user interface and a stable internet connection

6.3.2 SMART HOME INFRASTRUCTURE AND TECHNOLOGIES

SHE possesses a variety of ecosystems or in other words environments including the necessary infrastructure, technologies, and smart devices which gives the "smart identity" to the smart home. The established technologies transfer vital information to the interior of SHE as well outside the smart home transmitting information to the home holders. Technologies placed on the proper infrastructure enable security, automation, and energy sustainability provisions to be materialized and implemented in the SHE [14]. Moreover, smart home devices receive information from SHE ecosystems and executing home holders' orders.

Below, it is presented three (3) smart home ecosystems. Beginning with the Automation Ecosystem which encompasses three (3) infrastructures a) a Home Automation System, b) a Control System and c) a Home Automation Network and two (2) services a) Utilities Information Technologies, b) Business Information Services. Continuing with, Smart Home Security Ecosystem including a) Authentication mechanisms, b) Authorization mechanisms, c) Firewall, d) Intrusion Detection System (IDS), e) Residential property security services. Closing with, Smart-Home Sustainability Ecosystem including a) Resources' Information service, b) Home appliances interactive control, c) Household resources' management, d) Self-service payment service [15, 16].

A Home Automation System encompasses devices or groups of devices, which are providing diversified operations to the whole system. In this environment, emphasis is given to the devices' "behavior," in other words how devices intercommunicate, as well as to their operational category, more information will be provided in Section 6.2.3. IoT devices presently can possibly be associated with a system through sensors, actuators, processors, and transceivers.

A Control System encompasses a software control tool enabling manage information that is received or sent and vice versa from/to the SHE ecosystem. The administrator of this control-based tool is the home holder. The home holder has the total consciousness of the control tool taking data from the smart devices and performing orders to other devices or the total ecosystem. In general, devices which behave as sensors are sending information signals, and those that behave as actuators receiving information signals.

A Home Automation Network is an environment in which all SHE devices and control tools can exchange status and manage information. Appliances must be connected to internet in order to interact with other devices and the householders as well. Signals interacted among sensors and actuators with the assistance of processors, and transceivers form a communication "language" among devices.

Utilities Information Services is receiving continuous information from different social groups in order to automate home dweller's actions. Such information could be road traffic and rush hours, health/hospital information and epidemic prevention information, shopping lists from local shops such as supermarkets or grocery shops, municipal information regarding the local events of an area, etc. In addition, the service enables online purchasing and consulting services based on householders' needs.

Business information services take into account the home user's interests, interactive activities with other users, subscriptions to services, and are sending to householder relevant information. For example, weather forecast and weather conditions of an area, foreign exchange, relevant shopping products to their previous buying activity and the stock of the products, etc.

6.3.2.1 Smart-Home Security Ecosystem [14]

Authentication mechanisms encompass techniques that provide entity authentication and message authentication. In brief, entity authentication guarantees the authenticity of the entity, as well as message authentication justifies that the received order/information, derived from the right user. The authentication of entity happens after an identification process and a verification process, in addition message authentication uses a public key cryptosystems and digital signatures. More specifically, the entity wants to access the network and it is accepted or denied based on a specific network identifier. A network identifier consisted by proof by knowledge, proof by possession, and proof by property. The complexity level of the authentication mechanism could be defined by the user.

Authorization mechanisms include all the access restrictions that an authenticated entity has toward to sources and network services. Authorization mechanism can also decide the level of access that an entity have on a network services and resources in a SHE. The goal of this mechanism is to avoid threat-based sources in order to secure SHE. Based on bibliography, the authorization mechanisms categorized into (3) three categories a) server-based, b) peer-to-peer, and c) certificate-based authorization mechanisms.

Firewall is the first protection layer of the SHE. A firewall usually is implemented in the gateway of SHE and is scanning the network traffic from inside to outside of the SHE and vice versa. In case that an abnormal network traffic[1] is detected the firewall blocks the access inside and outside of the SHE. So, firewall provides guardiancy to SHE. The firewall techniques utilized so as to monitor the network traffic are packet filtering, proxy service, and stateful inspection. In packet filtering strategy, firewall analyses packets entering or leaving the private network, against a set of filters. After the analyzing step of traffic, the firewall accepts or discards the network traffic dependent on defined security rules.

Intrusion Detection System (IDS) is a device or application that monitors the network of SHE in order to identify any malicious action that is not detected by the firewall. Since IDS is the second protection layer of the SHE when a violation is identified it is sent to the Security Information and Event Management (SIEM) system in order to react and minimize threat consequences.

Residential property security services send information remotely to householders regarding the SHE current situation. The information resulted from various sensors delivered signals from smart-home devices, such as gas, smoke sensors, fire and intrusion alarms, frequency signals of water, energy and gas consumption and waste management services, etc.

6.3.2.2 Smart-Home Sustainability Ecosystem [16]

In the Smart Home Sustainability Ecosystem included four (4) resources' service technologies, briefly they are resources information service, home appliances interactive control, household resources management, self-service payment which are referred to various services regarding the monitor and orchestration of SHE resources (e.g., electricity energy, water, gas, etc.).

Resources' information service contains information for the prices, principles, policies of resources as well as resources' activity within the SHE. For instance, it contains information, for electricity and water price over various time period (e.g., hourly, daily, etc.), users' policy, the amount of consumed resources, the amount of remained resources, resources distribution within SHE, report for resources' balance, information history of smart home's activity.

Home appliances interactive control is a device or application that creates reports and advice home holders based on their regular use of resources. More specifically, after the analysis at the home holders' resources and needs, it creates a tailor-made program that guides the users to use, manage, and operating reasonably toward the SHE resources.

Household resources' management provides real-time information about the status and activity of SHE resources. For example, information on electricity and water, energy and water distribution in-home services, waste resources, etc. Based on this information, the data are evaluated, and a special individual plan is proposed to the owner with the short-term goal of reasonable and balanced use of resources.

Self-service payment service provides online payment toward to resources bills, such as electricity, water, etc. The payment can be made by a variety of channels such as, phone, SMS, website, self-service terminal, and other means.

6.3.3 SMART HOME DEVICES

Here it is illustrated a Table 6.1 which showcase the application areas of smart homes (see Section 6.2.6) with a number of selected devices, the category that are belonging to as well as their main provisions [17].

TABLE 6.1
Indicative Devices and Provisions of Application Areas

Application Areas	Devices	Device Category	Main Provisions
Smart Entertainment and Comfort	Lighting, temperature, and heating control	Home automation	Convenience, Remote control, Multitasking
	Smart fridge and refrigerator	Home automation	Multitasking, Disorder activities
	Smart cooking appliance	Home automation	Multitasking
	Smart speakers	Home automation	Remote control
	IP TVs	Home automation	Convenience
	Background music	Home automation	Convenience
Healthcare	Smart watch	Wearable	Fitness, Multitasking
	Fit Trackers	Wearable, Monitoring device	Fitness, Multitasking
	Implantable (e.g., pacemakers, defibrillators etc.)	Monitoring device	Wellness
	Smart insulin pump & smart continuous glucose monitoring	Monitoring & controlling	Wellness
	Indoor Air Quality Monitor	Monitoring devices	Quality conditions monitoring
	Connected inhalers for intelligent asthma monitoring	Monitoring devices	Wellness
Safety and Security	Smart alarm (stand-alone)	Physical Security	Safety
	IP Cameras & Facial Recognition	Cyber-Physical Security	Safety & Home assets' protection
	Smart Home Surveillance Cameras	Cyber-Physical Security	Monitoring & home assets protection
	Smart Locks	Physical Security	Monitoring & home assets protection
	Video Doorbell	Cyber-Physical Security	Safety & home assets protection
Energy Efficiency and Sustainability	Smart thermostat, smart lamps, smart plugs	Monitoring devices	Ambiance Control
	Smart irrigation, gas, electricity devices	Controlling devices	Controlling resources' consumption

Deepening into the SHE "world," the keyword toward smart devices is heterogeneity. To understand the diversity of smart devices not only toward to their provisions but also to their operations requires a top-down approach to their architectural layers.

A device is a composed object which has (3) basic layers. The external layer of the device which constitutes its physical characteristics, the software infrastructure included by the total capabilities of the device and the hardware infrastructure giving the ability to software to be implemented [10]. Many smart home devices do not contain hardware and software. Sensors, actuators, and devices communicate through connectivity protocols.

TABLE 6.2
Operational Categories of Smart Home Devices

Operational Categories

Smart Appliances
Sensors
Actuators

The smart home devices are classified into three (3) operational categories: sensors, actuators, and smart appliances (Table 6.2).

In detail, a) sensors are responsible to estimate SHE parameter and gather data (e.g., light, temperature, pressure, power, humidity sensors, etc.), b) actuators are taking actions based on sensors' information, and c) smart appliances are connected to internet and execute orders.

6.3.4 SHE Architecture

The implemented architectures in an SHE is varying, for that reason it is presented indicatively some architectures [10, 18].

6.3.4.1 Centralized Architecture

The centralized SHE architecture uses a computer system as main element for smart house controlling. This main element also called home gateway and represent the control system of the smart house in this architecture. The data comes from sensors, passes a control algorithmic procedure and then are sending to actuators. Moreover, data is exchanged from SHE environment to the external environment and vice versa with the intervention of the gateway. The exchanged data are focusing to communicate needs and taskforces of the householders. The data are passing the gateway after a control processing.

6.3.4.2 Distributed Architecture

The distributed SHE architecture has designed and implemented the software control tool of the SHE as a whole distributed computing system. The advantage of this kind of architecture is utilization of the smart devices resources to embed software components into the nodes of the home automation network. Nevertheless, the control system is place physically in the gateway, virtually and operationally it is distributed in the whole system.

6.3.4.3 Open Architecture Blueprint

In this kind of architecture, the devices with the information that are carrying are orchestrated by the home gateway via a cloud platform. The users are connected to the gateway through a local area network (LAN). Cloud service provides functioning services, remote access, management services, etc., to the home dwellers.

6.3.5 CONNECTIVITY AND PROTOCOLS

Communication protocols enable smart devices to be interconnected with each other and exchange information with safety. Protocols are separated to a) wireless communication, b) wired communication and c) hybrid communication.

Indicatively some protocols from the former category are Infrared (IR), Wi-Fi, Bluetooth, Thread, Zigbee, Z-wave, KNX. The protocols of the second category are separated between internal and external communication. Some indicative protocols for internal communication are I2C, SPI and for the external communication Ethernet, RS-232, RS-485, UART, USART, USB, UPB. In the third category protocols used for both wireless and wired communication e.g., UPnP, etc.

In a smart home, the most frequently used protocols are wireless communication protocols [19, 20]. Some wireless communication protocols are described below.

Wi-Fi is mainly used within smart homes, working under the IEEE 802.11 standard and under the Internet Protocol IPv6. Wi-Fi popularity emerges from its rapid transmission speeds and its capability to interconnect devices that are hundreds of meters away, through the Wireless Local Area Network (WLAN) network.

Bluetooth is a wireless technology standard used for exchanging data between mobile and fixed devices. Bluetooth connectivity characterized by rapid exchange of data and secure transfer of information. Devices' connectivity reaches short distances of 10 meters with the use of short-wavelength radio waves.

Thread is low-power mesh networking protocol for IoT products. Thread is an IPv6-based open standard that allows smart home appliances to safely and directly connected to the cloud.

Zigbee is usually used for home automation, working under the IEEE 802.15.4. standard. ZigBee reaches an average of 10 to 30 meters. ZigBee devices have a low cost, and lower power consumption comparing to other wireless communication protocols.

Z-wave is usually used for home automation. It is a proprietary standard intended to remotely control applications within the residential and business environments. Z-Wave provides a reliable, low-latency transmission of small packets of data up to 100 kbps[2]. Although with an outdoor range of 100 meters.

KNX is an open standard protocol that is used for automation for decades. It supports different types of communication media such twisted pair, radio frequency, power line, and IP/Ethernet. Also, KNX provides software tools to design, visualize, and troubleshoot home-automation systems and to implement smart home services.

6.3.6 APPLICATION AREAS

Application areas contain sets of devices that satisfies home users needs and preferences giving an advantageous and rewarding experience to householders. The most regular application areas are [10, 17]:

- Healthcare contains IoT devices which are used for visualizing and monitor various health services toward senescence, incapacity, health status, health activities and automated care activities. More specifically, these devices could monitor eating disorders, wellness, sleeping activity and disorders, activity

tracking, heart beats, insulin intaking, quality conditions monitoring (e.g., air, heat, humidity, etc.) monitor. On top of that, the abovementioned provisions represented graphically by appliances' interfaces and they are supported from remote location.
- Smart Entertainment & Comfort application area encompasses smart appliances which are used to orchestrate home environment automation. Some regular appliances services are related to smart lighting, temperature and heating control, home appliances' administration, identification and presentation, services toward entertainment, cooking preferences, voice-operated appliance control, intelligent appliance monitoring and control. More specifically, smart speakers, smart bulbs and lamps, smart cooking appliances, smart playground console, etc.
- Energy Efficiency and Sustainability application area encompasses smart appliances which are used to administrate SHE resources and preventing resources wastage. Householders are empowered to know through applications, with respect to the spending home assets and in this way the operating expenses of their SHE. Such application has the capacity dependent on householders' past information and resulting activities to make a customized vitality sparing system regarding householders' choices. Some usual appliances are related to lighting, temperature and heating control, resources management and reducing power (electricity and gas) and water wastage.
- Safety and Security application area encompasses smart devices which are protecting the SHE devices, infrastructure, and network from malicious attacks and subsequently smart home's assets and data. There are smart appliances such as smart locks, smarts alarms, smart home surveillance cameras, security applications, etc., that are responsible to monitor and control SHE remotely.

6.3.7 Key Security Concerns and Challenges

6.3.7.1 Security Concerns and Challenges

The more complex the smart home technologies become, the more security concerns in a SHE appears. Many security problems arise due to smart devices heterogeneity (see Section 6.2.3). More specifically, the diversified smart home architectures and the different connectivity protocols which smart devices' use in order to communicate within the same SHE, harden the implementation of a unique security plan. Moreover, many IoT devices lacking operational system (OS) and power resources (RAM, ROM, etc.), thus it is difficult to implement and install security mechanisms, as well as software updates. In addition, devices' existing vulnerabilities, poor configuration, and the utilization of default passwords are among the factors which are responsible for devices' security challenges. Last but not least, since the majority of smart devices have weak security mechanisms and secure encrypted communications are vulnerable and prone to malicious attacks that aim to steal their personal data.

6.3.7.2 Threat Landscape in SHE

SHE threats could be derived by internal and external actors. The former is referring to insider threats, such as home dwellers, visitors which act intentionally and unintentionally and the latter to third parties' vendors and malicious attackers.

An internal actor could be:

- A malicious attacker, which acts on purpose and aims to threaten the SHE, such as steal or damaging smart homes assets.
- A smart home dweller which harms SHE appliances and infrastructure by implementing for example unappropriated settings or wrong administration toward the assets, etc.
- A visitor that harms SHE assets accidentally.

An external actor could be:

- A malicious attacker that provoke data and privacy breaches, assets' damages, modification of security mechanisms, etc.
- A third-party vendor.

In a nutshell, both two (2) actors encompasses intentional and accidental threats. Moreover, irrespectively to the purpose the outcome remains the same, the violation and the disaster of the SHE.

6.3.7.3 Threats and Vulnerabilities [21]

Based on ENISA's report [22], threat categories are presented.

- Physical attacks can be caused both by the home users and malicious intruders toward smart home assets resulting in physical damages. Some physical damages are damages to physical appliances and infrastructure.
- Unintentional damages are damages that derived accidentally by home users or even smart home visitors and external parties. Some unintentional damages are:
  o Information leakage resulting from the vast number of networks that are interconnected with the smart home, alongside with the inappropriate security mechanisms, lacking encrypted mechanisms, etc. More specifically, is relating to the unintentional sharing of home user's information.
  o Erroneous use or administration of devices or systems are caused by the home dwellers and is correlated with the wrong utilization and administration of devices. As smart devices have wide range of capabilities, they become dangerous when it comes to misuse.
  o Using information from an unreliable source enables to malicious sources to get access to the smart home environment by compromising smart devices.
  o Unintentional change of data in an information system could be a deletion of vital information or a modification of SHE's functioning parameters.
  o Inadequate design and planning or a lack of adaptation allows to many threats to arise and attack to the smart home.
- Damages or loss (IT assets) can be caused accidentally or on purpose by a third-party.
  o Damage caused by a third-party is very similar physical attacks.

- o Loss from DRM (Digital rights management) conflicts happens when there are shortages in security management implementation, such as certifications, DRM policies, etc. These kind of shortages happen both in software and hardware level.
- o Loss of (integrity of) sensitive information can be caused by a lack of encryption mechanisms in storage repositories. This kind of attack is also caused by the shortage of security mechanisms and management mechanisms.
- o Loss or destruction of devices, storage media, and documents are usually stored on smart home's premises physically or digitally. The major threat is the data leakage of the devices and the loss of the device itself.
- o Loss of information in the cloud is a subset of the abovementioned threat, since cloud storage included in the total storage information of the SHE. The greatest threat on this type of threat is the data leakage, since it can provide vital information for the SHE daily regulations and the home holder as well.
- o Information leakage is one of the major SHE threats. Some reason can be found in the abovementioned categories.
- Outages in resources may cause the loss of important data interrupting the regular operations of SHE.
  - o Lack of resources/electricity might interrupt or even stop the regular functions of systems and appliances which are operating via power supply. The abrupt interruption in a device may cause functioning problems to the device itself.
  - o Internet outage do not allow to the home users to be connected with the SHE from external sources.
  - o Loss of support services is the abandonment of the smart home's operating services such as monitoring, security, analysis, cloud storage and management, etc. That causes a major impact to the home dwellers since the daily operations within SHE is interrupted or even stop functioning.
  - o Absence of personnel is referring occurs when the expertise users of the smart home are absent. This may conclude to the partial operational usage of the smart technologies.
  - o Network outages may be caused by resources outages, software errors, etc., as a result the user could not interact with the smart home infrastructure and devices.
- Eavesdropping/ sniffing/snooping attack is a passive attack, that incorporates a malicious piece of software that can be placed somewhere in the network and gather information. Malicious intruders can monitor the flow of data in and out of the home. Since this kind of attack is primarily looking to take advantage of householder's privacy, eavesdropping in SHEs is characterized as an attack of confidentiality.
- Masquerading usually is a complementary attack to other types of attacks, such as replay attacks (e.g., message modification, denial of service, malicious codes, etc.). In this specific malicious attack, the attacker act like an authorized user.

- Replay attack is an attack in which the malicious attacker acts secretly like a legal party targeting to take exchanged messages from two parties. Usually, the intruder gets into the secure smart home network blocks the messages that are exchanged, and then delays or fraudulently modifies the messages for his malicious purposes. The intruder can replicate valid service request sent from a smart device in the SHE environment and keep it.
- Distributed Denial of Service (DDoS) is a frequent attack in smart homes and is targeting to intervene to the normal traffic of smart home's server, device or network and overload it with unlimited flow of traffic. The attack is starting by using multiple compromised machines and overload them with malicious traffic and over-burden its services. The intruder sending a huge amount of malware to block the SHE traffic attempting to block the normal traffic to arrive its usual destination and finally misdirect it in order to achieve intruder's purposes. The purpose of this action is the complete control of intruder to the SHE assets and the elimination of householders control to SHE assets.
- Malicious codes are a piece of malicious software that intend to provoke problems into the SHE and exploiting its vulnerabilities. This kind of attack in smart homes network becomes more threaten, since householder's awareness.

6.3.8 SECURITY SOLUTIONS

Security solutions in SHE are a combination of cyber-physical and Cybersecurity good practices. Good practices are sets of prevention measures that minimize the possibility of cyber-threats and implemented at an infrastructure, a device and a network level. [smart-home 13]. Despite the different good practices which are implemented in each category type, the main prevention measure is the implementation of privacy by design approaches toward the infrastructure services, the devices, and the network of the SHE.

6.3.8.1 Design of SHE Architecture and Convergence Services

One remediation measure toward cyber-attacks is the implementation of the proper architecture and convergence interconnected services at the stage of the smart home design. Some of the most common tactics are a) lessening the number of external resources that are communicating with the smart home, b) utilizing the minimal types of smart home technologies and protocols, reducing the amount of the vulnerabilities that could affect the SHE, as well as assisting the implementation of a unique security plan, c) storage limitations toward CPU capacity and RAM capacity implemented by users, d) enlarging the local storage of data within the smart home, e) design of secure cloud-based smart home, f) classification of the operational framework into critical and noncritical in order to place into separated SHE, g) Choosing and utilizing secure communication systems, systems that allow local access and the cloud access is not mandatory, and h) in case of phone or device possess operating system (OS), to make sure that they run the most recent version available of its operating system.

6.3.8.2 Device Security Measures

Devices introduced in a SHE should follow all the reliable conditions in order to confer security within the smart home. These conditions are measures or settings that implemented by home users or by IoT constructors. Another step is the selection of secure and proper smart appliances by home holders. Security measures for devices are applying in all the three layers of a smart device: physical, software, hardware by the device's owners, and manufacturers. More specifically some of the security measures are, a) authentication and authorization mechanisms, b) implementation of Secure IP gateways, c) no fixed, default passwords, d) strong key implementation, e) utilization of encrypted communication.

6.3.8.3 Network and Communications Security Measures

Network security measures encompass remediation measures for reliable communication provision toward interconnected smart devices and infrastructure services. A secure smart-home network characterized by confidentiality, integrity, availability. Some of the network's prevention tactics are a) control the access points of the SHE such as smart homes gateway, control tools, home hub, smart TV, b) creation of white lists for the devices that do not derive from the SHE, c) remote management of devices activity, profile management, event management, d) authentication and authorization mechanisms in order to access the SHE's network.

6.3.9 SOCIAL AND ECONOMIC FACTORS

Provisions and the functions are basic factors for implementing IoT technology in a home, however, the adoption of smart home technologies is rigidly relying on social and economic factors.

Social factors for the adoption of smart technology are:

- the technical background of the householders in a way that they understand the proper technical infrastructure and equipment that they should implement.
- the occupation of householders
- the age of householders
- the income of householders
- the country, city even the region that the home is placed. Some countries are more technologically innovative involved than others.

Economic factors for the adoption of smart technology are varying some of them are:

- the financial status, as it is needed investment in order householders implementing smart home infrastructure, technology, and devices
- the consumption of resources, since the technologies that consume more resources (power, water, etc.) increases household expenses.

6.4 IOT APPLICATION DOMAIN #13: HEALTHCARE

6.4.1 BACKGROUND AND DRIVING FORCES

The aim of a well-functioning health system is to enhance the health status of individuals, families and communities; protect population against health threats as well as consequences of ill-health and; provide equitable access to people-cantered care [23]. Healthcare is a complex environment, wherein policies, facilities, technologies, drugs, information, and a full range of human interventions interact and is characterized by services, settings, and providers, as described below [24]:

- Services refer to the type of care delivered through the whole care process from promotion and prevention to diagnosis, rehabilitation and palliative care, as well all levels of care including self-care, home care, community care, primary care, long-term care, and hospital care.
- Settings describe services delivered by different types of facilities (clinics, health centers, district hospitals, dispensaries, pharmacies, etc.), institutions, and organizations.
- In addition, health providers can be classified as private or public, for-profit or not-for-profit, as well as by the varied trainings and scopes of practice that distinguish the profiles of providers, from nurses, primary care physicians, and specialists.

Healthcare systems are directly affected by the global financial crisis and get challenged from aging populations with multiple co-morbidities, emerging, and chronic diseases. In this demanding, complex, and changing environment, systems must continue to provide services that are safe, accessible, high quality, people-centered, and integrated to patients, persons, families, communities, and populations in general [24].

For that reason, new technologies that will result in enhanced services and will save and improve human lives is the main priority for the healthcare sector worldwide. A plethora of new technologies, ranging from IoT, wearable external and implanted medical devices (skin patches, insulin pumps and blood glucose monitors), order entry and administrative Information Systems (IS) to laboratory and operation theatre IS, Internet-based telemedicine, personal health records, asynchronous healthcare communication systems, and picture archiving communication systems have been implemented in the healthcare sector. They play an increasingly crucial role in the healthcare sector advancement, by providing an infrastructure to integrate people, processes, data, and technologies.

It has been reported that IoT will transform business in the next years and will lead to the next consumer technology that will provide benefits to life, society, and environment [25, 26]. The last years, IoT market in the healthcare is expected to grow at a compound annual growth rate (CAGR) of 30.8%, from $41.2 billion in 2017 to $158.1 billion by 2022 [27].

6.4.2 IoT in the Healthcare Sector

IoT has been defined as "a network of billions of interconnected devices or systems ('things') that can be remotely controlled over the Internet. These devices collect and exchange data that can be analysed and aggregated for use in monitoring, maintenance and improvement of processes, with the goal of delivering products and services to consumers." [28].

In the healthcare sector, IoT entails any ecosystem of connected medical technologies that support health and well-being services and can be used by different healthcare stakeholders in different ways, as described below (Figure 6.5):

- In-Home: IoT appears to be a sustainable, low-cost solution that can support care and well-being of elderly population and people with chronic diseases. These devices can monitor patients' health status and activities; can support tele-consultation and can alert healthcare professionals and/or other involved stakeholders automatically.

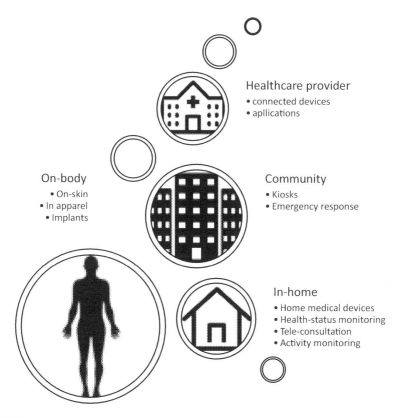

FIGURE 6.5 IoT ecosystem in healthcare—Based on [29]

- On-body: The human body is a source of several vital signs and biosensors are making it possible to design and build wearable medical devices that can monitor humans' health. They can be embedded in apparel, attached to skin, or even implanted under the skin.
- Community: For patients living in rural areas, where health services are not easily accessible, IoT solutions can be used to connect them with health professionals. This not only adds comfort to the healthcare process but also reduces the cost of services provision.
- Healthcare provider: Hospitals and healthcare providers are the largest users of IoT devices and applications that involve patient monitoring, X-ray machines, CT scans, and smart apps. They can be used to connect patients and healthcare professionals at remote places.

These devices are available to patients, citizens, and healthcare providers to monitor health data (e.g., diabetes, heart conditions); clinical data (e.g., blood glucose or heart rate); adherence data (e.g., medications), as well as consumer health data (e.g., physical activity). The data collected, can be accessible by health professionals that can track and give feedback to patients, in order to help them engage and make better health decisions in real time. They can also be used in emergency cases, where alerts are sent to health professionals, family, or emergency respondents. Some examples of healthcare IoT solutions include:

- Connected medical devices (e.g., MRI, CT scan, etc.) that collect and generate data and network with Healthcare Information Systems (HIS) for data visualization and analysis.
- Wearable clinical devices are certified and approved to be used devices by regulatory authorities. They can be used on-body or in-home by the patient and/or in healthcare setting. Examples of such devices are smart belts that detect falls, chest straps that record ECG, etc.
- Remote Patient Monitoring (RPM) devices are used to monitor patient health data outside of a traditional clinical setting. These data can then be sent to a healthcare professional, who can on his/her turn update care plans. In addition, reminders can be sent to patient. Common examples of RPM devices include voice apps that remind diabetes patients to take their insulin, while allowing their physician to monitor the disease.
- Point of Care (PoC) devices and kiosks are care-points that facilitate on-demand virtual monitoring of patients' vital data (e.g., thermometers, otoscope, dermatoscope, blood pressure cuff, etc.), tele-consultation, and tele-diagnosis [30, 31]. Point of Care devices and kiosks can be a great tool to treat and diagnose patients living in far off places.
- Clinical monitors record and store patients' data through digital methods (e.g., digital stethoscopes that can relay and save the heart sounds on a mobile app), so that they can be accessed whenever needed.
- Smart pills can support drug delivery, monitoring, and diagnosis. A smart pill with environmental sensors, a feedback algorithm and a drug release mechanism can support smart drug delivery and prevent accidental overdose. Moreover, it can record body temperature, pH and internal pressure, in

order to check how body is responding to medicine. In addition, a smart pill with built-in sensors can easily be swallowed and maneuvered to capture images of patients needed an endoscopy, measure fluids, and gases in the gut, etc. [32].
- Fitness wearables are devices (such as bands, wristwatches, necklaces, etc.) with built-in sensors that monitor patients' health and well-being, such as distance walked or run, calorie consumption, and in some cases heartbeat.

To further understand the current utilization of IoT in the healthcare sector, some published models/frameworks of such implementations are presented in Table 6.3.

6.4.3 Benefits of IoT in the Healthcare Sector

IoT provides various benefits to all processes (e.g., promotion and prevention to diagnosis, rehabilitation, and palliative care, as well all levels self-care, home care, etc.) and stakeholders (e.g., citizens, families, patients, communities, health professionals, caregivers, governments, organizations, etc.) involved in healthcare services delivery, such as clinical, operational and financial. More specifically, IoT supports and enhances healthcare services provision as presented below [27, 45–47]:

- Preventive care: Through activity and daily monitoring, as well as personalized plans, citizens can better manage behavioral and wellness information outside of appointments. The widespread access to real-time, high-fidelity data on each individual's health will reform health care by helping people live healthier lives and prevent disease.
- Diagnosis and treatment: IoT can collect and analyze intelligent and measurable data that support and enhance the speed and accuracy of diagnostics and target treatments more efficiently and effectively.
- Patient monitoring: IoT devices allow noncritical patients to be monitored in any place, by means of a smart medical device that can be wearable or ubiquitous.
- Medical devices management: Real-time location systems provide immediate or real-time tracking and management of medical equipment, potentially reducing device downtime.
- On-time reminders and alert: IoT devices collect and analyze intelligent and measurable data that are transferred to health professionals and caregivers for real-time tracking, while sending notifications and reminders to people about critical parts (medication and examination reminder, clinical alerts, etc.) via mobile apps and other linked devices. Reminders can also be sent to caregivers regarding important deadlines, such as medicine regimens and upcoming appointments.
- Emergency care: In event of an emergency, patients or even the application automatically can contact a healthcare professional, caregiver, emergency response teams or family.
- Research: IoT enables us to collect a massive amount of data that can be combined with other data; analyzed; transformed to information and knowledge that can support research activities.

TABLE 6.3
IoT Models/Frameworks in Healthcare

IoT application	Description
Cloud IoT-based healthcare framework for Electronic Healthcare Records.	The proposed framework supports patients to securely store health data; do self-assessment; monitor conditions and; share their information with healthcare providers. It also allows them to identify the best care at the optimal cost [33].
Intelligent Context-aware decision Support (ICADS) system.	The proposed system provides an effective basis for healthcare professionals, who can monitor the health status of patients; provide quality care services and reschedule and prioritize essential services [34].
Health informatics, IoT, and big data analytics framework	By integrating IoT, Complex Event Processing (CEP) and big data analytics technologies, the proposed solution can support real-time analytical processing of patient events from different sources [35].
Framework for managing systems of smart hospitals based on IoT.	This framework consists of three layers, namely physical, network and application layer. The physical layer collects data and sends them to the network layer. The network layer then processes and forwards these data to the application layer. Context awareness, which is a middleware between the network and application layers, analyzes the data received (from the physical) and transfers (to the application layer) only the required ones [36].
IoT-based remote treatment model	The proposed model virtually stores patient data and makes them accessible to healthcare professionals. It also supports professionals when delivering treatment by providing an intelligent clinical decision support system [37].
Sphere project	This project builds a generic platform that fuses corresponding sensor data to generate rich datasets that support the detection and management of various health conditions [38].
REACTION project	LinkSmart (within the REACTION European project) is a SOA-based middleware that monitors and manages the therapy of diabetic patients. It has also been used recently for a smart home application [39].
Smart e-Health Gateway	Fog assisted system architecture is proposed to cope with challenges in ubiquitous healthcare systems such as mobility, energy efficiency, scalability, and reliability issues. The proposed gateway provides features such as local storage, real-time data processing, embedded data mining, etc [40].
RFIDWSN hybrid monitoring system for smart health care environments	An IoT hybrid monitoring system that integrates RFID and WSN technologies to track location of healthcare assets (using passive and active RFID tags), and the location and health of patients (using an active wrist band that monitors skin temperature, heart-rate and movement) [41].
SilverLink	SilverLink uses object and human sensors for indicating user activities or health status collected from sensors. When a parameter changes, the system generates notifications / alerts to the emergency response team [42].
Fall detection system	A real-time fall detection system based on wearable sensors to detect the motion and location of the body was proposed and tested [43, 44].

IoT has the potential to lower costs, improve efficiency, and deliver better patient outcomes. It has been reported that it can save the healthcare industry $300 billion annually primarily through remote patient monitoring and improved medication adherence [48]. It also enables interoperability, machine-to-machine communication, information exchange, and data movement that make healthcare service delivery effective. To summarize, IoT can bring together people (citizens, patients, caregivers, and clinicians), data (patient or performance data), processes (care delivery and monitoring), and enablers (medical devices and mobile applications) to improve health care delivery.

6.4.4 Barriers

Despite the several benefits of IoT in the healthcare sector, there have been reported multiple barriers, such as security and privacy concerns, training, and data accuracy [25, 49] that are presented below:

- Data management: In the future, IoT will be extremely populated by a huge quantity of heterogeneous networked devices and sensors, which will be producing massive health data, as the human body is a dynamic system that changes its state continuously [50]. As the aim of IoT is also to extract valuable information from data, a novel data management approach is needed that will handle data in terms of variety, volume, and velocity [51].
- Scalability: By creating a smaller scale of IoT, with sensors on portable devices (e.g., smartphones) for data collection, all users can directly access medical services. This can then be scaled up to the entire hospital, so that patients and health professionals can use medical services, check updates, and health status by their smartphones. Finally, this model can be scaled up to the entire city, where, all data can be collected, processed and analyzed by smartphones through mobile apps and feedback will be sent seamlessly to patients to allow them to know their health status, medical exams results, etc. [50]. In this way, we can improve efficiency and quality of provided services, waiting times, as well as relationships between health professionals and patients [52].
- Complexity: The IoT is a diverse and complex network that needs multiple services, devices, increase of Internet bandwidth, etc. Thus, any failure or bugs in the software or hardware will have serious consequences [53].
- Compatibility: As different manufacturers get involved in IoT development, there is a need for common standards that will support devices' compatibility [54].
- Interoperability and standardization: Interoperability and standardization complexity lies in the fact that IoT is related to a wide variety of disciplines that are regulated by different affairs. The strict regulations mandated by medical standards increase the complexity even more [50].
- Security and privacy: Several surveys show that security is a top challenge for healthcare organizations planning, implementing, and adopting IoT solutions. It has been estimated that through 2022, half of all security budgets for IoT will go to fault remediation [55, 56]. If IoT security is compromised, health-related and other information can be accessed through unauthorized authentications and this can create risks to personal safety [50].

6.4.5 Security Issues of IoT in the Healthcare Sector

It has been re1`ported that there exist ten to fifteen million medical devices, with an average of 10–15 connected medical devices per patient bed in U.S. hospitals [57].

Security and privacy issues have become a growing concern in the healthcare sector, with ENISA mentioning that the most critical smart assets in the context of a smart hospital are the interconnected clinical information systems and networked medical devices [58].

According to a report published in 2019 [59], it has been identified that ninety-percent of healthcare organizations interviewed have some form of Cybersecurity vulnerability, with nearly half of them identifying IoT devices as their biggest threat. National Institute of Technology Standards (NIST) published a report namely "Considerations for Managing IoT Cybersecurity and Privacy Risks," in which three differences between medical IoT and traditional IoT are explained [60]:

- Many IoT devices interact with the physical world in ways conventional IT devices usually do not. NIST stated that IoT healthcare devices change physical systems and impact them in a different way. IoT sensor data should be managed effectively, as, e.g., when mitigating physical attacks on sensor technology, such as attacks performed through wireless signals that could cause sensors to produce false results. In addition, IoT network interfaces enable remote access to physical systems, as manufacturers, vendors, and other third parties may use remote access to IoT devices for management, monitoring, and maintenance purposes. This could put physical systems accessible through the IoT devices at much greater risk [60].
- Many IoT devices cannot be accessed, managed, or monitored in the same ways conventional IT devices can. They may also encounter one or more of the following challenges that affect Cybersecurity and privacy risk: (a) Lack of management features; (b) Lack of interfaces; (c) Difficulties with management at scale; (d) Wide variety of software to manage; (e) Differing lifespan expectations; (f) Unserviceable hardware; (g) Lack of inventory capabilities; (h) Heterogeneous ownership.
- The availability, efficiency, and effectiveness of Cybersecurity and privacy capabilities are often different for IoT devices than conventional IT devices. Because of the different ways in which medical IoT connects to the physical world and the information they store, transmit, and process, additional controls may need to be implemented to mitigate risks.

Additionally, as presented in Figure 6.6, ENISA identified and presented the following vulnerabilities of IoT devices in the healthcare sector that are related either to technical or organizational issues [58]:

- IoT devices are highly interconnected; therefore, security decisions made for a specific device can impact other systems or devices.
- IoT devices are spread everywhere in the hospital and as a result physical security is difficult to manage for all components.
- Medical devices are designed and built based on "intended use" cases, and not on "unintended use" or "abuse" cases. This might lead to systemic vulnerabilities and risks.
- IoT devices are deployed in a homogeneous way, which makes it easy for attackers to investigate possible attack paths. While device manufacturers and

Threats in IoT Smart Well-Being

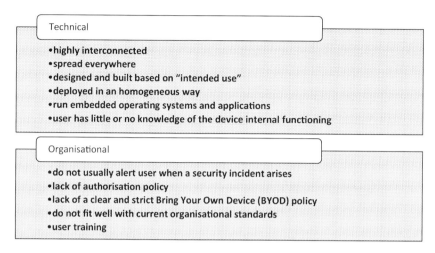

FIGURE 6.6 IoT devices vulnerabilities—based on [58]

security companies need to remove all vulnerabilities, criminals only have to find one. It is virtually impossible to patch all vulnerabilities for all devices.
- IoT devices run embedded operating systems and applications with little if any malware detection or prevention capabilities as well as encryption mechanisms.
- The user has little or no knowledge of the device internal functioning. This makes understanding potential threats as well as response activities in case of an incident very difficult.
- IoT devices do not usually alert user when a security incident arises. This might lead to delayed detection and response actions.
- In the smart hospital environment, there is as a lack of authorization policy. This can cause unauthorized users to gain access through an end device to a critical system.
- In a smart hospital environment, there is also lack of a clear Bring Your Own Device (BYOD) policy.
- IoT devices usually do not fit well with current organizational standards in the healthcare sector and it is difficult for IT departments to follow appropriate asset and change management processes.
- IoT devices users are not well trained on the usage of the apps and devices they use, as well as on security-related issues.

It appears that IoT in the healthcare sector is vulnerable to attacks with a report highlighting that eight out of ten healthcare organizations experienced a cyber-attack in the past year [59]. Several attacks have been identified in the normative literature, with some of them presented below:

- In 2017, a medical device vendor of pacemakers and other cardiac devices, patched its device software, as it was stated that its implantable devices were vulnerable to attacks that can harm patients' life [61].

- In similar lines, in 2016, Johnson & Johnson warned customers about a security vulnerability in one of its connected insulin pumps that hackers could exploit to overdose diabetic patients with insulin [62].
- Through MedJack, attackers could inject malware into medical devices, which could be spread across the healthcare network infrastructure. The data stolen was even used to track current drug prescriptions, enabling hackers to order medication online and then sell on the dark web [63].

It appears that IoT devices could be attacked in different ways that could harm consumers by (a) enabling unauthorized access and misuse of personal information; (b) facilitating attacks on other systems; and even (c) creating safety risks [64]. A cyber-attack on IoT in healthcare organizations could impact customer data (39%), patient safety (20%), and stolen intellectual property (12%). Security executives also are concerned about brand or reputational damage and operational downtime [59].

To tackle the aforementioned and other security-related issues several researchers proposed architectures and frameworks that are presented below:

- A secure and efficient authentication and authorization architecture for IoT-based healthcare has been proposed in 2015 [65]. The presented architecture (a) employs distributed smart e-health gateways that perform authentication and authorization on behalf of the medical sensors and; (b) relies on the certificate-based DTLS handshake protocol that is the main IP security solution for IoT. The authors developed and tested a prototype IoT-based healthcare system that proved to be more secure than the standard centralized delegation-based architectures. Furthermore, the impact of DoS attacks is reduced due to the distributed nature of the architecture.
- In 2016, a security framework for healthcare IoT was proposed [66]. This framework serves as a security guideline to design the system and as a validator to evaluate security in existing systems. In doing this, a security context associated with each piece of personal health information is created and when a piece of information is created, the associated security context should be generated automatically.
- To overcome IoT security challenges in the healthcare sector, authors proposed a holistic multi-layer approach [50]. For all IoT architecture layers identified (device, network, cloud, and human), several security measures, protocols or built-in capabilities are presented.
- A security framework that encrypts sensed physiological data with Light-Weight Encryption Algorithm and Advanced Encryption Standard cryptographic algorithms is proposed and evaluated [67].

6.4.6 Security Measures of IoT in the Healthcare Sector

As such, health structures are pointed out as potential targets, which highlights the need to enhance the protection of them. In order for healthcare organizations to prevent or at least reduce unauthorized access, use, disruption, deletion, corruption, etc.; to respond effectively, timely and efficiently and; minimize the impact of attacks to their network, information technology, and systems, it is important to take organizational and technical security measures, which are analyzed below.

With regard to organizational measures that will enhance Cybersecurity in healthcare organizations, it is crucial for healthcare organizations to assess cyber-risks. Cyber-risk assessments are used to identify, estimate, and prioritize risk to organizational operations, organizational assets, individuals, other organizations, and the Nation, resulting from the operation and use of information systems [68]. In addition, healthcare organizations should develop and incorporate IoT-specific laws, standards, plans, and policies that outline Cybersecurity measures and crisis management procedures.

Since the human factor is one of the major security threats in health domain, it is important that the personnel are aware of the basic IoT Cybersecurity related issues and their skills—both technical and behavioral—are improved [69]. In doing this, stakeholders that get affected by, or get involved in crisis management processes, can manage and cooperate effectively and timely on security planning, preparedness, response, recovery, and impact mitigation.

With regard to technical measures, it has been reported that healthcare organizations should adopt and implement different practices that will enhance IoT security, such as the following:

- Authentication: It ensures the validity of the claimed identities of the entities participating in communication (e.g., person, sensors, device, or application).
- Access control (authorization): Different access levels to devices, systems, and networks guarantee that individuals can only gain access to and perform operations on stored information and flows that they are authorized for.
- Availability: The availability security dimension provides reliable access to resources of interest to authorized people.
- Nonreputation: It refers to preventing an individual or entity from denying having performed a particular action related to data by making available proof of various network-related actions.
- Data confidentiality: It ensures that the data content cannot be understood by unauthorized entities.
- Data integrity: This security dimension ensures the correctness or accuracy of data.

In addition, network can be protected through the implementation of firewall, segmentation, and segregation techniques. Moreover, monitoring mechanisms should be employed, so as to support: (a) network protection from attacks; (b) detection of attacks and (c) response to attacks. Finally, a security by design approach would complete the aforementioned countermeasures, focusing on the Cybersecurity aspects for new IoT devices or systems that need to be planned and implemented already from the beginning, meaning the procurement, design, development, and maintenance phases.

6.5 CONCLUSIONS

This chapter presented the evolution, benefits, threats, and security measures in three interconnected domains, Smart Cities, Smart Homes, and in Smart Healthcare. All three domains are becoming part of the daily life of people all over the word, offering comforts, assistance, and automations among other. Nevertheless, most of these devices have many security flaws, exposing the safety, security, and many times sensitive and personal data of the user to malicious attackers. In the beginning, the concept of smart cities followed by the respective components have presented. An interesting part in this domain is the application, challenges, and opportunities introduced by the employment of IoT devices within smart cities. The socioeconomic benefits of smart homes were presented, providing significant information on the benefits but also of the significance of robust security so as to defend against attacks that will expose personal data to malicious attackers. Furthermore, it provided and interesting categorization of three possible architecture that a Smart Homeowner could choose from. Finally, it provided information about security challenges, threats, and measures. In the last part of the chapter, the healthcare domain was presented. More specifically we initially presented the way that different IoT-connected medical technologies support health and well-being services, as well as their benefits and barriers, with a focus on security issues, vulnerabilities and types of attacks. Finally, organizational and technical security measures that health sector should adopt to prevent; to respond effectively, timely and efficiently and; minimize the impact of attacks to their network, information technology and systems, are presented.

6.6 ACKNOWLEDGMENT

The work presented in this section has been conducted in the framework of CYBER-TRUST project, which has received funding from the European Union's H2020 research and innovation program under grant agreement no. 786698. The work reflects only the authors' view, and the Agency is not responsible for any use that may be made of the information it contains.

The work presented in this section has been conducted in the framework of SAFECARE project, which has received funding from the European Union's H2020 research and innovation program under grant agreement no. 787002.

NOTES

1. The traffic is called abnormal when it is detected with different parameters from those that homeowners had defined.
2. kilobytes per second

REFERENCES

1. S. P. Mohanty, U. Choppali, and E. Kougianos, "Everything you wanted to know about smart cities," *IEEE Consumer Electronics Magazine*, vol. 5, no. 3, pp. 60–70, 2016.

2. A. Singh, "The Internet of Things and smart cities," Apr. 10, 2019. [Online]. Available: https://www.iotevolutionworld.com/smart-home/articles/441882-internet-things-smart-cities.htm. [Accessed: Sep. 4, 2020].
3. H. Arasteh, V. Hosseinnezhad, V. Loia, A. Tommasetti, O. Troisi, M. Shafie-khah, and P. Siano, "Iot-based smart cities: A survey," in *16th International Conference on Environment and Electrical Engineering*, 2016.
4. ENISA, "Cybersecurity for Smart Cities," *ENISA*, 2015.
5. N. Sharma, M. Shamkuwar, and I. Singh, "The history, present and future with IoT," in *Internet of Things and Big Data Analytics for Smart Generation*, Poland: Intelligent Systems Reference Library, pp. 27–51, 2019.
6. H. Arasteh, V. Hosseinnezhad, V. Loia, A. Tommasetti, O. Troisi, M. Shafie-khah, and P. Siano, "Iot-based Smart Cities: A survey," in *16th International Conference on Environment and Electrical Engineering (EEEIC)*, Florence, Italy, 2016.
7. C. Lévy-Bencheton, E. Darra, G. Tétu, G. Dufay, and M. Alattar, "*Security and resilience of smart home environments: Good practices and recommendations*," in *ENISA*, Athens, Greece, 2015.
8. S. Tian, W. Yang, J. M. L. Grange, P. Wang, W. Huang, and Z. Ye, "Smart healthcare: making medical care more intelligent," *Global Health Journal*, vol. 3, no. 3, pp. 62–65, 2019.
9. K. Saravanan, E. G. Julie, and Y. H. Robinson, "*Smart cities & IoT: Evolution of applications, architectures & technologies, present scenarios & future dream*," in *Internet of Things and Big Data Analytics for Smart Generation*, Switzerland: Intelligent Systems Reference Library, Springer, pp. 135–151, 2019.
10. C. Badica, M. Brezovan, and A. Badica, "*An overview of smart home environments: Architectures, technologies and applications*," in *BCI*, Thessaloniki, 2013.
11. Fortune Business Insights, "Smart home market size, share & industry analysis, by product (Home monitoring/security, smart lighting, entertainment, smart appliances, and others (Thermostat, etc.)), and regional forecast, 2019–2026," *Fortune Business Insights*, 2020.
12. Markets and Markets, "Smart home market by product (Lighting control, security & access control, HVAC, entertainment, smart speaker, home healthcare, smart kitchen, home appliances, and smart furniture), software & services, and region—global forecast to 2024," *Markets and Markets*, 2019.
13. Mordor Intelligence, "Smart homes market—growth, trends, and forecast (2020—2025)," *Mordor Intelligence*, 2019.
14. A. Lazakidou, K. Siassiakos, and K. Ioannou, "*Wireless technologies for ambient assisted living and healthcare: Systems and applications*," in *Security in Smart Home Enviroment*, New York: IGI Global, pp. 170–191, 2010.
15. L. Moné, "IoT devices, sensors, and actuators explained," *LeanIX*, [Online]. Available: https://www.leanix.net/en/blog/iot-devices-sensors-and-actuators-explained. [Accessed: Apr. 24, 2020].
16. M. Li, W. Gu, W. Chen, Y. He, Y. Wu, and Y. Zhang, "Smart-home: Architecture, technologies and systems," *Procedia Computer Science*, vol. 131, pp. 393–400, 2018.
17. M. R. Alam, M. B. I. Reaz, and A. A. M. Mohd, "A review of smart homes—past, present, and future," *IEEE Transactions on Systems, Man, and Cybernetics, Part C (Applications and Reviews)*, vol. 42, no. 6, pp. 1190–1203, 2012.
18. C. Vassilakis, N. Kolokotronis, K. Limniotis, C.-M. Mathas, K.-P. Grammatikakis, D. Kavallieros, V. G. Bilali, S. Shiaeles, and J. Ludlow, "D2.1 threat landscape: Trends and methods," *Cyber-Trust Consortium*, 2018.

19. T. D. Mendes, R. Godina, E. M. Rodrigues, J. C. Matias, and J. P. Catalao, "Smart home communication technologies and applications: Wireless protocol assessment for home area network resources," *Energies*, vol. 8, no. July, pp. 7279–7311, Jul. 2015.
20. T. Serrenho and P. Bertoldi, "Smart home and appliances: State of the art—energy," *Publications Office of the European Union, Luxemburg*, 2019.
21. S. U. Rehman and S. Manickam, "A study of smart home environment and its security threats," *International Journal of Reliability, Quality and Safety Engineering*, vol. 23, no. 3, p. 9, 2016.
22. ENISA, "Threat landscape and good practice guide for smart home and converged media," *ENISA*, 2014.
23. WHO, "WHO," 2010. [Online]. Available: https://www.who.int/healthsystems/EN_HSSkeycomponents.pdf?ua=1. [Accessed: Apr. 10, 2020].
24. WHO, "WHO," 2020. [Online]. Available: https://www.who.int/topics/health_services/en/. [Accessed: Apr. 10, 2020].
25. European Commission, "Cross-cutting business models for IoT," 2017. [Online]. Available: https://op.europa.eu/en/publication-detail/-/publication/da41ca52-113a-11e8-9253-01aa75ed71a1/language-en/format-PDF/source-65168360. [Accessed: Apr. 10, 2020].
26. KPMG, 2018. [Online]. Available: https://assets.kpmg/content/dam/kpmg/pl/pdf/2018/06/pl-The-Changing-Landscape-of-Disruptive-Technologies-2018.pdf. [Accessed: Apr. 8, 2020].
27. MarketsandMarkets, "IOT healthcare market—global forecast to 2024," 2017. [Online]. Available: https://www.marketsandmarkets.com/PressReleases/iot-healthcare.asp. [Accessed: Apr. 4, 2020].
28. European Commission, "Internet of things," 2019. [Online]. Available: https://ec.europa.eu/digital-single-market/en/news/internet-things-brochure. [Accessed: Apr. 5, 2020].
29. Frost&Sullivan, "Role of IOT and wearables in healthcare," *Frost&Sullivan*, 2018.
30. AMWELL, "Telehealth kiosks," 2020. [Online]. Available: https://business.amwell.com/telemedicine-equipment/kiosks/. [Accessed: Apr. 13, 2020].
31. TELACARE, "What is a telemedicine kiosk and why is it the future of health care?" 2017. [Online]. Available: https://www.telacare.com/blog/what-is-a-telemedicine-kiosk-and-why-is-it-the-future-of-health-care. [Accessed: Apr. 12, 2020].
32. Alliance of Advanced Biomedical Engineering (AABME), "Smart pills enable convenient diagnostics and accurate therapy," 2017. [Online]. Available: https://aabme.asme.org/posts/smart-pills-enable-convenient-diagnostics-and-accurate-therapeutics. [Accessed: Apr. 17, 2020].
33. S. Tyagi, A. Agarwal, and P. Maheshwari, "*A conceptual framework for IoT-based healthcare system using cloud computing*," in *2016 6th International Conference Cloud System and Big Data Engineering (Confluence)*, 2016.
34. B. Manate, T.-F. Fortis, and V. Negru, "Optimizing cloud resources allocation for an internet of things architecture," *Scalable Computing: Practice and Experience*, vol. 15, no. 4, pp. 345–355, 2014.
35. C. Sheriff, T. Naqishbandi, and A. Geetha, "Healthcare informatics and analytics framework," 2015.
36. A. Pir, M. Akram, and M. Khan, "Internet of things based context awareness architectural framework for HMIS," 2015.
37. P. Chatterjee and R. Armentano, "Internet of things for a smart and ubiquitous eHealth system," 2016.
38. N. Zhu, "Bridging e-health and the Internet of Things: the sphere project," *IEEE Intelligent Systems*, vol. 30, no. 4, pp. 39–46, 2015.

39. A. Osello, "Energy saving in existing buildings by an intelligent use of interoperable ICTs," *Energy Efficiency*, vol. 6, no. 4, p. 707–723, 2013.
40. A. Rahmani, T. Gia, B. Negash, A. Anzanpour, I. Azimi, M. Jiang, and P. Liljeberg, "Exploiting smart e-health gateways at theedge of healthcare internet-of-things: a fog computing approach," *FutureGeneration Computer Systems*, vol. 78, no. 2, pp. 641–658, 2018.
41. A. Adame, A. Bel, A. Carreras, M. Meli-Segu, M. Oliver, and R. Pous, "Cuidats: An rfidwsn hybrid monitoring system for smart healthcare environments," *Future Generation Computer Systems*, vol. 78, no. 2, pp. 602–615, 2018.
42. SilverLink, "Smart home health monitoring forsenior care," *SilverLink*, 2016.
43. Y. Cheng, C. Jiang, and J. Shi, "*A fall detection system based onsensortag and windows 10 IoT core*," in *International Conference onMechanical Science and Engineering (ICMSE2015)*, 2015.
44. S. Gasparrini, E. Cippitelli, S. Spinsante, and E. Gambi, "A depth-based fall detection system using a kinect sensor," *Sensors*, vol. 14, pp. 2756–2775, 2014.
45. J. Tello and E. Barbazza, "*Health services delivery: A concept note*," WHO, 2015.
46. Louis Colombus, "2018 roundup of Internet of Things forecasts and market estimates," 2018. [Online]. Available: https://www.forbes.com/sites/louiscolumbus/2018/12/13/2018-roundup-of-internet-of-things-forecasts-and-market-estimates/#6edec69c7d83. [Accessed: Apr. 7, 2020].
47. J. Krawiec, J. Nadler, E. Tye, and J. Jarboe, "No appointment necessary: How the IoT and patient-generated data can unlock health care value," 2015. [Online]. Available: https://www2.deloitte.com/us/en/insights/focus/internet-of-things/iot-in-health-care-industry.html. [Accessed: Apr. 10, 2020].
48. C. Stern, "Goldman Sachs says a digital healthcare revolution is coming — and it could save America $300 billion," 2015. [Online]. Available: https://www.businessinsider.com/goldman-digital-healthcare-is-coming-2015-6. [Accessed: Apr. 5, 2020].
49. K. Taylor, M. Steedman, A. Sanghera, and M. Thaxter, "Medtech and the Internet of Medical Things | How connected medical devices are transforming health care," *Deloitte*, 2018.
50. B. Farahani, F. Firouzi, V. Chang, M. Badaroglu, N. Constant, and K. Mankodiya, "Towards fog-driven IoT eHealth: Promises and challenges of IoT in medicine and healthcare," *Future Generation Computer Systems*, vol. 78, no. 2, pp. 659–676, 2018.
51. S. Nandyala and K. Kim, "From cloud to fog and IoT-based real-time U-healthcare monitoring for smart homes and hospitals," *International Journal of Smart Home*, vol. 10, no. 2, pp. 187–196, 2016.
52. I. Hashem, V. Chang, N. Anuar, K. Adewole, I. Yaqoob, A. Gani, E. Ahmed, and H. Chiroma, "The role of big data in smart city," *International Journal of Information Management*, vol. 36, no. 5, p. 748–758, 2016.
53. M. Rathore, A. Ahmad, A. Paul, J. Wan, and D. Zhang, "Real-time medical emergency response system: Exploiting IoT and big data for public health," *Journal of Medical Systems*, vol. 40, no. 12, 2016.
54. S. Li, L. Da Xu, and S. Zhao, "The internet of things: A survey," *Information Systems Frontiers*, vol. 17, no. 2, pp. 243–259, 2015.
55. Gartner, "How IoT impacts data and analytics," 2018. [Online]. Available: https://www.gartner.com/smarterwithgartner/how-iot-impacts-data-and-analytics/. [Accessed: Apr. 12, 2020].
56. Bain & Ciompany, "Unlocking opportunities in the Internet of Things," 2018. [Online]. Available: https://www.bain.com/insights/unlocking-opportunities-in-the-internet-of-things/. [Accessed: Apr. 13, 2020].

57. Zingbox, "Threat report- medical devices," 2018. [Online]. Available: http://go.zingbox.com/rs/562-ZPO-907/images/Zingbox_2018_Annual_Threat_Report_Medical_Devices.pdf. [Accessed: Apr. 10, 2020].
58. European Union Agency for Cybersecurity (ENISA), "Smart hospitals security and resilience for smart health service and infrastructures," 2016. [Online]. Available: https://www.enisa.europa.eu/publications/cyber-security-and-resilience-for-smart-hospitals. [Accessed: 5, 2020].
59. Irdeto, "Cybersecurity survey: Protecting the Internet of Medical Things," 2019. [Online]. Available: https://go.irdeto.com/connectedhealth-himss-white-paper-irdeto-protecting-the-internet-of-medical-things/. [Accessed: Apr. 23, 2020].
60. National Institute of Technology Standards (NIST), "NISTIR 8228—Considerations for managing Internet of Things (IoT) Cybersecurity and privacy risks," 2019. [Online]. Available: https://nvlpubs.nist.gov/nistpubs/ir/2019/NIST.IR.8228.pdf. [Accessed: Apr. 10, 2020].
61. T. Spring, "St. Jude patches additional cardiac device," 2017. [Online]. Available: https://threatpost.com/st-jude-patches-additional-cardiac-device/123596/. [Accessed: Apr. 15, 2020].
62. J. Finkle, "J&J warns diabetic patients: Insulin pump vulnerable to hacking," 2016. [Online]. Available: https://www.reuters.com/article/us-johnson-johnson-cyber-insulin-pumps-e/jj-warns-diabetic-patients-insulin-pump-vulnerable-to-hacking-idUSKCN12411L. [Accessed: May 5, 2020].
63. L. Newman, "Medical devices are the next security nightmare," 2017. [Online]. Available: https://www.wired.com/2017/03/medical-devices-next-security-nightmare/. [Accessed: May 2, 2020].
64. Federal Trade Commission, "FTC staff report: Internet of Things: Privacy and security in a connected world," 2015. [Online]. [Accessed: May 5, 2020].
65. R. Moosavi, S. Gia, A. Rahmani, N. Ethiopia, V. Seppo, J. Isoaho, and H. Tenhunen, "*SEA: A secure and efficient authentication and authorization architecture for IoT-based healthcare using smart gateway*," in *Procedia Computer Science*, 2015.
66. O. Sangpetch and A. Sangpetch, "Security context framework for distributed healthcare IoT platform," in *Internet of Things Technologies for HealthCare*, Springer, 2016.
67. A. Bashir and A. Mir, "Secure framework for Internet of Things based e-Health system," *International Journal of E-Health and Medical Communications (IJEHMC)*, vol. 4, pp. 16–29, 2019.
68. NIST, "NIST glossary," 2019. [Online]. Available: https://csrc.nist.gov/glossary/term/RA. [Accessed: Dec. 10, 2019].
69. European Cybersecurity Organisation (ECSO), "ECSO healthcare sector report," European Cybersecurity Organisation (ECSO), Brussels, Belgium, 2018.

7 IoT Security Frameworks and Countermeasures

G. Bendiab, B. Saridou, L. Barlow, N. Savage, and S. Shiaeles

CONTENTS

7.1 Introduction 241
7.2 Major Cyber-Threats to IoT 242
    7.2.1 Malware 243
    7.2.2 Phishing..243
    7.2.3 Man-in-the-Middle (MitM)..244
    7.2.4 Distributed Denial of Service (DDoS) 244
    7.2.5 Botnets 245
    7.2.6 Injections..245
    7.2.7 Advanced Persistent Threats (APTs) 246
    7.2.8 Zero-Day Exploits..246
    7.2.9 Insider Threats...246
    7.2.10 IoT Device Hijacking 247
7.3 Operation of Firewalls on the Network Perimeter 247
    7.3.1 Firewalls Overview 247
    7.3.2 Types of Firewalls 248
        7.3.2.1 Packet Filter Firewall 248
        7.3.2.2 Stateful Packet Inspection 249
        7.3.2.3 Next-Generation Firewall (NGFW) 249
    7.3.3 Effectiveness of Firewalls in Limiting Threat Impact.............................249
        7.3.3.1 Ping of Death (PoD)...250
        7.3.3.2 SYN Flood 250
        7.3.3.3 Eavesdropping 250
        7.3.3.4 Java and ActiveX Attacks 251
    7.3.4 Firewalls Limitations 252
7.4 Operation of IPS in the Network..252
    7.4.1 IPS Overview 252
        7.4.1.1 Signature-Based Detection Method 253
        7.4.1.2 Anomaly-Based Detection Method..253
        7.4.1.3 Passive Network Monitoring..253
    7.4.2 Honey Pot for Intrusion Detection 254
    7.4.3 Effectiveness of IDS Operation in Limiting Threat Impact...............254
    7.4.4 IPS Limitations 256

7.5	Operation of Antivirus/Malware Detection Software		257
	7.5.1	Antivirus/Antimalware Software Overview	257
	7.5.2	Importance of Using Antivirus/Antimalware Software	258
	7.5.3	Limitations of Antivirus/Antimalware Software	260
7.6	Security Awareness of Individuals		260
	7.6.1	Security Awareness Training and Its Importance	261
		7.6.1.1 Password Management and Authentication	261
		7.6.1.2 Email/Phishing and Social Engineering	261
		7.6.1.3 Removable Media	262
		7.6.1.4 Social Media Usage	262
		7.6.1.5 Physical Security	262
	7.6.2	Limitations	263
7.7	Existence of Update/Patching Policy		263
	7.7.1	Security Vulnerabilities and Exploits	264
	7.7.2	Update/Patching Policy	265
	7.7.3	Effectiveness of Update/Patching Policy in Limiting Threat Impact	266
	7.7.4	Limitations	267
7.8	Existence of Logging/Alerting Policy		267
	7.8.1	Logging/Alerting Policy Overview	267
		7.8.1.1 Logging Management	267
		7.8.1.2 Alerts Management	268
	7.8.2	Effectiveness of Alerting/Logging Policy in Limiting Threat Impact	269
	7.8.3	Limitations	269
7.9	Conducting Periodic Security Checks of Security Mechanisms		270
	7.9.1	Overview	270
	7.9.2	Effectiveness of Running Periodic Security Checks in Limiting Threat Impact	271
		7.9.2.1 Insecure Software/Firmware and IoT	271
		7.9.2.2 DDoS Attacks	271
		7.9.2.3 Botnets	272
	7.9.3	Limitations	272
7.10	Existence of Security Policy to Install/Update Equipment/Software		273
	7.10.1 Security Policy Overview, Objectives		273
	7.10.2 Security Policy & General Good Practices for New Software Installation		273
	7.10.3 Effectiveness of Security Policy		274
		7.10.3.1 Malware, Malicious Software, Email Malware	275
		7.10.3.2 Use of Nonapproved Hardware	275
	7.10.4 Limitations		276
		7.10.4.1 Memory Scraping	276
		7.10.4.2 Physical Damage/Theft/Loss	276
		7.10.4.3 Drive-by Download Attacks	277
7.11	Existence of Backup Policies		277
	7.11.1 Backup Policy Overview, Objectives		277
	7.11.2 Types of Data Backup Policies		278
	7.11.3 Good Practices for an Effective Backup Policy		279

7.11.4 Effectiveness of Backup Policy in Limiting Threat Impact 279
  7.11.5 Limitations 280
7.12 Conclusions 281
References 283

7.1 INTRODUCTION

Internet of Things (IoT) devices are quickly changing the way people access services and utilize internet infrastructure. They are in every part of our daily lives, whether at work or home, we are constantly connected. With this ever-evolving connection, users can find themselves at risk from a variety of both malicious and accidental threats [1]. In fact, cyber- threats may affect the functioning of modern society by causing electrical blackouts [2], disrupting the normal functionality of IoT devices and computer networks, and making them unavailable for legitimate users, and causing large theft of valuable sensitive information like financial credentials and medical records [2]. They can even cause failure of medical and military equipment and breaches of national security secrets. The latest threat landscape shows that cybercriminals have improved their tactics to the point where they become extremely difficult to detect and remediate. They are now successfully attacking computers/devices and networks every 39 seconds, according to a recent study by the University of Maryland[1].

It is evident that cyber-threats are seen as one of the top challenges facing IoT devices and edge-computing frameworks. The information security community warned that even a single unsecured device connected to a network could cause a potential threat to the whole network and serves as a point of entry for a wide variety of hacking attempts [3]. Another worrying trend is the continuing improvement of the advanced persistent threats (APTs), which present a very real and urgent menace to businesses and IoT systems. Unlike other hacking techniques, APTs attacks are targeting specific organizations, by using sophisticated levels of expertise and significant resources, for achieving a very specific goal such as extracting or destroying extremely high-valuable data [4]. Security best practices for defense against those potential cyber-threats include basic but extremely important countermeasures like patching systems, closing unused ports, periodic security audits, and threat assessment, employing protective software and devices such as firewalls, intrusion detection systems, antivirus, and proxy servers/NATs, etc.

This chapter looks at the basic security frameworks and countermeasures that can be used to ensure the integrity and security of IoT infrastructure. It begins with a brief description of the most common cyber-threats for IoT devices. Then, it gives you an understanding of different security measures, either technical or nontechnical, that can be applied to protect from cyber-threats, and how these countermeasures can alter the threat exploitability level. There are a large number of security measures that can be applied to protect the IoT ecosystem; however, this chapter focuses only on a subset of security measures that are relevant for securing IoT devices. This includes firewalls, IDS/IPS, antivirus, system patching, logging/alerting policies, periodic security checks and assessment, individual awareness, backup policies, and security policies that can be used for acquiring and installing new/update equipment and software.

7.2 MAJOR CYBER-THREATS TO IOT

A cyber-threat refers to a hypothetical malicious act that might intentionally exploit a vulnerability to gain unauthorized access to control a system device and/or a network, damage, disrupt, or steal an information technology asset, or any other form of sensitive data of corporations, governments, and individuals [1]. Cyber-threats come from different malicious actors, which could be an individual that creates attack vectors using their own software, a criminal organization with large number of skilled employees developing and executing more sophisticated attacks, organized terrorist group, business competitor, or even a nation state that is trying to learn another country's national secrets.

Most IoT devices run on low power and limited computational capabilities in terms of processing and storage. This means that it will be challenging for IoT device manufacturers to design and incorporate complicated security protocols in these devices [5]. This makes these devices easy targets for hackers, which can even use them as tools for massive cyber-attacks. For example, attackers can exploit vulnerabilities in IoT devices to steal their data or reprogramming the devices to facilitate other attacks such as Distributed Denial of Service (DDoS) attacks. To shed light on the current situation, the OWASP[2] Internet of Things Project released the Top 10 in IoT 2018, a vulnerabilities list regarding the most serious IoT threats for 2018 [6]. These include:

1. **Weak, Guessable, or Hardcoded Passwords**: In fact, most IoT devices are not configured to allow users to change default passwords, and that leaves them vulnerable to a host of password attacks. Further, Many IoT products are often released with insecure firmware that contains backdoors.
2. **Insecure Network Services**: Vulnerable network services (unpatched software, open ports, insecure communication channels, etc.) can potentially lead to these devices being compromised and utilized as a part of a larger botnet.
3. **Insecure Ecosystem Interfaces**: Common interfaces used to communicate with IoT devices are web interface, cloud, back-end APIs, and mobile interfaces. Vulnerabilities in these interfaces, such as weak encryption, lack of authentication/authorization, can lead to compromising the device, or worse, the overall network.
4. **Lack of Secure Update Mechanism**: With the lack of any secure update mechanism in place, there is no guarantee about the security of the IoT devices. Thus, IoT device manufacturers should provide periodic security updates/patching of their devices.
5. **Use of Insecure or Outdated Components**: Using outdated and insecure software, from insecure customizations of the operating system to using non-approved third-party hardware or software components could lead to compromising the overall security of the IoT device and create an entry point for potential cyber-attacks.
6. **Insufficient Privacy Protection**: Insecure storage of personal data and processing of personal information generated by IoT devices can lead to many privacy issues and disclosure of this sensitive data without the user's permission.
7. **Insecure Data Transfer and Storage:** The data breach incidents that occur on a daily basis are mainly caused by insecure storage of user's personal

information on the device or in the ecosystem that is used insecurely, improperly, or without permission.
8. **Lack of Device Management**: Inappropriate management of the connected IoT devices, by relying on old and nonreliable methods, can compromise the entire network security as they are interacting with the network and have access to it.
9. **Insecure Default Settings**: The default device configurations (e.g., passwords) on IoT devices are often insecure. Thus, not changing default settings before using the devices makes them vulnerable to potential cyber-attacks.
10. **Lack of Physical Hardening**: IoT devices should be protected against physical attempts by malicious users to extract sensitive information that can later be exploited to launch remote attacks or gain remote control of the device.

Security researchers reveal that due to all these security vulnerabilities, IoT devices have become an increasingly attractive target for cyber-attackers, with different kind of attacks, to secretly control the device, damage the device, gain unauthorized access to the data on the device, and/or otherwise affect the device's operation without permission. Further, compromised devices provide hackers with an entry point for penetrating other devices/machines on the same network and therefore, compromise all the network. According to security reports [6–9] common cyber-threats and attacks against IoT devices include:

7.2.1 MALWARE

Malware is probably the most common threat to any system, including individual user's systems and networks. It refers to any piece of software that was intentionally designed to perform malicious tasks such as stealing data, perturbing the normal functionality of a computer/device, or a network, or taking control of a system [10, 11]. It is usually written by groups of hackers for monetary gain, either by spreading the malware themselves or selling them on the Dark Web. Malware has actually been a threat to individuals and organizations since the early 1970s when the Creeper virus first appeared. After that, a wide variety of types of malware have been developed, including computer viruses, worms, trojan horses, bots, adware, spyware, exploits, keyloggers, and ransomware [12]. Recently, malware has been determined by security reports as the most frequently encountered cyber-threat, with high evolution in terms of sophistication and diversity [10].

A recent security report by AV-TEST institute[3] affirms that the AV-TEST analysis systems recorded over 144.91 million new malware samples in 2019 and over 38.48 million new samples in the first trimester in 2020. Another report by Webroot[4] found that over 93% of malware was obfuscated and has the ability to continuously change its code to avoid detection by security mechanisms.

7.2.2 PHISHING

Phishing is a kind of cyber-threat in which cybercriminals try to steal sensitive information (e.g., login credentials and credit card numbers) or install malware on the victim devices using deceptive e-mails, websites, or text messages [13]. Phishers typically employ social engineering techniques to masquerade themselves as trusted entities to gain the trust of the victims and duping them into opening malicious emails ("malspam"), or text messages ("spoofs") that imitate a person or a trusted organization (e.g., bank, government office, etc.) [14]. The recipient is then tricked into clicking a malicious link, which can lead to damaging losses in terms of identity theft, sensitive intellectual property and customer information, and even national-security secrets [14]. It can also lead to locking a system as part of a ransomware attack or installing a malware that can be used by cybercriminals for conducting further attacks like DDoS.

Along with the phishing attacks that address general Internet population, several other variants of spear phishing, often known as whale phishing attacks have been involved in last years [15]. This kind of phishing attacks is targeting at specific individuals, organizations, or businesses by using clever tactics to fool even experienced security professionals. Spear phishing attackers usually perform prior researches before launching their attacks in order to boost the chances that the victim will carry out all the actions necessary for infection [15]. A new report by FBI noticed a dramatically increase of spear-phishing attacks against multiple industry sectors [16].

7.2.3 MAN-IN-THE-MIDDLE (MITM)

The Man-In-The-Middle (MITM) attack is one of the most well-known threats in computer security [17], where a malicious actor positions himself as a relay/proxy between two endpoints, targeting the actual data that flows between them. An attacker intercepts the connection between a client and a server, behaves like a server for the client and client for the server in order to inject false information and intercept the data transferred between the two endpoints. This could be achieved through interfering with legitimate networks or creating fake networks that the attacker controls. Then, compromised traffic is stripped of any encryption in order to steal, change, or reroute that traffic to the attacker's destination of choice such as phishing websites [17].

In the context of IoT devices, security reports[5] have found that 98% of all IoT data traffic is not encrypted, which make it an easy target to MitM attacks with catastrophic consequences. For instance, attackers may use MitM to fake temperature data from a monitoring device in order to force a piece of machinery to overheat and therefore, cease production, which can lead to physical and financial damage to the operating organizations.

7.2.4 DISTRIBUTED DENIAL OF SERVICE (DDoS)

A Distributed Denial-of-Service (DDoS) attack is one of the most challenging threats on the Internet. It refers to malicious attempts to disrupt normal traffic of a target service, server or network and make it unavailable for legitimate users. This can be accomplished by overwhelming the victim or their surrounding infrastructure with a flood of Internet traffic. In order to be more effective, DDoS attack involves multiple compromised (i.e., infected with malware) online computer systems or IoT devices which form a botnet. The attacker is able to direct the machines (also called bots) in the botnet by sending updated instructions via remote control. Each bot will contribute to the DDoS attack by sending requests to the victim, potentially causing the target server or network to overflow capacity.

Security reports confirm that IoT networks can both amplify and be the targets of DDoS attacks. In this context, security incidents confirm that the bigger security challenge is that security weaknesses of these devices can be easily exploited by hackers to form large botnets (zombie armies), and therefore launch massive DDoS attacks. According to the A10 Networks latest report,[6] which detected nearly 6 million DDoS attacks in the fourth quarter of 2019, Mirai continues to be the malware of choice for botnets and WD-Discovery become the third most common source of DDoS after SNMP (Simple Network Management Protocol) and SSDP (Simple Service Delivery Protocol).

7.2.5 Botnets

The word "botnet" comes from the combination of "robot" and "network" and is used to describe a network, or "army," of compromised devices, including computers, IoT devices and smartphones. This army of "bots" is controlled through a Command & Control scheme (C&C) by the "bot herder." Botnets are used for large-scale malicious activity, including DDoS attacks, phishing, spamming, and data harvesting. The attacker will typically gain and maintain access into a device (turning it into a *zombie machine*) through a Trojan horse program that comes as an email attachment or web download [18].

Despite large efforts (and budgets) made by companies and the security community to mitigate the ever-growing use of botnets, attackers continue to come up with new ways to obfuscate their activity. To achieve this, they will usually employ malware morphing to avoid detection by antivirus software, strong cryptography for their communication, exploitation of weak security infrastructure or sophisticated social engineering attacks [19]. An older study by P&G, which was run for a 6-month period against its devices on a global scale, revealed that 3000 out of its 80000 PCs were infected [20].

7.2.6 Injections

Injection attacks refer to a wide range of attack vectors in which attackers inject malicious input into a program or query or inject malware onto a system in order to execute remote commands. This fake input will be interpreted as a part of a command or query and executed, generating wrong results. Some of the more common types of injections are SQL Injection (SQLi), Cross-Site Scripting (XSS), shell injection, code injection, XPath Injection, and XML external entity (XXE) injection [21]. Injections are amongst the most popular and dangerous attacks that can lead to data theft, loss of data integrity, DDoS attacks, as well as full compromise of the target system. In fact, there are numerous free tools that help amateur hackers to easily launch this kind of attack with potentially serious consequences. For instance, over 420,000 websites around the world were attacked with an SQL injection in 2014, which allowed Russian hackers to steal more than 1.2 million identifiers and passwords [22].

7.2.7 ADVANCED PERSISTENT THREATS (APTs)

An APT is a broad term used to describe the threat of highly sophisticated and targeted cyber-attacks, in which experienced groups of hackers with significant resources gain access to a specific system and remain undetected for a long period of time, with potentially destructive consequences [23]. To achieve their purposes, APT groups use advanced attack techniques, including advanced exploits of zero-day vulnerabilities, highly targeted spear-phishing, and other advanced social engineering techniques. The Ukraine power grid attacks in December 2015 is an example of highly successful APTs that caused power outages for more than 250,000 customers, which take a long time to recover [24].

GhostNet, Stuxnet, DeepPanda, APT28, APT34, and APT37 are more examples of destructive APTs attacks. According to security reports, the Stuxnet worm that was deployed to attack Iranian nuclear facilities went undetected for almost 2 years. This attack destroyed numerous centrifuges in Iran's Natanz uranium enrichment facility by causing them to burn themselves. Over time, this dangerous computer worm has been exploited by other groups to target critical infrastructures including water treatment plants, power plants, and gas lines [25]. These incidents indicate the urgent need for effective and robust security mechanisms to respond to the new threats.

7.2.8 ZERO-DAY EXPLOITS

The term "Zero-day" may refer to a recently discovered vulnerability or an exploit for a vulnerability that hackers can use to attack systems, with "zero" representing the number of days since the software vendor discovered the new vulnerability or its exploit, and started releasing a patch to fix it. The term "zero-day exploit" describes the malicious code that was written by criminal hackers and spies to exploit a "zero-day vulnerability." Exploits can go unobserved for many years and are often sold on the black market for large sums of money. These threats are extremely dangerous because only the attacker is aware of their existence, so no security patches available to fix these vulnerabilities and block its corresponding zero-day exploits.

Security reports confirm that zero-days are becoming more common, with the large number of new zero-day threats originating every day, mainly because of the emergence of the large market for buying and selling zero-day vulnerabilities and corresponding exploit kits.

7.2.9 INSIDER THREATS

Insider threat is one of the most common threat factors for the IT systems and infrastructure of companies. This threat originates from users (e.g., employees, contractors, or any other business associates) with legitimate access to an organization's asset, who use that access either maliciously or unintentionally, to cause data breaches and/or harm to the organization [26]. The results of such an attack can be catastrophic and costly to the business. Insider attacks are particularly more dangerous than outsider attacks because an insider already has direct access to the organization and its network and does not need to hack in through the outer perimeter; thus, they are harder to defend against. It is also more challenging to detect insider threats because they can easily evade existing defenses and it is very difficult to distinguish between a legitimate user's normal activity and a potentially malicious activity, especially for those with elevated levels of access [26].

A recent study by SANS institute[7] (www.sans.org/), about advanced threats, found major gaps in insider threat defense mechanisms driven by the lack of visibility into a baseline of normal user behavior and poor management of privileged user accounts. Another new study by the Pomeron institute,[8] sponsored by ObserveIT and IBM, noticed that the average annual cost of insider threat was $11.45 million in 2020, with an increase of 31% compared to 2019.

7.2.10 IoT Device Hijacking

One of the most common security threats to IoT devices is device hijacking, where attackers attempt to take control of the target device. This kind of attack is quite difficult to detect because the attacker does not change the basic functionality of the victim device. In the case of weak security defense, device hijacking could be extremely effective, producing DDoS attacks that can cripple IT infrastructures and systems as it takes only one compromised device to potentially re-infect all other devices in the network and remotely perform its malicious activities. As we rely more and more on IoT devices in our daily activities, those attacks can become more disruptive or even dangerous.

In summary, it can be concluded that very skilled cybercriminals exist in great numbers and that all of them are an imminent threat to the IT infrastructures. Even the most amateur hacker can easily find all the tools they need online, at virtually no cost, to traverse highly secure systems at will [19]. This has been proved by the increasing number of incidents reported annually. Thus, securing your system and network in an appropriate manner is critical. This chapter will investigate various security tools and procedures that can be used to defend against those attacks, which manifest themselves in a variety of forms. The security tasks that will be described in this chapter deals with devices and policies including firewalls, intrusion-detection systems, antivirus/antispyware, patching policies, logging/alerting procedures, periodic security checks, individual awareness, and secured backup policies.

7.3 OPERATION OF FIREWALLS ON THE NETWORK PERIMETER

7.3.1 Firewalls Overview

A firewall is one of the most fundamental devices used to implement network security and it constitutes the first line of defense against malicious traffic and intrusion attempts. According to Marcus Ranum, "A firewall is a system or group of systems that enforces an access control policy between two networks" [27]. This network security tool can be either a hardware or a software positioned strategically between the internal network and the external untrusted networks, and sometimes internally to create segregation between different network zones, e.g., separating different departments of a business [28]. The primary purpose of this device is to filter traffic and control its flow via one or more methods. It is also relied upon for important logging and audit functionality with their logs allowing security administrators to identify most forms of threats/misuse of the network, leading to the creation/amending of rulesets to counter these issues [29]. The methods of protection vary greatly, with the cheapest and most basic form being the enforcement of a security policy or ruleset. With this method, network packets will be inspected to see if their characteristics match that of the pre-configured ruleset [30]. More precisely, the packet header is checked against the security policy/ruleset configured on the firewall, which will either allow or deny packets from accessing services and hosts on the internal network. The most common methods of protection that might be used in a particular firewall implementation are [19]:

- Packet filtering
- Stateful packet filtering
- User authentication
- Client application authentication

The basic firewall operation includes filtering incoming packets based on specific parameters such as packet size, source IP address, protocol, and destination port.

7.3.2 TYPES OF FIREWALLS

Firewalls can be categorized by what they protect, the two types being a network-based firewall, and a host-based firewall. With network-based firewalls generally existing as hardware and guarding entire networks. And host-based solutions often appearing as software to protect individual hosts [29]. Packet data can be represented differently throughout different levels of the network, due to the packets being reformatted several times to convey to the protocol where to send the packet. This has led to the creation of a range of filtering methods that can be implemented on a firewall, these methods vary in cost and functionality with each combatting threats to different degrees. Each type has different capabilities and features.

7.3.2.1 Packet Filter Firewall

The cheapest and most basic type of firewall is a Packet Filtering Firewall [31]. This form of filtering is stateless meaning that each incoming packet is inspected individually with no context, meaning the firewall will not know whether the packet is part of a traffic stream. This firewall operates primarily on the network and transport layer, inspecting source, and destination data from each packet's header, dropping packets that conflict with existing rulesets [27]. For example, if a firewall has a rule to block Simple Mail Transfer Protocol (SMTP) packets, then any packets with the destination port 25 shall be dropped. Only those packets that match the preconfigured rules are allowed through [19]. This type of firewall works by filtering packets based on packet size, protocol used, source IP address, and many other parameters. Due to the limited amount of checks being carried out by the filtering method, these firewalls are very easy to configure and inexpensive [31]. In fact, some operating systems, such as Windows 10 and Linux, include built-in packet filtering capabilities [19].

Firestarter, McAfee Personal Firewall, Norton Personal Firewall, and Outpost Firewall are examples of the most common firewall products that are based on the packet filtering method. These firewalls are very effective on a simple network with only a few servers running a small number of services (perhaps a Web server, an FTP server, and an e-mail server), however, with a wide area network that is connecting multiple sites in geographically diverse regions, configuring packet filtering rules can become quite complicated, which may result in blocking legitimate network service or allowing malicious network services [27].

7.3.2.2 Stateful Packet Inspection

Another type of firewall that can be implemented is known as a stateful packet inspection (SPI). Unlike packet filtering firewalls, this protection method can track the state of a connection, whether this is at the initial creation of the connection, the data transfer, or the termination state [32]. This understanding of the data stream allows for both the continued blocking of unwanted external connections to the network, whilst also allowing authorized internal hosts to reach out and maintain a two-way connection [27]. This functionality can be established in two ways, one being the opening of return ports, the second being the examination of TCP flags.

Those capabilities give the firewall the ability to identify the context in which a specific packet was sent. This makes these firewalls far less susceptible to ping floods and SYN floods, as well as being less susceptible to spoofing attacks [19]. Most popular SPI firewall products include SonicWALL, Linksys, Wolverine, and Cisco.

7.3.2.3 Next-Generation Firewall (NGFW)

A Next-Generation Firewall (NGFW) is the more advanced and effective solution for network security as NGFWs combine knowledge of both rulesets and policies, with an understanding of threats relating to the application-level protocols and applications used in the network [33]. This enriched knowledge base makes it able to detect and block packets that are deemed to be malicious, along with a variety of attacks. This type of firewall can implement a range of additional security controls on a network. This includes deep packet inspection, where not only basic packet data is inspected, but the identification of what applications are being used for the data stream. When paired with relevant and easily updated intelligence, this allows it to detect possible exploits and attacks at a greater level of accuracy [33].

If, however, this detailed inspection misses a threat such as advanced malware (APTs), the firewall will also have a form of an intrusion prevention system, allowing for both the sandboxing and the fast remediation of the threat before entering the network. This is a popular option for commercial networks as one security solution can provide both detection and remediation. Most popular NGFW products include Forcepoint NGFW, SonicWall, Cisco Firepower NGFW, and Sophos XG.

7.3.3 Effectiveness of Firewalls in Limiting Threat Impact

A firewall is absolutely an essential security requirement for every network. It defends the insider (i.e., trusted) network from attacks that come from untrusted outsiders and ensures that only authorized traffic passes into the network under control [19]. In fact, all network traffic entering or leaving the internal network passes through the firewall, which examines each packet and blocks those that do not meet the specified security benchmarks. In order to work effectively, the firewall must inspect both incoming packets entering the network in an attempt to detect potentially malicious attacks, but it must also inspect outgoing packets to limit threats such as malware communication and data exfiltration [33]. Thus, organizations should use the most appropriate firewall solutions to prevent potential security risks. In this context, a properly configured firewall around your network and devices can help to prevent a large number of attacks that may include backdoors, many variants of DDoS attacks, macros, remote logins, spam, virus, and port-scanning attacks. Generally, firewalls can help prevent the following well-known intrusions:

7.3.3.1 Ping of Death (PoD)

A PoD is the most primitive form of DDoS attacks in which an attacker sends malformed or oversized packets using a simple ping command to a computer or a service [19]. Sending an oversized ping can crash the target computer or service because most operating systems, including Windows, Mac, Unix, Linux, as well as network devices like printers and routers, cannot handle a ping larger than the maximum IP packet size, which is 65,535 bytes [19]. When the victim system attempts to reassemble the oversized fragments, a memory overflow could occur and lead to various problems including crash. This attack was particularly effective because the attacker can easily spoof his identity. Moreover, the attacker does not need detailed knowledge of the target machine, except its IP address. To avoid PoD attacks, and its variants, Firewall can be used to block all ICMP ping messages or selectively block fragmented pings [34], and drop any oversized packet.

7.3.3.2 SYN Flood

SYN flood is another form of DDoS attack that exploits the handshake process of a TCP connection by repeatedly sending SYN requests faster than the targeted system can process them, causing network saturation [19]. This kind of DDoS attacks is able to overwhelm all available ports on the target machine, making it unavailable to legitimate traffic. Attackers may also spoof their IP addresses on each SYN packet they send in order to avoid detection and make their identity more difficult to discover. Stateful packet inspection or SPI firewalls can easily stop SYN floods by examining all the packets from a given source. Thus, a large number of SYN packets from a single IP without corresponding SYN/ACK packets would be classified as abnormal and be blocked [19, 35]. Some advanced firewalls can block open and closed ports if a suspect IP address is probing them. These firewalls can also be configured to alert administrators if they detect connection requests across many ports from only one host [36].

7.3.3.3 Eavesdropping

An eavesdropping attack, also known as sniffing or snooping attack, basically refers to all kinds of attacks like stealing e-mail passwords, messages, files, data, information over the network connection by listening on the connection [37]. Generally, it is a piece of software that can simply be sitting somewhere in the network path and capturing all the relevant network traffic for later analysis. An attacker can insert the software onto a compromised device by direct insertion or by a virus or other malware. The most common active eavesdropping attack is the man-in-the-middle (MitM) attacks. This kind of attacks takes advantage of unsecured network communications to access data as it is being sent or received by its user. Eavesdropping is difficult to detect because the network transmissions will appear to be operating normally. Personal firewalls or host-based firewalls with VPNs and antivirus are effective in detecting and preventing passive network eavesdropping attacks, by using packet filtering and configuring the firewalls to reject any packet with a spoofed address.

7.3.3.4 Java and ActiveX Attacks

ActiveX controls are essentially small programs that could have access to the entire system if installed, this feature can be used for malicious purposes. Like other malicious programs, they could install malware, generate pop-ups, log your keystrokes and passwords, and do other malicious things. Firewalls are also very effective in preventing Java and ActiveX attacks targeting web servers [35]. For instance, CISCO IOS routers and the PIX firewalls[9] can be used together to define policy information about what URLs that users can or cannot access. In this case, the PIX firewall will check each user web access request against the predefined policy before permitting or blocking it. This may also help in preventing users from accessing known malicious and phishing websites by configuring the firewall to block data from certain locations (i.e., IP addresses), applications or ports.

Firewalls can also prevent overlapping fragments, Teardrop attacks, fragmented ICMP, and LAND attacks. In addition, all firewalls log the various activities that occur on the internal network. Produced logs can provide valuable information about security incidents, and therefore, help in determining the source of an attack, methods used to attack, and other data that might help either locate the perpetrator or at least prevent a future attack using the same techniques [19]. In fact, reviewing the firewall logs to check for abnormal activities should be a part of every network administrator routine. Also, reviewing the normal network activity over a period of time can create a baseline (e.g., average number of incoming and outgoing packets per hour, minute, and day) that helps administrators noticing when any abnormal activity may occur [19, 33].

Moreover, using internal firewalls on the top of perimeter firewalls provides an extra layer of security and significantly reduce the impact of insider threat by separating insiders from assets [38]. Internal firewalls are usually located between two parts of the same network, or between two separate organizations that share the same local network to identify east to west threats. The term "East to West" refers to the internal traffic to the local network, which may include traffic from server to server, server to client, client to server, client to client, or between IP subnets or routed VLAN interfaces but not leaving the insider network. Whereas the term "North to South" designs the traffic going from the local network to the outside world (i.e., Internet), and vice versa [39]. Internal firewalls can prevent the lateral movements of threats within the internal network and protection from potential insider threats. It can also minimize the attack surface by using micro-segmentation, which divides the network into granular zones that are secured separately [38]. This will increase the attacker's breakout time and give more time to respond to the attack.

7.3.4 FIREWALLS LIMITATIONS

By establishing a strong perimeter defense between inside and outside of your network, a properly deployed firewall is considered the foundation for your network security. However, firewalls alone cannot protect the network from all potential external and internal threats. In fact, firewalls can be used as an integral part of a larger security policy to safeguard your network [19, 40]. They should always be the first line of defense, in conjunction with other security measures that protect the entire network from both outsider and insider threats, such as intrusion detection systems and antivirus [19, 29]. Firewalls only help as a perimeter to block unauthorized data transmissions, but they do not provide anti-virus, anti-malware, or anti-spyware capabilities [35]. For instance, firewalls alone, cannot protect against completely new threats and insider intrusions, especially when an insider collaborates with an outside adversary [33]. Also, firewalls are not able to protect against viruses and infected files since it may not be possible to scan all incoming traffic [19, 29].

Moreover, most firewalls cannot deal with attacks at a relatively low level such as firmware manipulation, memory scraping, remote firmware attacks, nonapproved hardware, Transfer of malware or malicious commands via sensors, false sensor data injection attacks, hardware Trojan attacks, back off attack, information leakage via sensors, MAC Layer Jamming, malicious mobile nodes, selective forward attack and physical layer DoS attacks [19, 29]. Generally, firewalls cannot assist in reducing the risk related to those attacks or the consequences of successful attacks and in many cases, firewalls themselves can become the target of those attacks such as "Pulse Wave" DDoS attacks [41]. This could open the whole network for further attacks, and therefore not provide services to the legitimate users for which this network has been built. NGFWs and advanced firewalls may add transport layer security by acting as proxies. However, they cannot mitigate cases for accessing services with insufficient security which are outside the network perimeter.

Like any other software program, firewalls have some vulnerabilities that attackers may exploit. For instance, missed security patches of firewalls can be exploited by malicious attackers to compromise them [37]. Therefore, it is critical to check firewall activity periodically and ensure that it is fully correctly configured and has no vulnerabilities.

7.4 OPERATION OF IPS IN THE NETWORK

7.4.1 IPS OVERVIEW

An Intrusion Prevention System (IPS) is a security tool that is capable of not only detecting malicious activities, but also to take preventive actions to secure both the host and the network against potential threats that would normally pass through a traditional firewall device. It works in inline mode to provide protection from malicious attacks in real-time [42, 43]. Most IPS security systems also perform IDS (Intrusion Detection System) functions. They are available in two main categories, Host-based Intrusion/Prevention Detection Systems (HIDS/HIPS) and Network-based NIDS/NIPS. HIDS/HIPS is commonly used to analyze the activities on a particular machine, while the NIDS examines network traffic flows to detect and prevent intrusion threats [44]. It continuously monitors the network traffic, looking for possible malicious and unauthorized inputs aimed at compromising the basic network security [43, 45], and taking automated actions to stop them by sending alerts to the administrator (i.e., like in the case of an IDS), dropping the malicious traffic, blocking traffic from the source address, or terminating the connection [46].

An IPS is originally developed for addressing requirements of lacking in most firewalls by applying more checks on the passing traffic and analyzing the whole packets, both header and payload. Hence, it is recommended to place the IPS behind the firewall to ensure that only legitimate traffic passes to the IPS for further investigation. In this context, several intrusion detection techniques have been proposed by the research community, however signature-based detection and anomaly-based detection are the most used by the majority of IPSs.

7.4.1.1 Signature-Based Detection Method

Signature-based detection technique refers to a database of uniquely identifiable patterns (or signatures) in the code of each threat. When a new threat is identified, its corresponding signature is created and stored in a database of signatures [46]. Then, these signatures are compared against the network traffic passed through the IPS to identify possible attacks. This type of detection techniques is very effective at detecting known threats but largely ineffective at detecting previously unknown threats, obfuscated threats, and many variants of known threats [45]. Which leaves a security hole for all those kinds of threats. Signature-based detection technique is mainly adopted by the well-known open-source IPSs Snort (www.snort.org/) and Suricata (https://suricata-ids.org/). Cisco has also several models of intrusion-detection, like Cisco IDS 4200 Series, which are designed to be sensors throughout the network.

7.4.1.2 Anomaly-Based Detection Method

Anomaly-based detection, as its name indicates, focuses on identifying unexpected or unusual patterns of activities [47]. An IPS using this detection method builds a pattern (or a profile) of normal activity by monitoring the characteristics of a typical activity over a period of time. Then, current network activity is compared against the predefined profile to detect deviations. Every significant deviation of the current activity from the pre-defined profile is qualified as malicious [48]. This technique is very useful for detecting unknown and obfuscated threats; however, it generates the highest number of false alarms, known as false positives, due to the inability to capture the behavior drifts with time. This means that a large number of normal activities are likely to be considered as malicious [48]. Bro-IDS (https://zeek.org/) is an example of NIPSs that use this detection method.

7.4.1.3 Passive Network Monitoring

Most NIDSs, like Bro IDS and Suricata, implement passive network monitoring techniques to identify malicious behavior. Passive network monitoring captures network traffic that circulates through a network and analyzing it, without altering the traffic [19]. It uses security threshold parameters to identify if the activity is acceptable or may be malicious and generate alarms for the security administrators [19].

Unlike active monitoring, passive monitoring generates less strain on network resources because the interval between testing is greater.

7.4.2 HONEY POT FOR INTRUSION DETECTION

The most complete IDS/IPS solution should have a perimeter IDS/IPS working in conjunction with a perimeter firewall. Ideally, an IDS/IPS should be placed on each major server with a honeypot solution [19]. Honeypot is a variant of standard IDSs but with more focus on tracking intruders and collect information about their activities [49]. The main idea is to setup a honeypot system that simulates a real valuable server or even an entire subnetwork, fooling intruders into thinking it is a legitimate target and therefore, they will be attracted to the honeypot rather than to the real system [19]. Once the attackers compromise the honeypot system, the software can closely track their activities and assess their behavior [49]. Figure 7.1. illustrates the honeypot concept.

Honeypot systems often use hardened operating systems and are usually configured so that they appear to offer attackers exploitable vulnerabilities [49]. For instance, it may represent itself as an enterprise database server that storing consumers' information; however, it is actually separated and closely monitored [19]. Because legitimate users do not have access to the honey pot system, any attempts to communicate with it will be viewed as a possible attack. Inspecting and logging this activity can greatly help enhance security by providing insight into the level and types of threat a real network infrastructure may encounter [49].

7.4.3 EFFECTIVENESS OF IDS OPERATION IN LIMITING THREAT IMPACT

IDSs/IPSs are considered as one of the fundamental components of a typical security architecture, which gives visibility into the activities of a network, enabling timely

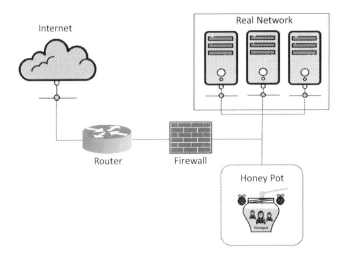

FIGURE 7.1 Honeypot implementation

detection and response to suspicious activities [45]. These security tools complement a firewall by providing a thorough inspection of both the header and content of all packets traversing the network [42]. From the point of view of a security policy, IDSs/IPSs are the second line of defense against cyber-attacks after the firewalls. In fact, these security tools are highly effective in protecting the entire network (NIPS) and machines (HIPS) against a wide variety of cyber-attacks. Using signature or anomaly-based detection techniques, a properly configured IPS can monitor the working of routers, firewall, key servers, hosts, and files in your network, catch intruders, and take actions in real time to thwart anomalies that firewall, or antivirus software may miss. In addition, by using the extensive attack signature database, IPSs ensures quick and effective detection and prevention of any attack based on malicious traffic, with a low risk of raising false alarms [50]. Specifically, IPS can effectively detect and stop:

- Many types of DoS/DDoS attacks that attempt to make a server, service, or network unavailable such as "**Smurf Attacks**," which perform DDoS attacks by flooding a system with a large number of Internet Control Message Protocol (ICMP) packets using IP broadcast address [51]. IPS can also prevent Ping of Death DoS attack, SYN Flood Attacks, and IP Fragmentation Attacks. In an IP Fragmentation Attack, malicious actors exploit the datagram fragmentation mechanism to overwhelm the network, by confusing the target system as to how TCP/UDP datagrams should be reassembled. For instance, attacks may send overlapping fragments with nonaligned offsets [19].
- SSL Evasion attacks that exploit Secure Sockets Layer (SSL) and its successor, Transport Layer Security (TLS) to hide malicious content, evade detection, and bypass critical security controls like firewalls, secure web gateways, Unified Threat Management (UTM), and anti-malware solutions [51]. Especially, Android malware is now successfully using SSL to hide their routines and to evade detection. In this context, a recent study by Gartner[10] found that more than 50% of the network attacks targeting enterprises are using SSL encryption.
- Port Scanning Attacks that attempt to find open ports on specific hosts, by sending client requests to a range of server ports on a host [51]. Attackers take advantage of the available tools such a Nmap[11], NetCop,[12] and Netcat[13], to find open ports, which present potential vulnerabilities that attackers can exploit to gain illegitimate access to a network or a system.
- OS Fingerprinting Attacks that attempt to identify the operating system of a specific target and learn which exact vulnerabilities to exploit [51]. These attacks could be passive or active, where active fingerprinting is easier and return accurate information. However, attackers prefer passive fingerprinting because they are more difficult to detect. Most of these attacks are performed with the Nmap tool.
- Buffer Overflow Attacks that attempt to exploit buffer overflow vulnerabilities, by putting extra data into a fixed-length buffer than the buffer can handle. The inserted extra information can overflow into adjacent memory space, and therefore, corrupting the execution path of an application by overwriting parts

of its memory. Successful buffer overflow attacks usually result in a system crash and give attackers possibility to run malicious actions.
- SMB (Server Message Block) relay attacks, which is a kind of MitM attacks that exploits Windows vulnerabilities to capture SMB authentication requests, and relay them to another host [51]. SMB is a transport protocol used in Windows Active directory domains for sharing files and printers and accessing remote services. This kind of attack is very dangerous because any party that has access to the network can capture traffic, relay it, and get illegitimate access to critical servers.
- Sinkhole attack, which is a kind of sensor network routing attack [52]. This attack occurs when a malicious node in a Wireless Sensor Network (WSNs) attracts network traffic of its neighbor's by advertising fake routing updates and performing data forging or selective forwarding of data passing through it [53]. It could be carried out by either a compromised node in the network or by introducing a fake one. It can be also accomplished using wormhole attacks. It can be exploited by cybercriminals for launching further attacks like acknowledge spoofing attack and drops or altered routing information [52]. Sensor nodes usually cannot deal with this attack because of their resource constraints. Therefore, using an IPS can help identifying patterns in network packets that may indicate a Sinkhole attack.

IPSs can also effectively prevent ARP Spoofing attacks [51], gateway interface (CGI) attacks and preserve the privacy of users as IPS records the network activity only when it finds a malicious activity, and also protects the user sensitive information in the computer network against attacks from intruders [42]. Further, IPSs are able to detect certain types of malicious requests to lower-levels of Software Defined Networking (SDN) components, for which the intention is to be passed to higher levels and modify characteristics of the SDN. This is accomplished by applying protocol and/or identity rules; the reaction component of the IPS may assist in mitigating the attack.

7.4.4 IPS Limitations

IDSs/IPSs are essential components of any organization security, but they are just one part of a security plan that must include other security measures. Like most technologies, these security tools have numerous limitations and lack certain capabilities that should be covered by other security tools [54, 55]. Most important limitations of IDS/IPSs are:

- **Encrypted network traffic**: NIPS cannot analyze encrypted network traffic as the payload information is not available; thus, intruders can use encrypted data to bypass the IDS and slip into the network, which leaves the network vulnerable until the intrusion is discovered [54]. This is a very challenging problem, especially when more than 940% of modern malware are encrypted, according to a recent study by WebRoot.[14] In this context, attackers are using a number of techniques to evade NIPSs such as slow scans, fragmentation techniques,

address spoofing/proxying, pattern changes, and TCP/IP attacks that aim at confusing the NIDS [56].
- **Frequent false alarm**: While IPSs are very effective in detecting abnormal activities, especially with the signature-based detection technique, they have a significant issue with the number of false alarms, which could be important, especially in large networks and with the anomaly-based detection technique [47, 56]. In many cases, false alarms are more frequent than actual threats, and consume important time and resources in responding to them [56].
- **Limited visibility**: Like firewalls, most IPSs focus on the perimeter attack surface, starting with the firewall, which only provides protection of "North to South" traffic that either enters/leaves the local network from/to the outside world [19]. However, it does not consider the east-west traffic, which refers to the internal traffic to the local network, knowing that only 20% of discovered threats are from the north-south monitoring.
- **Resource limitations**: NIDSs can suffer from issues with resource limitations, especially in analyzing large volumes of data, in near real-time [54]. For instance, in order to effectively detect a wide range of attacks, the NIDS must maintain state for a large number of TCP connections, which may require a significant amount of memory on the NIDS host. Also, many research studies found that NIDSs, especially Snort, may fail in high-speeds; losing more packets as the speed and volume of traffic increase, which have a significant impact on detection accuracy of the NIDS [47, 55].

In addition, IPSs cannot mitigate issues related to traffic sniffing, spoofed IP addresses, MAC congestion, MAC Jamming, physical layer DoS attacks. Moreover, low-level host-based attacks like firmware manipulation, memory scrapping false sensor data injection, information leakage via sensors, malware transfer between sensors and remote firmware attacks are all outside the scope of IPSs. IPSs cannot assist in reducing the risk related to these attacks or even the consequences of successful attacks. Actually, since IPSs are critical components of a network, they themselves are potential attack target [56]. For instance, attacks may exploit the fact that NIDSs will drop packets on heavily loaded links to flood a link whilst trying, to compromise a particular asset in the network [47, 56]. Improperly configured NIPSs are vulnerable to many types of DDoS attacks such as invalid data and TCP/IP stack attacks, which may cause an NIDS to crash [56].

7.5 OPERATION OF ANTIVIRUS/MALWARE DETECTION SOFTWARE

7.5.1 Antivirus/Antimalware Software Overview

The most obvious defense against viruses and malware on the device and Critical Information Infrastructure (CII) is the antivirus program (also called anti-malware Software). This software is capable of tracking and preventing all kinds of known malware including viruses, worms, trojans, spyware, phishing attacks, rootkits, spam, and other malicious programs circulated by cybercriminals for their malicious purposes [19]. Antivirus usually scans computing/portable media devices, networks, and incoming e-mails for known malware files. It usually works in a very traditional way, which is based on signatures, also called signature-based detection technique [57]. This technique uses a database called "virus dictionary," which store a list of known malware files. The virus dictionary is usually stored in a small file, often called ".dat" file (short for data file) [19]. This file is periodically updated by the Antivirus vendor by including signatures (patterns) of newly discovered malware. When the antivirus scans a file, it takes a snippet of that code and compares it to the files in their virus dictionary [57]. If the files match, the antivirus considers that file as malware, stops it from running, and moves it into the quarantine folder. Then, the user can choose to delete it automatically or keep it if he is confident that it is a false detection. This technique works only if the ".dat" file is updated, and only for known malware [19].

Another technique used by antivirus engines for malware detection is based on the system monitoring for certain types of behaviors that may indicate malicious activities such as attempts to writing to a hard drive's boot sector, alter the system registry, etc. [19]. This approach is also called behavior-based detection approach. This approach uses a dynamic analysis of malware's behavior to create patterns (profiles or baselines). Then, the current activity of the system is analyzed for suspicious activities, where any attempts to perform actions that are abnormal like attempts to disable security controls, or installing rootkits will be treated as malicious, or at least suspicious [12]. The main advantage of this approach is that it is not affected by evasion techniques like obfuscation, and it can capture the polymorphic and metamorphic strains of malware [58]. Moreover, this technique is entirely independent of the source code of the malware [12].

Antivirus scanners also search for Terminate and Stay Resident (TSR) program files that remain in memory until it is needed and then performs some function. Some legitimate programs do this such as some special virus scanner programs, however, it could be also a sign of a malware [12, 19]. Antivirus programs often use additional methods to enhance the malware detection such as identifying critical files to the system (e.g., registry, the boot.ini, etc.) and monitor any attempt to modify these files [19]. In this case, the antivirus sends a notification to the user who should permit or deny the modification. Recently, most antivirus providers are incorporating predictive analysis and artificial intelligence into their antivirus software to be able to detect new malicious programs (i.e.,, zero-day malware).

There are a large number of antivirus packages available for individual computers/devices and networks, however, McAfee, Norton, AVG, Norton, and Kaspersky are the most widely known and used. While Spyshelter, SpyHunter, and TotalAV, are the best free malware detection tools that provide protection against spyware.

7.5.2 Importance of Using Antivirus/Antimalware Software

Antivirus software protection is an essential part of any security strategy to detect, prevent, and take actions against malicious software on the device and the CII. With nearly 30% of cyber detected security breaches involving successful malware infections, it is evident that employing updated virus-scanning software will greatly help

IoT Security Frameworks and Countermeasures 259

to avoid a large number of malware attacks, or at least minimize their effects [59]. In fact, security experts compare a system without antivirus software to a house with an open door or unprotected door that attracts intruders and burglars into your home. Similarly, Unprotected computers or networks will attract all kinds of malware attacks to the system. Thus, an antivirus will act as a protected or closed-door against all the malicious intruding malware. More specifically, antivirus is highly effective against:

- **Malware and their transmission**: antivirus or antimalware can detect any potential malware and remove it before it can harm the system. This includes virus, spyware, rootkits, trojans, worms, adware programs, mobile devices malware, and other online cyber-threats [59, 60].
- **Malicious spam sites and ads**: antivirus can block malware spam attacks that can infect a system through pop-up ads and other spam sites [59]. These spam attacks are capable of installing spyware to hijack the users' browser and steal sensitive information (e.g.,, financial credentials, social security numbers, passwords, credit card numbers, and other important data), compromising their privacy, or even causing substantial financial losses.
- **Browser extension attacks**: browser extensions are small pieces of programs executed in the browser context in order to provide additional capabilities and enrich the browser functionalities [60]. Malicious extensions are generally used by attackers to get unauthorized access to the user private information. Antivirus program can detect these malicious extensions in your browser and block them [60].
- **Phishing attacks**: Good antivirus software can stop phishing emails reaching your inbox, by scanning e-mails and attachments. In this context, some antivirus scanners examine e-mails on the e-mail server before downloading it to the destination system. Others can scan received e-mails and attachments on the destination computer system before passing them to the e-mail program [19]. Further, most commercial network antivirus scan e-mails on the server before sending it on to the target system.
  Moreover, more sophisticated anti-phishing protection software can detect and block, in real-time, malicious URLs, files, and spoofing attacks.
- **Injection attacks:** robust antivirus programs can prevent some kinds of injection attacks such as malicious code injection, which exploit vulnerabilities in a program to inject malicious code into that program [21].

Furthermore, antivirus can ensure protection from removable devices, by automatically scanning any connected device to your computer. It can automatically put suspicious programs on these devices in virtual containers in order to prevent them from accessing the computer's data and other resources [19]. Further, modern antimalware programs can ensure the protection of critical system objects and resources against OS-level malware attacks (e.g., BIOS malware) that attempt to compromise OS and obtain root privilege. They can also prevent some application layer DoS/DDoS attacks that involve malware infection.

7.5.3 LIMITATIONS OF ANTIVIRUS/ANTIMALWARE SOFTWARE

Although the effectiveness of antivirus software in mitigating many potential cyberthreats, users should not consider it to be universal protection against all cyberthreats. Instead, users should consider an antivirus program to be one part of a comprehensive security plan (i.e., endpoint protection). This security tool has also some limitations and lacks certain security protection capabilities.

The main limitation of the existing antivirus programs is that they can only protect against malicious programs that are known, and only if kept updated. Thus, cybercriminals are using zero-day exploits to avoid detection and bypass antivirus [19, 57]. An example of zero-day exploits is "**EternalBlue**",[15] known as the most enduring and damaging exploit of all time. This zero-day vulnerability was realized in 2017 by a group of hackers called Shadow Brokers. Whilst a patch was quickly developed by Microsoft to fix this vulnerability, cybercriminals have been successively exploiting it to launch a worldwide cyber-attack by the dangerous "**WannaCry ransomware**" crypto worm. Investigations estimated that the attack affected more than 200,000 computers running the Microsoft Windows operating system, across 150 countries, with overall damage costs reached billions of dollars. With over 350,000 new malicious programs released every day, zero-day exploits become a challenging problem to the Cybersecurity community.

In addition, antimalware programs are not able to mitigate most network-level threats and those related to hardware/sensor-level, including flooding attacks, MAC spoofing, identity spoofing, network, transport and physical layers DDoS attacks, bypassing network visualization, powerline jamming, traffic diversion, and traffic sniffing, SSL evasion attacks, attacks related to hardware modification and manipulation, side-channel attacks, Sybil attacks, remote file inclusion, and many others.

Overall whilst anti-malware programs have some limitations, they still a vitally important part of both home and business networks security. Constant developments are being made to further secure systems, with the introduction of next-gen solutions utilizing artificial intelligence to detect malware [12, 58], and application awareness features to reduce the number of false positives reported. This ongoing development and the introduction of anti-malware coming pre-packaged with operating systems shows that these solutions will continue to be utilized to counter the relentless stream of malicious software developed daily.

7.6 SECURITY AWARENESS OF INDIVIDUALS

Individuals are the weakest link in information security, a fact frequently exploited by cybercriminals and fraudsters to cause operational disruption, loss of critical data, and brand damage. According to a security survey, more than 78% of security professional believe that employee negligence for security practices is the biggest threat to endpoint security, where more than 90% of organizations involved in the survey have experienced at least one insider threat per month [61]. One of the very famous examples of this particular type of cybercrime is the massive data breach of the Anthem health insurance company, in 2015. In this incident, hackers got administrative access to the company's database for over a month through a phishing email that was sent to their employees. The attackers exploited compromised credentials of a database administrator in the company to steal 78.8 million customers records that contain medical IDs, social security numbers, income data, employment history, etc., and sold them on the dark web [62]. Thus, security awareness of individuals has become one of the most important investments a company can make in order to reduce Cybersecurity risk related to user behavior.

7.6.1 Security Awareness Training and Its Importance

Security awareness is a formal process for training and educating employees and other individual users on how to keep safe against potential threats and avoid situations that might put their sensitive data at risk. As with monitoring, security awareness training requires to be a continuous process [63]. Security reports warn that human error is the cause of up to 95% of Cybersecurity breaches, and with simple awareness training courses, this number can be dramatically reduced.

In this section, we only consider the awareness of individuals and employees of a generic profile, because they are the weakest link in information security. A generic profile is relevant to the use of information systems (i.e., applications, production, and consumption of information). In pertinent cases, we consider the situation where individuals with a generic profile are also device owners, hence they assume responsibility regarding the management of their own devices. Users with this profile not only should learn the latest threats, but they should also understand the impact their actions can have on the overall security of the organization. Key training topics of such a process typically include:

7.6.1.1 Password Management and Authentication

Passwords management is one of the most common in any information awareness program. Commonly used and simple passwords can be easily guessed by cybercriminals and give them access to a large range of accounts. A study[16] found that 59% of end-users use the same password for every account. According to security reports, stolen or weak passwords are still the most common reason for data breaches [63]; thus, end-users should be aware of the security risks related to password theft. With security awareness training, end-users will learn how to create strong, secure passwords, and then store and utilize them properly. Also, using two-factor authentication will provide extra layers of security to protect the integrity of their accounts.

7.6.1.2 Email/Phishing and Social Engineering

Users are constantly exposed to sophisticated phishing and other social engineering attacks. Most of these attacks, including phishing, are using techniques for gaining trust and therefore, persuade users to provide their sensitive data or perform an action to benefit the attacker [63]. These types of attacks are effective because many people are unaware of what social engineering attacks are and the threat they pose. While there is no definite way to prevent these attacks, security awareness training that spread phishing awareness can provide an effective strategy to limit the chances of successful attacks [63]. Current figures clearly reflect the need for awareness of phishing attacks, Proofpoint[17] data demonstrations that, 99% of phishing attacks required human interaction to succeed. The security company suggests that training end-users to recognize potentially harmful emails and reporting suspicious ones, phishing attacks can be dramatically reduced.

7.6.1.3 Removable Media

Removable media presents a useful, cost-effective storage solution that is available in various size abilities and form factors, with different transfer speed capabilities (e.g., USB drivers, SD cards, optical media, smartphones, legacy media, music players, etc.) [64]. These removable media consider another important security awareness topic because they are used daily by individuals and employees to copy/past sensitive data to/from them [63]. In fact, if not managed properly, they can expose your system and the whole company network to many risks related to data security, malware infection, and copyright infringement [64]. These removable media can carry malware that may cause many security problems for a user or an organization. In this context, TechAdvisory.org[18] reports that more than 25% of malware is spreading today through USB devices. Therefore, end-users and companies need to learn the best practices that can reduce the risks associated with using portable devices.

Users should also understand the main risks of using removable media as a primary storage solution, as they are susceptible to media failure, loss, theft, or damage. It is therefore really important for users to learn the importance of storing sensitive, important, and confidential information safely and securely on the organization's file servers [64].

7.6.1.4 Social Media Usage

Despite the benefits of social media to organizations, there were a lot of cases of information leakage, malware, identity theft, financial loss, espionage, and sabotage through such use. Deploying social media by employees arises many security risks related to the inappropriate release, leakage or theft of information strategic to the organization, and exposure of organization networks and systems to viruses and malware due to human error, phishing scams, sophisticated attackers (i.e., APT). For instance, cybercriminals can use information shared on social media to create targeted email attacks (i.e., spear phishing) and cause reputational harm for organizations. With the proliferation of social engineering attacks and scams on social media, training end-users and employees the basis of safe social media usage is becoming crucial for reducing risk of the potential leverage that hackers can gain from this access to their personal network.

7.6.1.5 Physical Security

Physical security describes security measures that can be used to control access to a building, facility, and resources from unauthorized people, including attackers and accidental intruders, such as employees who may not be aware of the restricted area [63]. Raising awareness about physical security among end-users and encouraging them to take an active stance in defending their assets and workplace is the most effective way to combat the whole spectrum of physical security threats. For instance, simple awareness of the risks of leaving documents, unattended computers, and passwords around the office space or home can reduce security risks.

In addition to the security awareness training, a method for users to report suspicious activity should also be created, which could be a web form or other tools that can be integrated into the established intelligence cycle [63]. Also, understanding the threats that users may face online, the threat actors, and the motivation of online criminals can effectively help users to better understand the threats they pose and report them.

7.6.2 Limitations

Security awareness and training of individuals/employees can significantly help in reducing the impact of potential security threats; however [63], it could not handle the impact of most network-level threats, such as botnets, flooding attacks, traffic sniffing, DDoS attacks, and network intrusions. Individuals could only report service outages or performance degradations, which usually stem from this kind of attacks, as security-related infrastructure problems. In addition, trained personnel cannot identify most low-level attacks such as MAC Spoofing and MAC layer jamming. However, they can detect and report powerline jamming attacks, which try to impact the performance of power line communication (PLC) network.

Furthermore, training and raising security awareness cannot prevent zero-day attacks that exploit unknown/undisclosed vulnerabilities. With a large number of zero-day identified every day, it is a clear and ongoing risk that must be tackled by other means, as whilst employees can be trained to the point of knowing what a zero-day attack is, they cannot realistically be trained to ensure that they can notice/detect brand new exploits [19]. Similarly, raised awareness of Man in the Middle Attack can have limited impact as it can be very hard for nontechnical users to detect the threat of this kind of attacks before and after compromise [65]. Trained users can also understand Formjacking cyber-attacks [66], their forms and impacts, however, it can be near impossible for them to detect a malicious web form before entering their data, and once data is entered the attackers can utilize this immediately with no obvious notification to the victim user.

7.7 EXISTENCE OF UPDATE/PATCHING POLICY

Update[19]/Patching[20] is one of the crucial activities involved in keeping devices and companies' networks secure. Leaving machines and devices unpatched makes them vulnerable to various cyber-attacks. According to the Ponemon Institute[21], most than 57% of data breaches are directly attributed to attackers exploiting a known vulnerability that has not been patched. Thus, it is important, especially for organizations, to have a policy for update/patch management to ensure keeping components (i.e., hardware, software, and services) in their IT infrastructure up to date with the latest updates and patches.

Before, describing update/patching poly and its effectiveness in reducing the impact of potential threats, first, we need to explain security vulnerabilities and exploits.

7.7.1 SECURITY VULNERABILITIES AND EXPLOITS

Security vulnerabilities are weak spots that are usually present in a network, an individual device, or software systems [67]. They are commonly found in more complex software systems (i.e., operating systems), firmware, protocols (e.g., TCP/IP protocol), and ciphers. Unlike "cyber-threats," which may involve an outside element, vulnerabilities are open doors in a system or network asset (e.g., computer/device, database, application, service, etc.) that cybercriminals may exploit to access a target system [19]. An example of the most common security vulnerabilities is SQL injection [21], which allow attackers to get illegal access to SQL database, by tricking an application into sending unexpected SQL commands. Successful SQL injections allow attackers to steal sensitive data, spoof identities, and perform many other harmful activities [68]. Based on where the vulnerability exists, security vulnerabilities can be divided into different broad categories including:

- **Network Vulnerabilities:** these are weaknesses or flaws in a network's software, or hardware in the network, which expose it to possible intrusion by outside parties [67]. Examples include outdated or unpatched software, insecure Wi-Fi access points, etc.
- **Operating System and Software Vulnerabilities:** these are errors in the code or the logic of operation within a particular operating system or application software, which can be exploited by hackers to gain access to an asset (e.g., server) the OS is installed on [67]. CodeRed[22] and Slammer are examples of worms that exploited vulnerabilities in the Windows OS.
- **Human Vulnerabilities:** vulnerabilities could also be created by individuals, especially when configuring software, hardware, social media, privacy, and security settings, etc. [19]. For instance, poorly selected and guessed passwords can be easily hijacked by malicious actors. Also, misconfigured Internet services, such as turning on JavaScript in web browsers, enable JavaScript-based attacks when accessing untrusted and malicious web sites.

Attackers and malicious actors continuously search for vulnerabilities by using automated scanning tools that examine the web for weak spots they could exploit. After identifying a vulnerability, attackers use an exploit, which could be a piece of software, or a sequence of commands, specifically designed to leverage the detected vulnerability for malicious activities [67]. Exploits depend on failures and mistakes, such as unpatched systems and out-of-date software, to achieve their goals [19, 67]. Exploit kits are one of the most popular methods, to automatically launch exploits against installed vulnerable applications such as Adobe Flash Player and Java Runtime environment.

7.7.2 Update/Patching Policy

As mentioned before, unpatched software/firmware can make devices/networks a target of exploits and malware access. Update/patching is the process responsible on identifying, acquiring, testing, and installing missing patches (code changes) on running operating systems, applications, software, and firmware tools, on a computer/device or a network of computers/devices. This helps to keep systems updated on the latest patches and updates, repair security holes (i.e., software vulnerabilities) that have been discovered and fixed, or even remove bugs (i.e., functionality problems) [67, 68]. Software updates can also add new features to the system and remove outdated ones. With automated updates/patching management solutions, companies can keep their endpoints updated with the latest patches, fix bugs and possible security vulnerabilities, and enhance them with new features. In general, patch management life cycle consists of the following five stages:

- **Updating vulnerability details from software vendors**: software vendors release patches to fix vulnerabilities identified after the release of software. Thus, a patch management solution should keep up to date with the latest released patches/updates [19].
- **Scanning the network for vulnerabilities**: companies can use automated tools to automatically scan their systems/networks and identify which endpoints contain vulnerabilities and need to be patched. Examples of these tools are Vuls,[23] Archery,[24] Microsoft attack surface analyzer,[25] and Nessus.[26]
- **Identifying patches for vulnerabilities**: after identification of vulnerabilities, they should be analyzed in order to find missing and appropriate patches.
- **Deploying patches and validation**: after identification of the missing patches, they should be downloaded from the vendor site and deployed. In this context, automated patch download ensures to automatically deploy and validate patches based on the deployment policies, without any manual interference [68]. This will help to deploy new patches immediately and therefore, minimize the time that systems are vulnerable to the associated software flaws.
- **Generating Status Report**: once the patches are correctly deployed, reports on the status of the automated patch management tasks are updated. Reports should expose situations that require an immediate return to the analysis phase, such as a failure in deployment [67]. This will greatly help in monitoring the patching progress in the company.

Currently, it becomes imperative that security plan for each company involves a proper update/patching policy to automatically apply updates/patches to groups of tagged endpoints at planned times. The policy should define how and when each patch is to be undertaken, who is responsible for executing it and how patch verification will be performed. Also, it should effectively handle regular patch notifications that come from multiple sources, and remotely deploy updates for running operating systems. With a clear and effective update/patching policy, a great surface of cyber-attacks, especially malware would be avoided, or their effects would at least be minimized.

7.7.3 EFFECTIVENESS OF UPDATE/PATCHING POLICY IN LIMITING THREAT IMPACT

Clear and effective update/patching policy is the best defense against malicious cyber-attacks. It ensures that all pieces of software ranging from security solutions to everyday tools are patched and up to date, and therefore, they do not introduce major security holes within a system or a network [67]. In this context, Edgescan report[27] confirmed that an average of 50 new critical vulnerabilities are discovered every day, where the average window of exposure is 69 days. The report highlighted that biggest cases of security incidents are results of unpatched systems such as the famous WannaCry ransomware attack, in May 2017 [69].

This ransomware attack had been successfully devastated systems of the UK's National Health Service (NHS) in May 2017, along with hundreds of other businesses worldwide. This was achieved by successfully exploiting a vulnerability in a range of Microsoft operating systems, using the famous EternalBlue exploit (CVE-2017-0143)[28]. The ransomware affected over 200,000 systems across the NHS, costing approximately £92 million in initial customer damages and remediation. Investigations found that victim systems were outdated and running operating systems that were no longer supported by security patches.

Routinely updated and patched operating systems and software is the only real safeguard against various critical cyber-attacks such as Ping of Death (PoD) DDoS attack, buffer overflow attacks, some data forging attacks, exploit kits, identity spoofing, remote firmware attacks, and firmware manipulation. They can mainly assist in:

- Reducing the exploitability of application layer DoS/DDoS attacks, IoT-based DDoS attacks, and botnets, since measures such as blacklisting attacker sites, applying rules to filter malicious requests, or applying rate/size/time limits could alleviate the problem.
- Disabling and limiting access to insecure network services and upgrading existing services with better access control mechanisms. The impact of access to these services may also be reduced [67].
- In some cases, they can assist in replacing old versions of software lacking capabilities for transport-level encryption with new ones that include this capability. They can also add new security mechanisms or replacing weak implementations with stronger ones [67].
- Removing multiple vulnerabilities that pave the way to network intrusion, including insecure services, inappropriate configurations, lack of access control mechanisms, SSL weak cipher suites, etc [68].
- Updating malware detection programs (e.g., Antivirus, IDSs/IPSs) with new signature database that will detect new realized malware (Zero-day malware). New versions of software or firmware with enhanced self-defense could be also installed and malware that installs modified firmware could be blocked [68].

IoT Security Frameworks and Countermeasures

7.7.4 LIMITATIONS

Despite the effectiveness of the update/patching policy in limiting the exploitability and technical impact of a wide range of threats, it cannot effectively contribute to the mitigation of some network and low-level attacks such as flooding attack, MAC Congestion Attack, MAC Spoofing, physical layer DoS attacks, powerline jamming, traffic sniffing, selective forward attacks, side-channel attacks, and physical damage, theft, or loss. Also, such a policy cannot protect against some insider threat categories including careless users, disgruntled employees, and malicious insiders. Clear update and patching policies can help to mitigate these threats to an extent by limiting the technical vulnerabilities and exploits that can be utilized by malicious insiders. However, there are limitations to this approach, as determined insiders could still misuse their system privileges to circumvent defenses and achieve their goals. This can only be limited by increased logging and alerting to deter the insider threat.

Another limitation of updating/patching policy is the use of nonapproved software by careless employees, which may prefer to make their own choice of what software to utilize at work. This kind of software will generally not be on the security team's radar, meaning they will not be checking for vulnerabilities, exploits, or patches for this software.

7.8 EXISTENCE OF LOGGING/ALERTING POLICY

7.8.1 LOGGING/ALERTING POLICY OVERVIEW

Monitoring and logging Cybersecurity events provide potential Indicators of Compromise (IOC) that can lead to Cybersecurity threats if not addressed quickly and effectively. Thus, it is important for companies to have a clear and effective logging/alerting policy that can identify potential IOC at an early stage, investigate them electively and take appropriate actions to mitigate, or at least minimize the technical impact of Cybersecurity threats [70]. It generally refers to the process of monitoring, recording, and storing details about the execution of a program including operating systems, custom software, applications, etc. The main goal is to track and reporting (i.e., alerting) errors along with the related data in a centralized way [71], by grouping all aspects of Cybersecurity monitoring, logging and alerting together in one framework. Figure 7.2. shows the main steps of a logging/alerting process.

7.8.1.1 Logging Management

The term logging can refer both to the process of event logging or to the log files that record the logged events. For Cybersecurity monitoring, events logs can be used to help with detection of Cybersecurity threats and/or their investigation [70]. They

FIGURE 7.2 Main steps of logging/alerting process

keeps track of security-related activities that have been attempted or performed on a system, a network, or application that processes, transmits, or stores confidential information [71]. They are composed of log entries, where each entry contains information related to a specific event that has occurred within a system or network [71]. In this context, four main types of event logs can be used for security monitoring and advanced detection of potential cyber-threats [70]:

1. **System logs (syslog):** These log files contain events that are logged by operating systems such as Windows, Linux, and Mac. It may record information about system errors, warnings, device drivers, start-up messages, abnormal shutdowns, system changes, etc.
2. **Network logs:** this type of event logs, such as emails, firewalls, VPN, and NetFlow logs, help to monitor the network activities and identify issues in the network before they become real threats.
3. **Technical logs (techlogs):** this kind of event log files may include HTTP proxy logs, DNS, DHCP, and FTP logs, web and SQL server logs.
4. **Logs from Cybersecurity and logging tools:** events logs can be generated by a variety of security software such as antimalware (i.e., antivirus), firewalls, intrusion detection, and prevention systems, from network devices - such as routers, switches, etc. [71], and other relevant security management tools such as sandboxing and virtual execution engines.

Most log files are saved in a plain text format, which minimizes their file size and allows them to be viewed in a basic text editor [71].

7.8.1.2 Alerts Management

Alerts give timely awareness to problems in running systems and applications, so security administrators can resolve the problems quickly. Thus, it is important to identify a list of prioritized security alerts to quickly investigate possible problems and recommendations for how to remediate attacks. Security alerts usually provides a brief summary of an incident, when and where it occurred, and a description of the offender or suspect, if applicable. Alerts management policy should specify what should be monitored, when to trigger an alert and defined the order of priority in which alerts will be treated. This is done by assigning a severity to each alert, which could be high, medium, and low. Each alerts management policy can define its proper levels of severity based on the local requirements. A well-tailored alerting policy can lead to faster responses by security teams allowing for any potential damages and network propagation to be prevented or limited.

Effective logging/alerting policy help to protect confidential information and through careful trend analysis, identify security incidents and policy violation (e.g., invalid logical access attempts), and therefore establish meaningful improvements to security management programs [71]. However, systems that process large volumes of log sources and related data need an appropriate logging management policy to ensure efficient handling and response to all reported security incidents. The policy should define the necessary requirements and recommendations for log management activities, including log generation, storage, transmission, analysis, and disposal of all generated logs [71]. Such a policy should identify security breaches and alert the system or the network administrators to take the appropriate mitigation actions in real time.

7.8.2 Effectiveness of Alerting/Logging Policy in Limiting Threat Impact

The existence of a logging/alerting policy can identify security incidents, often in real time, and alert the system/network administrators to take appropriate action after either an attempted or a successful attack. However, the effectiveness of such a policy greatly depends on the response of the system/network administrators. The reaction of administrators should be in a reasonable time to mitigate the technical impact of the threat, and in some cases decrease the threat exploitability.

For instance, application layer DoS/DDoS attacks may be identified by a logging/alerting policy from event logs of the device itself, or if the DoS was successful. However, they cannot be mitigated by a logging/alerting policy alone. The actions of the network/system administrators may limit or mitigate the consequences of an application layer DoS. Similarly, exhaustion of resources due flooding attacks may be logged and identified by a logging/alerting policy but cannot be mitigated by this policy alone. The mitigation of these attacks highly depends on the actions of the network/system administrators. In addition, effective analyses of logs and alerts can help to identify insecure network services and disable them. Exploitation of such services could also be detected from the actions of the compromised system (if its behavior diverges from the baseline behavior).

Logging/alerting policy could identify insider violations (i.e., insider threats) because accidental or malicious actions are carried out by users on systems or the network. Whether the user is authorized or not, their actions could trigger alerts depending on what normal behavior threshold has been established. Whilst this security measure may be limited by system administrators and other privileged employees, the existence and awareness of the logging practice may be enough to discourage insiders from intentionally misusing systems and tools for personal gain. Well-established logging and alerting policy can alert nonapproved hardware and software (e.g., USB devices, that have not been issued by the business), which present a real risk, especially to companies as it opens a potential access route for both malware and remote access tools into the company network. Physical damage, theft or loss of a device/asset may also be identified by the lack of generated logs. However, mitigation and technical impact of these incidents depends on the actions of the system/network administrators and the nature of data stored on the device.

Logging mechanisms can report to security administrators when untrusted devices have been connected to the network. They can also report actions and effects of malicious software on the infected devices (e.g.,, logs from IPSs).

7.8.3 Limitations

A logging/alerting policy cannot deal with threats related to the lower communication layers (i.e., physical and MAC) like MAC Congestion, MAC layer jamming, MAC spoofing, selective forward attack, sinkhole attack, spectrum sensing, hardware Trojan, side-channel attack and malicious mobile nodes, which operate at the physical or MAC layer and may not be identified by a logging mechanism on the device itself. In addition, a logging/alerting policy cannot detect malicious modifications of the hardware, unless they happen as the device is active, then any changes can be logged and reported to the system/network administrators. Regarding botnets, a logging/alerting policy may identify systems that are part of a botnet only if their behavior differs significantly from the baseline behavior. Systems that are a victim to botnet DDoS attacks could be identified from the logs of network devices and the actions of the network/system administrators may limit or mitigate the consequences of a potential DDoS attack.

Moreover, poorly configured logging/alerting policy can raise many privacy issues and leakage of personal information through log files. Other than that, some attacks targeting private data breaches could be impeded through fear of agent identification and prosecution. Thus, log files should be appropriately configured and secured to avoid leakage of sensitive data and trigger appropriate incident response actions in such a case.

7.9 CONDUCTING PERIODIC SECURITY CHECKS OF SECURITY MECHANISMS

7.9.1 Overview

Conducting periodic security checks and reviews of security mechanisms can be defined as the systematic checking, evaluation and update of data, devices, networks, and their components [72]. Security checks involve activities that measure, assess, and test procedures on a regular basis that could potentially present security gaps within an organization. These measures are usually adopted as a part of a company's general multi-stage security management process, which combines risk assessment, testing and reviewing, risk mitigation and operational security [72]. This continuous process requires careful planning, calls for many actors to collaborate and whose effort will prove its worth in the long-term. In fact, the existing complexity of corporate information systems can impede companies from seeking security audit services at first [73]. Nevertheless, the rise of security breaches and their disastrous economic consequences led to the establishment of several acts, requiring organizations to undertake security operations for their systems [74].

For instance, the Federal Information Security Management Act (FISMA), which is a US federal law enacted in 2002, expects US state agencies to undertake periodic risk assessment for identifying potential vulnerabilities and perform continuous monitoring of security controls [75]. According to this law, Information Security Frameworks (ISFs) in organizations require periodic assessments and security checks that act as preventive measures, rather than a post-incident investigation. This approach also reflects the constant change of the technology landscape as the European Union Agency for Cybersecurity (ENISA) states "a one-off certification is limited when considering online or cloud services" [76].

The deliverables of security checks and assessment include a detailed report with a prioritized list of vulnerabilities, vulnerability analysis, proactive ways to eliminate

IoT Security Frameworks and Countermeasures 271

these threats and recommendations for future tests of the systems. Security evaluators can also set up automated periodic tests (as opposed to manual ones) that comprises both scanning (e.g., through dedicated tools) and reporting processes (e.g., an email sent to an administrator) [77]. Because the slightest missed vulnerability can prove catastrophic, the level of detail in each security assessment is remarkably high and it is common for an assessment to produce a plethora of information, e.g., many false positives and negatives. In other words, security tests might report potential risks that represent no actual problems, and this is even more crucial when these tests are performed on a regular basis. Consequently, it is left to the administrator to decide how to deal with this information and carry on with a strategic plan [72, 77].

7.9.2 EFFECTIVENESS OF RUNNING PERIODIC SECURITY CHECKS IN LIMITING THREAT IMPACT

Continuous check and monitoring of security mechanisms and procedures represent a cyclical step-by-step process that facilitates actors to gain a deeper understanding of their systems. By regularly performing checks, examiners not only protect their infrastructure, but also gain insights when they compare their current strategies to the performance of past countermeasures [78]. This process is considered an essential measure in security mechanisms, as it has been proven to be effective against many types of threats. Running periodic security checks is largely effective against the following threats:

7.9.2.1 Insecure Software/Firmware and IoT

Fundamentally, the software/firmware of IoT devices must be protected in the same way that code must be protected in general-purpose computer systems. One of the most effective measures in keeping devices and their firmware secure is by conducting regular testing and patching processes. Some software or firmware are configured to automatically run checks for updates; in other cases, vendors will publish and push updates once they discover vulnerabilities or features that enhance performance [79]. A survey conducted by Symantec[29] showed that most compromised machines were exposed to malicious activities because they were left unupdated from already known security threats [80]. Another recent study regarding botnet armies underlined the relationship between infected machines and a lack of diligent update behavior from users [81].

Despite efforts to educate users and corporations, there are still major actions related to insecure software/firmware and malpractices. From the manufacturer's side, one example would be certain IoT devices that are not "updatable," meaning their OS and/or software cannot be updated by design. What is more, some of them will run under an outdated OS with vulnerabilities that are already known [82].

7.9.2.2 DDoS Attacks

Periodic checks may identify traces of exploitation in log files and/or prominent measures to apply to firewalls/antivirus-antimalware systems/applications to confront application layer DoS attacks (e.g., rate limiting, size limiting, etc.). Many defending mechanisms have also been developed, such as pushback, an added functionality in routers that drops suspicious packets and prevents their traffic in upstream routers [83]. Identification and labeling of packets as malicious are products of continuous traffic monitoring and pattern recognition of what constitutes "normal" network traffic behavior. IPSs that are based on Network Behavior Analysis (NBA) are heavily used in the battle against DDoS attacks by discarding their traffic.

Running regular security checks on IPSs helps general performance and health of network flow, as well as update pattern behavior on detection systems to eliminate false alarms [84]. This is particularly important since attackers constantly come up with new innovative ways to send DDoS attacks. One example is the **Slowloris attack**, in which the attacking machine causes the host to crash by sending requests at an extremely slow pace.

7.9.2.3 Botnets

Another threat in the mitigation process, where conducting periodic tests has proven crucial, is the presence of botnets. As botnet activity can go unnoticed, running regular checks on devices becomes more crucial than ever. Conducting security checks is considered a key measure to defend systems against bot armies. Unpatched antivirus systems, vulnerable firewalls, and unautomated checks make organizations extremely attractive to hackers who aim for low-hanging fruit [85]. European Union Agency for Cybersecurity (ENISA) also recognizes the effectiveness of conducting periodic software audits and network penetration tests in the fight against botnets. It also cautions against operating on unpatched software and recommends that companies improve their patch management practices [86].

In addition, network security checks can assist in identifying vulnerabilities and minimize the attack landscape by replacing insecure network services with secure ones, limiting access to insecure network services, or blocking malicious requests.

7.9.3 LIMITATIONS

Despite the many benefits of running regular tests, there are many areas where this tool is not sufficient to provide security in mechanisms and procedures. For instance, periodic checks cannot mitigate potential threats related to the lowest layers of the OSI model (i.e., physical and MAC) like physical DDoS attacks jamming and tampering, which take place on wireless media. This type of attack cannot be dealt with the traditional checking and testing of procedures described in the previous section. In contrast, and according to [87], defending a network of nodes against jamming involves a variety of methods, such as spread-spectrum, priority messages, lower duty cycle, region mapping, and mode change. Popular defense mechanisms against tampering attacks include tamper proofing and hiding, even though their success depends on many factors [87]. Bearing this in mind, it is fair to say that periodic security checks provide limited protection and insight against these types of attacks. For instance, after a tempering attack, it is not uncommon for a node which has stopped working to go unnoticed despite running network checks.

Further, the problem of malicious mobile nodes could not be handled by conducting periodic checks on security controls and mechanisms. Mobile malicious node attacks are very dangerous and need enhanced detection mechanisms to minimize the

IoT Security Frameworks and Countermeasures 273

damage they can cause. Also, periodic checks cannot deal with low-level attacks like malicious modification of hardware, hardware Trojan, and false sensor data injection.

7.10 EXISTENCE OF SECURITY POLICY TO INSTALL/UPDATE EQUIPMENT/SOFTWARE

7.10.1 Security Policy Overview, Objectives

Security policy refers to "a written document in an organization outlining how to protect the organization from threats, including computer security threats, and how to handle situations when they do occur" [88]. For an organization to determine potential threats, it will need to keep an updated list of all its assets and define how they could possibly be linked to malicious activity. After successfully identifying assets and threats, the next step in the security policy is to describe the right measures that protect systems and/or avoid exposure to those threats [19, 88]. In general, a security policy may include different aspects of the organization, such as software, network, equipment, the physical building, as well as potential malicious activity coming from organization members (with privileged information or access to physical systems) or people outside the company environment, such as hackers, competitors, activists, etc [89].

After extensive analysis of the above, security policies should also include the likelihood that these threats appear. Basically, a security policy needs to satisfy several requirements. More specifically, it should:

- Appropriately protect the confidentiality and integrity of people and information assets.
- Set the rules for expected behavior by employees, customers and other users, system administrators, management, and security personnel.
- Provide adequate authorization that enables security personnel and network administrators to monitor, probe, and investigate incidents.
- Define and authorize actions associated with the consequences of a violation
- Define the organizational consensus baseline stance on security and helps make staff aware of the views of the organization and senior management.
- Aid in creating an environment that minimizes risk, and aid in remaining compliant to the regulations and legislation that applies to the organization [89].

In addition, because of the policy's impact in future decisions, it is important that the policy reflects the company's existing environment, goals, and personnel to act as an advancement factor rather than an impediment [89].

7.10.2 Security Policy & General Good Practices for New Software Installation

Installation of new software (i.e., operating systems, office applications, financial applications, applications development, etc.) is a basic task in all companies around the world. However, if this task is not sufficiently controlled, it can lead to potential security risks and legal issues for the company. For instance, the company may support additional costs for unwanted software/equipment. In addition, installation of nonapproved software can increase the number critical security vulnerabilities that are added to a system by allowing the installation of malware, such as rootkits and Trojan horses, without the user knowledge [90]. Thus, it is crucial for companies to set up a security policy to address all security issues relevant to new software installation and deployment on a computer system. In this context, the **ISO 27001 control A.12.6.2** [90] has been proposed the following good practices that can be included in a typical security policy regarding new software installation and deployment:

- Employees are not allowed to download software from the Internet or bring software from home without authorization. This can help to prevent using of nonapproved software by employees and limit the number of vulnerabilities on the network.
- When an employee detects the need for using a specific software application, a request needs to be transmitted to the IT department. This request can be stored by the IT department as a record or as evidence.
- The IT department shall determine if the organization has a license of the software requested in order to ensure legal use of the software and guarantee that it is always patched and up to date.
- If there is license, the IT department notifies the employee and will proceed to install the software on the computer of the user who requested it.
- If there is no license, a responsible party must assess whether the requested software is necessary for the performance of the duties of the employee. For the evaluation, the financial feasibility of the software purchase must also be analyzed when the software costs money.
- If the software costs money, an analysis should be made as to whether there is another similar tool on the market that is cheaper or even free (Total Cost of Ownership must be calculated). In this context, top management should participate in the decision on the acquisition of new software
- Once the decision has been made, the IT department will proceed to include the software in their inventory and will install the software [90].

7.10.3 Effectiveness of Security Policy

The increase in the number and type of cyber-threats that can put companies at risk pushes them to lock down and enhance the security of their networks. One aspect of the network's overall security is a good software/equipment installation/update policy. When it comes to installing or updating software, setting up an appropriate security policy can help to minimize the risk of loss of program functionality, the exposure of sensitive information contained within computing network, the risk of spreading malware in the company network, and the legal exposure of running nonapproved and unlicensed software. Particularly, an effective security policy can help to mitigate and limit the impact of the following cyber-threats:

7.10.3.1 Malware, Malicious Software, Email Malware

An effective security policy can help to mitigate and limit the impact of malware by preventing employees from accidentally causing malware to spread throughout the organization. More specifically, security policy will prove effective if it restricts end-users of devices from installing nonapproved applications from the Internet. Since employee devices are connected directly to the World Wide Web, organizations should dedicate time and effort to train employees and contractors for activity that does not put IT systems at risk [19]. According to this, users should be made aware of potential threats and learn about the responsibilities or restrictions of installing unauthorized software. Nonapproved software should be identified by authorized tools and prevented from execution. In general, software should be tested according to updated lists of recognized programs and services that are malware-clean, back-door-clean and that adhere to security standards [90, 91].

Another issue that needs to be addressed in security policies is filtering Internet access and emails; that is, blocking access to known malicious sites to prevent employees from downloading malware that can spread throughout the organization. This will prevent employees from downloading malware that can spread throughout the organization network [91]. According to [92], as of 2020, 94% of malware is delivered via email. Malware emails can cause security breaches through macro viruses or malicious links. Good practices such as email filters can block email containing malicious attachments and prevent malware activities. It was also found that hackers prefer targeting high ranking employees to steal data, because of their privileged access rights to systems [91].

7.10.3.2 Use of Nonapproved Hardware

An effective security policy can help to mitigate and limit the impact of threats related to nonapproved hardware. In fact, hardware-related threats have been present for a long time in the security landscape with attackers exploiting existing technology to perform new types of attacks. New advancements, such as IoT, smart devices and smartphones require a thorough analysis of potential risks and call for integrated security policies within organizations. Setting up procedures and checks for acquiring and installing or upgrading hardware can considerably reduce the probability that nonapproved hardware is used. For example, attackers will often try to insert modified hardware into organizations to achieve their goals, and it is an organization's obligation to block off any device/hardware that does not comply with certain characteristics described in their security policy.

Modification or extension of existing hardware is a common type of attack, where the attacker extends or modifies external and internal interfaces of hardware. The distinction between external and internal interfaces refers to the device casing. For example, interfaces that are accessible without opening/tampering the casing are called external interfaces, such as USB ports. In contrast, if the casing needs to be removed, leaving the hardware inside intact, then it is referred to as an internal interface, such as pins that expose a JTAG interface [93]. A good example of modification/extension of hardware is the **Cottonmouth-1,**[30] which "extends an existing USB cable in a noninvasive way and supports over-the-air attacks." This hardware implant "will provide a wireless bridge into a target network as well as the ability to load exploit software onto target PCs." Other examples include the FireWire plug or a PCIe device. Finally, in order to construct an effective policy against nonapproved equipment, it is important to remember that hardware can be "impacted by threats in a twofold way: a) the hardware itself is a physical asset to users (based on value and function) which can be impacted, and at the same time, b) can be modified to impact other asset types, such as user health and property [93]."

In this context, an effective policy against nonapproved software/equipment can prevent and reduce the impact of risks related to "back off attacks" that are mainly caused by hardware trojans or infected operating systems/drivers. Also, guaranteeing that network software/infrastructure is malware-clean and adheres to security standards can reduce the probability that insecure network services are active, reducing thus the attack surface including MAC Congestion, MAC spoofing, malicious mobile nodes, sinkhole attack, network intrusions, physical and application layers DDoS attacks and many others.

7.10.4 LIMITATIONS

Although security policy and procedures regarding new/updated equipment and software exhibit many benefits against certain types threats, it is unable to defend against certain types of threats like memory scraping, traffic sniffing, buffer overflow, drive-by download attacks, physical damage, theft, or loss.

7.10.4.1 Memory Scraping

This type of attack uses a malicious script that parses data stored briefly in the memory banks of specific Point-of-Sale (POS) devices. More specifically, this technique "captures the data stored on the card's magnetic stripe in the instant after it has been swiped at the terminal and is still in the system's memory" [94]. The collected data can either be sold in the underground black markets, or they can be used directly to create cloned copies of the cards. Cloned card holders can then use them to shop directly in stores [95]. In the infamous Target data breach in 2013, attackers stole financial information and personal data of more than 110 million customers. The attackers were able to steal card data by installing memory-scraping malware on the checkout POS devices of the Target stores [95].

In general, memory scraping is the most critical threat to POS systems. A recent study reviewed and extracted features from 22 malware families. Their analysis showed that memory scraping behavior consists of three different stages of behavior: a) infection and persistence, b) process and card data search, and c) data exfiltration. The results showed that this type of attack is still quite immature, as the code rarely includes sophisticated techniques to avoid static or dynamic malware analysis [95].

7.10.4.2 Physical Damage/Theft/Loss

Security policies regarding software and hardware updates can provide limited protection from physical damage, theft, and loss. Unexpected damage, such as extreme weather conditions (e.g., storms, floods, earthquakes, etc.) can have devastating results on hardware, equipment, and physical assets of an organization. In addition,

IoT Security Frameworks and Countermeasures 277

physical damage can be caused by criminal activity or acts of mischief, such as terrorism, explosion, riots, smoke, and civil commotions [19, 96]. There are certain cases where security policies can do little to prevent damage. For instance, in cases of theft, if offices, data centers, or sites where computer hardware is kept are not sufficiently secured or left unattended, it will then be quite easy for criminals to gain access and break in. Sometimes, criminals can also gain access by masquerading as suppliers, for example, a technician, a cleaner, or a utility company representative [96]. This type of "attack" is related to social engineering and it is fairly easy for employees to be tricked, especially when criminals disguise themselves as maintenance professionals, which gives them both access to the site and the physical equipment area (e.g., the electrical/network wiring rack).

Unfortunately, physical security tends to be overlooked by most organizations; taking countermeasures against hacking and sensitive data breaches are considered a higher priority, which leaves an opportunity for attackers to improve their methods over physical and remote access of systems.

7.10.4.3 Drive-by Download Attacks

Drive-by download attacks is another threat that cannot be dealt with by security procedures describing new updates and software installations. A drive-by download attack is a common attack among cybercriminals in which an automated download of software is installed on a device without the user's consent. Downloading malware can happen in one of two ways:

- The user has authorized the download but is not aware that the download includes a malicious program, for instance, an unknown or counterfeit executable program, ActiveX component, or Java applet.
- The user has not authorized the download and is not aware that the download has been installed on the device, for instance, a virus, spyware, malware, or crimeware.

Essentially, the download can be initiated in various ways, such as an email attachment, a malicious link online, an advertisement pop-up window [97], etc.

7.11 EXISTENCE OF BACKUP POLICIES

7.11.1 Backup Policy Overview, Objectives

A backup policy is one of the most critical functions in any security policy; once an organization experiences data loss and the inability to protect its data, its longevity and reputation in the market are compromised. Often, when a business experiences physical damage, the irreplaceable costs do not come from the physical loss of the equipment, but rather from the data that was store inside that hardware. The most common reasons why organizations will experience some form of data loss:

- Accidental deleting files or formatting drives
- Viruses and Malware

- Damage or loss of computers and hard drives
- Power failures
- Fire or natural disasters.

In general, a backup policy helps avoid this risk by providing recovery, restoration and retrieval of data that is lost or compromised. It is essentially a pre-defined recovery scheme that provides specific guidelines related to questions like "who," "what," "where," "when," and "how," regarding the data of the organization. Where the question "**who**" describes the person, who is authorized to access, schedule, and conduct the backups, "**what**" refers to what data need to be backed-up, "**where**" defines the location of the backup copies located either on the premises or at a remote location (e.g., cloud), "**when**" defines the frequency of data backup; whether the data is backed up daily, weekly, monthly, etc., and the question "**how**" describes the hardware resources or software required or recommended for performing the backup and the type of backup e.g.,, incremental, differential [19], etc.

To summarize, the ongoing availability of data is critical to the well-being of an organization, and units responsible for providing and operating administrative applications need to ensure all information is adequately backed up.

7.11.2 Types of Data Backup Policies

There are various types of data backup and recovery, each designed to address different risks, vulnerabilities, and storage needs. Effectively backing up files, networks, servers, and other assets begins with addressing the capabilities of a network and selecting the proper type of backup for the circumstances. Thus, it is imperative for organizations and individuals to understand the differences among the main types of backup policies to ensure full protection of data while allocating their resources in an optimal manner. The three main types of backup policies available include [19]:

1. **Full backups**: This type of backup policy copies the entire data set every time a backup is initiated. As a result, they provide the highest level of protection. However, most organizations cannot perform full backups frequently because they can be time-consuming and take up too much storage capacity.
2. **Incremental backups**: Back up only the data that has been changed or updated since the last backup. This method saves time and space for storing but can make it harder to perform an entire restore. Incremental is a common form of cloud backup because it tends to use fewer resources.
3. **Differential backups**: These backup policies are similar to incremental backups because they only contain data that has been altered. However, differential backups back up data that has changed since the last full backup, rather than the last backup in general. This method solves the problem of difficult restores that can arise with incremental backups [98].

There is no specific ideal backup strategy. However, the selected backup strategy will depend on the organization needs and should be periodically tested by restoring the backup data to a test machine.

IoT Security Frameworks and Countermeasures

7.11.3 Good Practices for an Effective Backup Policy

Setting up a backup strategy is an essential measure for organizations to protect their assets from data loss. In this light, a backup schedule needs to be planned carefully to ensure full protection of data while assuring continuity of operations. Organizations who are in the process of designing an effective backup plan should consider the following practices:

- **Document all procedures**: a backup regime must be followed regularly and accurately with all procedures being clearly documented. Logs should be accessible to anyone following the backup procedures.
- **Use of high-quality media**: It is important to choose reputable, high-quality media from vendors that perform ongoing tests and qualifications.
- **Regularly clean and maintain backup equipment**: Data storage devices must be cleaned regularly as part of the backup schedule. For instance, debris on the tape head can cause backup errors.
- **Store media securely onsite and offsite**: Secure offsite storage provides additional protection against flood, fire, and theft. Onsite storage options should be carefully examined. For example, even though fireproof safes provide additional protection, archived data should be periodically inspected to ensure that no obvious damage or corruption has occurred due to environmental or other types of damage.
- **Rehearse data recovery and disaster recovery procedures**: It is essential for organizations to rehearse recovery procedures regularly. Doing a test restore on a regular basis both ensures the pre-planned procedures can be followed during an incident [99].

7.11.4 Effectiveness of Backup Policy in Limiting Threat Impact

Best practice for backup and recovery help to mitigate or minimize the effect of various threats such as the effect of contaminating the software/firmware of devices with malware through restoration of "clean" backup. For instance, backups can be used as the last line of defense and the most effective countermeasure during the mitigation process of ransomware attacks. Today, ransomware is one of the top Cybersecurity threats in all sectors, such as government, manufacturing, retail, finance, healthcare, etc., with attacks being held even in countries with access to the most advanced security technologies. In addition, ransomware is not restricted to computers, but it can also attack servers, mobile, and cloud systems. When it comes to costs, monetary damage caused by hackers exceeds the amount of the ransom, as organizations have to cover costs resulting from downtime, data loss, network/systems restoration, and reputation [100, 101]. Despite efforts, there is no security measure that can truly protect systems from ransomware. Users need to combine a variety of measures, such as antivirus systems, firewalls, IDS, use of authorized software, visiting reputable websites, etc., to help decrease the probability of a ransom attack [100]. To keep systems secure, organizations can choose among the three types of backup for their data: a full backup, an incremental backup, or a differential backup [19]. However, organizations need to constantly update their strategies, as hackers are able to encrypt or block access to backup data as well.

Backups are also very effective against physical damage to data storage equipment that can come from a variety of sources. Either a mechanical failure, extreme weather conditions, or an act of terrorism, this type of damage cannot be usually predicted or controlled by organizations. Theft and loss are two other ways that enterprises fall victim of sensitive data loss. Other than external threats, such as identity theft or hacked customer databases, sensitive data manipulation can also be the result of an insider's act of mischief. For instance, an employee causing leaking or corrupting files as a form of revenge. However, it is not uncommon for companies to experience data loss due to human error, for example, an employee erasing important files by mistake. These problems have led companies and organizations to protect their assets by setting up advanced strategies, including preventive measures such as environmental sensors, monitoring devices, and disaster recovery plans, such as backup policies [102].

Furthermore, effective backup policies can help in reducing the impact of missing or weak implementations of security mechanisms, network intrusion and poor physical security through restoration of "clean" and secure backups. In the case of false sensor data injection threat, if the injection has resulted in tampering with the data, a clean backup could be restored. However, if additional data (other than the falsified ones) have been added to the dataset, restores are not adequate [19].

7.11.5 Limitations

Although a backup policy exhibits many benefits against certain types of threats, it is unable to defend against certain other types of threats like malicious manipulation of firmware used to control a device's hardware. This kind of software presents security risks as it can be manipulated to execute malicious code. More specifically, USB devices have been well known to have been used for malicious purposes by hackers' sites, spy on the users, and exfiltrate data [19, 64]. One example is the Bad USB firmware hack that exploits computer components to connect to malicious famous attacks involving heavy manipulation of firmware was the **Stuxnet** case, in which attackers used a USB device to attack Iran's International Atomic Energy Agency. The malware exploited four zero-day flaws to attack the centrifuges and tear them apart [103]. Ever since those attacks, investigation analysis on reputable devices reveal new cases of firmware manipulation or find vulnerabilities that represent new opportunities for attackers.

Once an attacker has managed to compromise the firmware, the device is now ready to be used as an attack vector. Some of the most common malicious techniques include:

- **Altering Boot Process**: By compromising the system firmware or firmware within the Trusted Platform Module, attackers can establish persistence by disrupting the secure boot process of a system.
- **Option ROM Attacks** Option ROM attacks are often used as a part of an initial infection or to spread malicious firmware from one component to another.

IoT Security Frameworks and Countermeasures

Option ROM is generally employed by components to get the acceptable firmware during the boot process.
- **Direct Memory Access (DMA)**: DMA attacks are a common method in firmware-based attacks. DMA gives components direct access to system memory without going through the OS.
- **Processor Level Exploits**: Attacks such as Row hammer can allow an attacker to flip bits in areas of RAM in order to escalate privileges.
- **Disabling Devices:** Lastly, firmware can be used to disable a device temporarily or permanently. These techniques can be used against virtually any device including servers [104].

Backup policies are powerless in the battle against firmware manipulation. This is because up until now backup strategies have only been implemented as post-exploitation measures and mitigation actions. In contrast, firmware manipulation methods are used by hackers as attack vectors. To overcome this threat, security experts note the need to reinvent firmware by separating it from hardware vendors and instead start seeing declarative firmware as "the best chance we have of real bottom-up security [105]."

Furthermore, Backup policies cannot deal with security threats related to insufficient transport layer encryption. This security threat, which has been ranked ninth in the OWASP Top 10 risks in recent years, continues to represent a serious vulnerability in network security today, caused by applications that do not take any measures to protect network traffic [6]. Most applications use secure communication channels based on standard encryption techniques and protocols, such as SSL and TLS. However, they frequently fail to properly authenticate, encrypt, and protect the confidentiality and integrity of sensitive network traffic by using weak algorithms, expired or invalid certificates, or do not use them correctly. This leaves sensitive data and session IDs exposed and may result in complete compromise of devices, applications, and user accounts. This threat becomes increasingly prevalent since the use of IoT devices in our everyday life has exploded in the last decade. Still, many IoT devices fail to employ transport encryption in their data transmissions [106]. Thus, organizations that seek to protect their systems from potential attacks should apply network encryption mechanisms and only choose to work with highly secure devices/applications.

7.12 CONCLUSIONS

IoT devices have become an integral part of our society with more and more devices connected to the global network, making everything "smart." However, the increasing spread of these devices in our daily life creates many security and privacy challenges. It is clear that weakly secured or completely unsecured smart devices introduce numerous security risks including MitM, data and identity theft, device hijacking, DoS/DDoS, Botnets, malware, firmware manipulation, and tampering, spoofing, sinkhole, and many other serious cyber-attacks. From the discussions made throughout this chapter, it can be concluded that IoT security should not only consider the device security, but all elements involved in the IoT ecosystem; including mobile applications and interfaces, operating systems and software, cloud, network interfaces, encryption protocols, and physical security. In fact, there is a need for different security countermeasures to cover all security aspects related to the IoT ecosystem and ensure that the information and services are being exchanged in a safe environment.

In this context, security features should be incorporated into each IoT solution during the design phase. This can be achieved by large collaborations between all those who participate in developing IoT devices including IoT manufacturers, IoT connectivity architects, IoT platform developers, IoT application developers, IoT service developers, and IoT experience designers. Each actor has a responsibility and a moral obligation towards the community to deploy sufficient protection measures. In fact, building hardware that incorporates hardened security features into IoT devices is a critical missing piece that must be addressed. In addition to implementing security at the design phase, regular software/firmware updates/patching is another crucial security measure that can help to protect the IoT ecosystem, repair security holes (i.e.,, software vulnerabilities) that have been discovered and fix, or even remove bugs. It can also enhance IoT devices with new features. Thus, it is recommended that IoT device manufacturers promote security updates/patching and vulnerability management.

Since most cyber-attacks use compromised devices to probe deeper into a network, stopping potential attacks begins with early detection by using security components like firewalls, NIDS, and antivirus. These security solutions can take many forms and detect many different types of attacks. In this context, adding firewall and IDS capabilities to embedded devices is critical to providing early warning of a cyber-attack. This enables security administrators to take appropriate actions in order to block attacks, quarantine compromised systems, and protect their networks, in real-time. In addition, user/employee's awareness and training encourages them to be aware of the vulnerabilities that the IoT devices may experience. This could be another effective way to combat the whole spectrum of IoT security threats. For instance, when selecting an appropriate IoT device, consumers should require that the vendors have properly defended the device against common attacks.

It is also important to have monitoring systems in place when an event occurs. Once the event has been detected, a responsive action must be triggered to prevent any malicious use of the device. Effective logging/alerting policy can log in real-time, abnormalities in the data and alerting the system/network administrators to take appropriate action. Periodic security checks and reviews of security mechanisms also play a central role in protecting IoT networks from cyber-attacks and ensure that the appropriate level of security is being maintained. Moreover, it will keep security administrators on top of the newest security risks. Existence of good software/equipment installation/update policy will help identify security ambiguities, mitigate the risks, and put precautionary measures in place. Finally, reliable backup policies help to provide consistent and efficient methods for recovering data that is lost or compromised.

In addition to all discussed countermeasures, the entire communication channel from the sensors to the service providers must be secure by using robust encryption

IoT Security Frameworks and Countermeasures

protocols along with strong authentication methods. This ensures that the devices are communicating with known and trusted entities. In summary, various security countermeasures against all possible attacks and threats are indispensable to protect the IoT infrastructure, without disrupting operations, service reliability, or profitability. Properly implementing all the discussed security controls will certainly enhance the overall security of the IoT ecosystem, improve the quality of life and provide benefits to consumers and enterprises.

NOTES

1. https://eng.umd.edu/news/story/study-hackers-attack-every-39-seconds
2. https://owasp.org/
3. https://www.av-test.org/en/statistics/malware/
4. https://mypage.webroot.com/rs/557-FSI-195/images/2020%20Webroot%20Threat%20Report_US_FINAL.pdf
5. https://start.paloaltonetworks.com/unit-42-iot-threat-report
6. https://www.a10networks.com/marketing-comms/reports/state-ddos-weapons/
7. https://www.ibm.com/security/threat-detection-analysis
8. https://www.observeit.com/2020costofinsiderthreat/
9. https://www.cisco.com/en/US/docs/security/pix/pix30/user/guide/pixugint.html
10. https://www.venafi.com/
11. https://nmap.org/
12. http://www.netcopweb.com.br/
13. http://netcat.sourceforge.net/
14. https://www.webroot.com/us/en
15. https://www.avast.com/c-eternalblue
16. https://securityboulevard.com/2018/05/59-of-people-use-the-same-password-everywhere-poll-finds/
17. https://www.proofpoint.com/us
18. https://www.techadvisory.org/
19. **Update**: is a new version of a software program with enhanced functionality, new features and/or bug fixes.
20. **Patch**: is a fix of a known problem with an operating system or software program.
21. https://workflow.servicenow.com/it-transformation/ponemon-vulnerability-response-study/
22. https://pcantivirusreviews.com/Top-Threats-Removal/Code-Red/
23. https://vuls.io/
24. https://www.archerysec.com/
25. https://www.microsoft.com/en-us/download/details.aspx?id=58105
26. https://www.tenable.com/?tns_languageOverride=true
27. https://www.edgescan.com
28. https://www.cvedetails.com/cve/CVE-2017-0143/
29. https://securitycloud.symantec.com/cc/#/landing
30. https://nsa.gov1.info/dni/nsa-ant-catalog/usb/index.htm

REFERENCES

1. H. Taylor, "What are cyber threats and what to do about them," Available: https://prey-project.com/blog/en/what-are-cyber-threats-how-they-affect-you-what-to-do-about-them/. [Accessed: Jun. 17, 2020].

2. M. S. Abu, S. R. Selamat, A. Ariffin, and R. Yusof, "Cyber threat intelligence—Issue and challenges," *Indonesian Journal of Electrical Engineering and Computer Science*, 2018. doi: 10.11591/ijeecs.v10.i1.pp371–379.
3. NSA, "*Best practices for keeping your home network secure*," 2011. Available: https://www.nsa.gov/Portals/70/documents/what-we-do/cybersecurity/professional-resources/csi-best-practices-for-keeping-home-network-secure.pdf, [Accessed: Jun. 03, 2020]
4. R. Brewer, "Advanced persistent threats: Minimising the damage," *Network Security*, 2014. doi: 10.1016/S1353-4858(14)70040-6.
5. M. Safaei Pour, E. Bou-Harb, K. Varma, N. Neshenko, D. A. Pados, and K. K. R. Choo, "Comprehending the IoT cyber threat landscape: A data dimensionality reduction technique to infer and characterize Internet-scale IoT probing campaigns," *Digital Investigation*, 2019. doi: 10.1016/j.diin.2019.01.014.
6. OWASP, "Internet of Things (IoT) top 10 2018," *OWASP Internet of Things Project*, 2018. Available: https://wiki.owasp.org/index.php/OWASP_Internet_of_Things_Project#tab=IoT_Top_10. [Accessed : Sep. 20, 2020].
7. ENISA, *ENISA Threat Landscape Report 2017*. 2018.
8. S. Corporation, "2016 internet security threat report," 2016.
9. A. Mosenia and N. K. Jha, "A comprehensive study of security of internet-of-things," *IEEE Transactions on Emerging Topics in Computing*, 2017. doi: 10.1109/TETC.2016.2606384.
10. A. Afianian, S. Niksefat, B. Sadeghiyan, and D. Baptiste, "Malware dynamic analysis evasion techniques: A survey," *ACM Computing Survey*, vol. 52, no. 6, 2019. doi: 10.1145/3365001.
11. B. Ali, M. Shiaeles, S. Bendiab, and G. Ghita, "MALGRA: Machine learning and N-gram malware feature extraction and detection system," *Electronics*, vol. 9, no. 1777, 2020. doi: 10.3390/electronics9111777.
12. D. Uppal, V. Mehra, and V. Verma, "Basic survey on malware analysis, tools and techniques," *International Journal of Advanced Computer Science*, vol. 4, no. 1, pp. 103–112, 2014. doi: 10.5121/ijcsa.2014.4110.
13. J. Hong, "The state of phishing attacks," *Communications of the ACM*, 2012. doi: 10.1145/2063176.2063197.
14. B. B. Gupta, A. Tewari, A. K. Jain, and D. P. Agrawal, "Fighting against phishing attacks: state of the art and future challenges," *Neural Computing and Applications*. 2017. doi: 10.1007/s00521-016-2275-y.
15. Z. Benenson, F. Gassmann, and R. Landwirth, "Unpacking spear phishing susceptibility," in *Lecture Notes in Computer Science (including subseries Lecture Notes in Artificial Intelligence and Lecture Notes in Bioinformatics)*, 2017. doi: 10.1007/978-3-319-70278-0_39.
16. F. S. Diego, "FBI warns public that cyber criminals continue to use spear-phishing attacks to compromise computer networks," *FBI*. Available: https://archives.fbi.gov/archives/sandiego/press-releases/2013/fbi-warns-public-that-cyber-criminals-continue-to-use-spear-phishing-attacks-to-compromise-computer-networks. [Accessed: Jun. 30, 2020].
17. M. Conti, N. Dragoni, and V. Lesyk, "A survey of man in the middle attacks," *IEEE Communications Surveys and Tutorials*, 2016. doi: 10.1109/COMST.2016.2548426.
18. S. S. C. Silva, R. M. P. Silva, R. C. G. Pinto, and R. M. Salles, "Botnets: A survey," *Computing Networks*, 2013. doi: 10.1016/j.comnet.2012.07.021.
19. C. Easttom, *Network Defense and Countermeasures: Principles and Practices*. Pearson Education, USA, October, 2013.
20. G. Anthes, "Software: The eternal battlefield in the unending cyberwars, Computerworld," *P&G*, 2009. Available: https://www.computerworld.com/article/2523720/software--the-eternal-battlefield-in-the-unending-cyberwars.html. [Accessed: Apr. 14, 2020].

21. T. Pietraszek and C. Vanden Berghe, "Defending against injection attacks through context-sensitive string evaluation," in *Lecture Notes in Computer Science (including subseries Lecture Notes in Artificial Intelligence and Lecture Notes in Bioinformatics)*, 2006. doi: 10.1007/11663812_7.
22. Outpost24, "TOP 10 of the world's largest cyberattacks, and to prevent them," *Outpost24*, 2018. Available: https://outpost24.com/blog/top-10-of-the-world-biggest-cyberattacks. [Accessed: Jul. 14, 2020].
23. Kaspersky, "What is an Advanced Persistent Threat (APT)?" 2020. Available: https://www.kaspersky.com/resource-center/definitions/advanced-persistent-threats [Accessed: Jul. 12, 2020].
24. M. Assante, "Analysis of the cyber attack on the Ukrainian power grid," 2016.
25. McAfee, "What is stuxnet?" *McAfee*, 2020. Available: https://www.mcafee.com/enterprise/fr-fr/security-awareness/ransomware/what-is-stuxnet.html#:~:text=Share%3A, used to automate machine processes. [Accessed: Jul. 14, 2020].
26. V. Koutsouvelis, S. Shiaeles, B. Ghita, and G. Bendiab, "*Detection of insider threats using artificial intelligence and visualisation*," in *Proceedings of the 2020 IEEE Conference on Network Softwarization: Bridging the Gap Between AI and Network Softwarization, NetSoft 2020*, 2020. doi: 10.1109/NetSoft48620.2020.9165337.
27. H. M. Lynch, "Firewall fundamentals," *Information Systems Security*, vol. 9, no. 5, pp. 1–11, 2000. doi: 10.1201/1086/43312.9.5.20001112/31374.6.
28. CISCO, "What is a firewall?" Available: https://www.cisco.com/c/en/us/products/security/firewalls/what-is-a-firewall.html. [Accessed: May 23, 2020].
29. H. Abie, "An overview of firewall technologies," *Telektronikk*, 2000. doi: 10.1.1.20.9215.
30. D. W. Chadwick, "Network firewall technologies," *NATO Science Series III: Computer and Systems*, 2001.
31. Y. H. Cho, S. Navab, and W. H. Mangione-Smith, "Specialized hardware for deep network packet filtering," in *Lecture Notes in Computer Science (including subseries Lecture Notes in Artificial Intelligence and Lecture Notes in Bioinformatics)*, 2002. doi: 10.1007/3-540-46117-5_48.
32. M. G. Gouda and A. X. Liu, "*A model of stateful firewalls and its properties*," in *Proceedings of the International Conference on Dependable Systems and Networks*, 2005. doi: 10.1109/DSN.2005.9.
33. K. Neupane, R. Haddad, and L. Chen, "*Next generation firewall for network security: A survey*," in *Conference Proceedings - IEEE SOUTHEASTCON*, 2018. doi: 10.1109/SECON.2018.8478973.
34. H. J. Wen and J. H. M. Tarn, "Internet security: A case study of firewall selection," *Information Management and Computer Security*, 1998. doi: 10.1108/09685229810227658.
35. R. Zalenski, "Firewall technologies," *IEEE Potentials*, 2002. doi: 10.1109/45.985324.
36. F. Muhammad Arifin, G. Andriana Mutiara, and I. Ismail, "Implementation of management and network security using endian UTM firewall," *IJAIT (International Journal of Theoretical and Applied Information Technology)*, 2017. doi: 10.25124/ijait.v1i02.874.
37. A. Chopra, "Security issues of firewall," *International Journal of P2P Network Trends and Technology*, vol. 22, no. 1, pp. 4–9, 2016. doi: 10.14445/22492615/ijptt-v22p402.
38. K. Brancik and G. Ghinita, "*The optimization of situational awareness for insider threat detection*," in *CODASPY'11 - Proceedings of the 1st ACM Conference on Data and Application Security and Privacy*, 2011. doi: 10.1145/1943513.1943544.
39. E. G. Amoroso, *Cyber Attacks Protecting National Infrastructure PREFACE*. Butterworth-Heinemann, Elsevier; USA, 2011.
40. K. Scarfone and P. Hoffman, *Guidelines on Firewalls and Firewall Policy* NIST Special Publications, USA, 2009.

41. I. V. Chugunkov, L. O. Fedorov, B. S. Achmiz, and Z. R. Sayfullina, "*Development of the algorithm for protection against DDoS-attacks of type pulse wave*," in *Proceedings of the 2018 IEEE Conference of Russian Young Researchers in Electrical and Electronic Engineering, ElConRus 2018*, 2018. doi: 10.1109/EIConRus.2018.8317090.
42. "A survey on IPS methods and techniques," *International Journal of Computing Science Issues*, 2016. doi: 10.20943/01201602.3843.
43. F. Alsakran, G. Bendiab, S. Shiaeles, and N. Kolokotronis, "Intrusion detection systems for smart home IoT devices: Experimental comparison study," *Communications in Computer and Information Science*, 2020. doi: 10.1007/978-981-15-4825-3_7.
44. D. Stiawan, A. H. Abdullah, and M. Y. Idris, "*The trends of Intrusion Prevention System network*," in *ICETC 2010 - 2010 2nd International Conference on Education Technology and Computer*, 2010. doi: 10.1109/ICETC.2010.5529697.
45. N. Chakraborty, "Intrusion detection system and intrusion prevention system: A comparative study," *International Journal of Computing Business Research*, 2013.
46. PLOALTO Network, "What is an intrusion prevention system?" Available: https://www.paloaltonetworks.com/cyberpedia/what-is-an-intrusion-prevention-system-ips. [Accessed: Apr. 27, 2020].
47. Y. Tayyebi and D. S. Bhilare, "A comparative study of open source network based intrusion detection systems," vol. 9, no. 2, pp. 23–26, 2018. doi: 10.1016/S0169-5347(00)02077-2.
48. V. Jyothsna, V. V. Rama Prasad, and K. Munivara Prasad, "A review of anomaly based intrusion detection systems," *International Journal of Computer Applications*, 2011. doi: 10.5120/3399-4730.
49. R. Selvaraj, V. M. Kuthadi, and T. Marwala, "Honey pot: A major technique for intrusion detection," *Advances in Intelligent Systems and Computing*, 2016. doi: 10.1007/978-81-322-2523-2_7.
50. V. Kumar, "Signature based intrusion detection system using SNORT," *International Journal of Computer Applications and Information Technology - IJCAIT*, vol. I, no. Iii, pp. 35–41, 2012. [Online]. Available: http://ijcait.com/IJCAIT/index.php/www-ijcs/article/view/171.
51. P. Borkar, "IPS security: How active security saves time and stops attacks in their tracks," *Exabean*, 2019. Available: https://www.exabeam.com/ueba/ips-security-how-active-security-saves-time-and-stop-attacks-in-their-tracks/.
52. H. Shafiei, A. Khonsari, H. Derakhshi, and P. Mousavi, "Detection and mitigation of sinkhole attacks in wireless sensor networks," *The Journal of Computer and System Sciences*, 2014. doi: 10.1016/j.jcss.2013.06.016.
53. A. Rehman, S. U. Rehman, and H. Raheem, "Sinkhole attacks in wireless sensor networks: A survey," *Wireless Personal Communications*, 2019. doi: 10.1007/s11277-018-6040-7.
54. A. Sarmah, "Intrusion detection systems: definition, need and challenges," *Information Security*, 2001.
55. S. B. Ambati and D. Vidyarthi, "A brief study and comparison of, open source intrusion detection system tools," *International Journal of Advance Computational Engineering and Networking*, no. 110, ISSN: 2320–2106, 2013. [Online]. Available: http://www.iraj.in/journal/journal_file/journal_pdf/3-27-139087836726-32.pdf.
56. S. Schupp, "Limitations of network intrusion detection," 2000. [Online]. Available: https://www.giac.org/paper/gsec/235/limitations-network-intrusion-detection/100739.
57. M. Gilbert, D. Haag, and T. S. Hottes, "Antivirus software?" *The BC Medical Journal*, 2011.

58. R. Sihwail, K. Omar, and K. A. Z. Ariffin, "A survey on malware analysis techniques: Static, dynamic, hybrid and memory analysis," *International Journal on Advanced Science, Engineering and Information Technology*, vol. 8, no. 4–2, pp. 1662–1671, 2018. doi: 10.18517/ijaseit.8.4-2.6827.
59. B. Potter and G. Day, "The effectiveness of anti-malware tools," Computer Fraud & Security, 2009. doi: 10.1016/S1361-3723(09)70033-8.
60. P. Picazo-Sanchez, J. Tapiador, and G. Schneider, "After you, please: browser extensions order attacks and countermeasures," *The International Journal of Information Security.*, 2019. doi: 10.1007/s10207-019-00481-8.
61. Code24, "Unpreditable humans still the weakest link in data security," *Code24*, 2015. Available: https://blog.code42.com/wp-content/uploads/2015/12/code42-unpredictable-humans.png. [Accessed: Aug. 12, 2020].
62. C. Riley, "Insurance giant Anthem hit by massive data breach," *CNN BUSINESS*, 2015. Available: https://money.cnn.com/2015/02/04/technology/anthem-insurance-hack-data-security/. [Accessed: Aug. 12, 2020].
63. P. Marshall, "Building an information security awareness program," *Journal of Government Information*, 2002. doi: 10.1016/j.jgi.2003.12.014.
64. G. Messina, "What is removable media?" *Infosec*, 2019. Available: https://resources.infosecinstitute.com/category/enterprise/securityawareness/removable-media/#gref. [Accessed: Aug. 28, 2020].
65. Y. Chen, K. Ramamurthy, and K. W. Wen, "Impacts of comprehensive information security programs on information security culture," *Journal of Computer Information Systems*, 2015. doi: 10.1080/08874417.2015.11645767.
66. W. Rash, "You need to protect your website against formjacking right now," *PC Magazine Australia*, 2019.
67. A. W. Rufi, "Vulnerabilities, threats , and attacks," *Network Security. 1 2 Companion Guide (Cisco Network Acadamic)*, 2006.
68. H. Okhravi and D. Nicol, "Evaluation of patch management strategies," *International Journal of Computational Intelligence Theory and Practice*, vol. 3, no. 2, pp. 109–117, 2008.
69. L. Vaas, "Top 10 most exploited vulnerabilities list released by FBI, DHS CISA," *SOPHOS*, 2020. Available: https://nakedsecurity.sophos.com/2020/05/15/top-10-most-exploited-vulnerabilities-list-released-by-fbi-dhs-cisa/.
70. J. Creasey, "Cyber security monitoring and logging guide," p. 60, 2015 [Online]. Available: http://www.crest-approved.org.
71. K. Kent and M. Souppaya, *Guide to Computer Security Log Management*. NIST Special Publications, USA, 2006.
72. D. Landoll, *The Security Risk Assessment Handbook: A Complete Guide for Performing Security Risk Assessments*, 2nd. ed, 2011.
73. J. Yarden, "Why you should perform regular security audits," *TechRepublic*, 2005. Available: https://www.techrepublic.com/article/why-you-should-perform-regular-security-audits/. [Accessed: Jul. 09, 2020].
74. A. N. Craig, S. J. Shackelford, and J. S. Hiller, "Proactive Cybersecurity: A comparative industry and regulatory analysis," *The American Business Law Journal*, 2015. doi: 10.1111/ablj.12055.
75. E. Hulitt and R. B. Vaughn, "Information system security compliance to FISMA standard: A quantitative measure," *Telecommunication System*, 2010. doi: 10.1007/s11235-009-9248-8.

76. D. L. Marnix Dekker, C. Karsberg, and M. Lakka, "Schemes for auditing security measures: An overview," 2013. [Online]. Available: http://knjiznica.sabor.hr/pdf/E_publikacije/Schemes_for_auditing_security_measures.pdf.
77. H. F. Cervone, "Computer network security and cyber ethics (review)," *Portal: Libraries and the Academy*, 2007. doi: 10.1353/pla.2007.0017.
78. T. G. of the H. K. S. A. Region, "Security risk assessment & audit," *Office of the Government Chief Information Officer*, 2017. Available: https://www.govcert.gov.hk/doc/ispg-sm01_en.pdf.
79. C. Tankard, "The security issues of the Internet of Things," *Computer Fraud & Security*, 2015. doi: 10.1016/S1361-3723(15)30084-1.
80. Symantec, "Internet security threat report 2013." [Online]. Available: https://www.insight.com/content/dam/insight/en_US/pdfs/symantec/symantec-corp-internet-security-threat-report-volume-18.pdf.
81. M. Khan, Z. Bi, and J. A. Copeland, "*Software updates as a security metric: Passive identification of update trends and effect on machine infection*," in *Proceedings - IEEE Military Communications Conference MILCOM*, 2012. doi: 10.1109/MILCOM.2012.6415869.
82. M. Hypponen and L. Nyman, "The Internet of (Vulnerable) Things: On Hypponen's law, security engineering, and IoT legislation," *Technology Innovation Management Review*, 2017. doi: 10.22215/timreview1066.
83. J. Ioannidis and S. Bellovin, "Implementing pushback: Router-based defense against DDoS attacks," *Network and Distributed System Security Symposium*, 02, 2002. doi: 10.1007/s13398-014-0173-7.2.
84. "Managing information security," *Kybernetes*, 2011. doi: 10.1108/k.2011.06740caa.012.
85. M. Bailey, E. Cooke, F. Jahanian, Y. Xu, and M. Karir, "*Al survey of botnet technology and defenses*," in *Proceedings - Cybersecurity Applications and Technology Conference for Homeland Security, CATCH 2009*, 2009. doi: 10.1109/CATCH.2009.40.
86. D. Plohmann, E. Gerhards-Padilla, and F. Leder, "*Botnets: Measurement, detection, disinfection and defence*," 2011.
87. A. D. Wood and J. A. Stankovic, "Denial of service in sensor networks," *Computer (Long. Beach. Calif).*, 2002. doi: 10.1109/MC.2002.1039518.
88. S. Banks, "Security policy," *Computers Security*, 1990. doi: 10.1016/0167-4048(90)90058-2.
89. J. J. Fay and D. Patterson, "The importance of policies and procedures," in *Contemporary Security Management*, pp. 495–522, Butterworth-Heinemann, Elsevier, USA, 2018. doi: 10.1016/B978-0-12-809278-1.00024-4.
90. A. J. Segovia, "Implementing restrictions on software installation using ISO 27001 control A.12.6.2," *ISO*, 2016. Available: https://advisera.com/27001academy/blog/2016/02/08/implementing-restrictions-on-software-installation-using-iso-27001-control-a-12-6-2/. [Accessed: Sep. 18, 2020].
91. E. Meyer, "Defending your data: Securing against internal and external threats," *AI & Data Analytics*, 2016. Available: https://inform.tmforum.org/features-and-analysis/2016/03/defending-your-data-securing-against-internal-and-external-threats/. [Accessed: Sep. 19, 2020].
92. J. Fruhlinger, "Top cybersecurity facts, figures and statistics for 2020," *CSO*, 2020. Available: https://www.csoonline.com/article/3153707/top-cybersecurity-facts-figures-and-statistics.html. [Accessed: Sep. 19, 2020].
93. ENISA, "Hardware threat landscape and good practice guide," 2017. [Online]. Available: https://www.enisa.europa.eu/publications/hardware-threat-landscape.

94. J. Hizver and T. C. Chiueh, "An introspection-based memory scraper attack against virtualized Point of Sale systems," in *Lecture Notes in Computer Science (including subseries Lecture Notes in Artificial Intelligence and Lecture Notes in Bioinformatics)*, 2012. doi: 10.1007/978-3-642-29889-9_6.
95. KrebesOnSecurity, "Cards stolen in target breach flood underground markets," 2020. Available: https://krebsonsecurity.com/2013/12/cards-stolen-in-target-breach-flood-underground-markets/. [Accessed: Sep. 19, 2020].
96. J. Oglevie and P. Rooney, "Physical security," in *Electric Power Substations Engineering*, J. D. McDonald (eds.), CRC Press, Boca Raton, p. 304, 2004.
97. M. Cova, C. Kruegel, and G. Vigna, "*Detection and analysis of drive-by-download attacks and malicious JavaScript code*," in *Proceedings of the 19th International Conference on World Wide Web, WWW '10*, 2010. doi: 10.1145/1772690.1772720.
98. B. Beard and B. Beard, "Differential backups," in *Beginning Backup and Restore for SQL Server*, Bradley Beard, Apress, USA, 2018. doi: 10.1007/978-1-4842-3456-3.
99. B. Gischel and B. Gischel, "Data backup," in *EPLAN Electric P8 Reference Handbook*, Hanser Publications. EPLAN, KG, Germany, p. 551, 2011.
100. J. R. Youngblood, "Ransomware," in *Business Theft and Fraud*, CRC Press, Taylor & Francis Group, Boca Raton, FL, 2016.
101. S. Mansfield-Devine, "Ransomware: Taking businesses hostage," *Network Security*, 2016. doi: 10.1016/S1353-4858(16)30096-4.
102. S. Liu and R. Kuhn, "Data loss prevention," *IT Professional*, 2010. doi: 10.1109/MITP.2010.52.
103. J. P. Farwell and R. Rohozinski, "Stuxnet and the future of cyber war," *Survival (Lond)*, 2011. doi: 10.1080/00396338.2011.555586.
104. Eclypsium, "Anatomy of a firmware attack," *Eclypsium*, 2019. Available: https://eclypsium.com/2019/12/20/anatomy-of-a-firmware-attack/. [Accessed: Oct. 28, 2020].
105. M. Shuttleworth, "ACPI, firmware and your security," 2014. Available: https://www.markshuttleworth.com/archives/1332. [Accessed: Sep. 21, 2020].
106. I. Andrea, C. Chrysostomou, and G. Hadjichristofi, "*Internet of Things: Security vulnerabilities and challenges*," in *Proceedings - IEEE Symposium on Computers and Communications*, 2016. doi: 10.1109/ISCC.2015.7405513.

8 Cyber-resilience

E. Bellini, G. Sargsyan, and D. Kavallieros

CONTENTS

8.1 Introduction 291
8.2 Cyber-resilience 295
    8.2.1 Risk vs Resilience 295
    8.2.2 Cyber-resilience Frameworks 297
    8.2.3 Toward a Cyber-resilience Holist View 304
8.3 Emerging Cyber-resilience Strategies Approaches 309
    8.3.1 Network-Based Approach (Prevention) 309
    8.3.2 Moving Target Defense (Protection) 311
        8.3.2.1 Identity-Based Randomization 312
        8.3.2.2 Non-Identity Randomization 312
        8.3.2.3 Infrastructure Randomization Technique 312
        8.3.2.4 Traffic Configuration Randomization Technique 312
    8.3.3 Game Theory Approach (Protection) 313
    8.3.4 Risk Perception and Epidemiological Based Approach
           (Absorption) 314
    8.3.5 Probabilistic and Adapting Modelling (Adapt) 316
8.4 Use Cases 317
    8.4.1 Transport Infrastructure 317
    8.4.2 Financial System 319
    8.4.3 Telecommunication 319
    8.4.4 Energy Sector 320
    8.4.5 Emerging Technologies 321
8.5 Recommendations 323
8.6 Conclusions 324
8.7 Acknowledgment 324
References 325

8.1 INTRODUCTION

Our environment is becoming saturated with computing, sensing, and communication devices that interact among themselves, as well as with humans: virtually everything is enabled to generate data and respond to appropriate stimuli (Internet of Everything—IoE). According to ERICSSON Mobility report [1], the number of connected devices is expected to exceed the number of mobile phones, and so is anticipated to revolutionize the way we do business, communicate, and live. This Cyber Physical World (CPW) convergence results in humans being deeply immersed in the information flows from the physical to the cyber world, and vice versa [2]. In fact, the cyber layer is even more the core of many critical functions of the society. Cyber infrastructures include environments that support advanced data communication (networks), acquisition (sensors), storage (hardware), management (database), mining (algorithms), computing and information processing services distributed over the Internet beyond the boundaries of a single institution.

The services being offered via platforms for enabling the IoE vision are becoming highly pervasive, ubiquitous, and distributed; machines, objects, and services become more intelligent and create a large-scale decentralized pool of resources interconnected by highly dynamic networks. Moreover, the lines between Information Technology (IT) and Operational Technology (OT) are blurring. As more and more objects are connected, communicate, and interact with each other, there has been a surge in the number of endpoints and potential ways for cyber criminals to gain access to networks and infrastructure systems. OT is becoming increasingly accessible, with threat vectors now extended to base-level assets as sensors. The issue here is that Cybersecurity is usually managed following an IT-driven approach. IT is fluid and has many moving parts and gateways, making it highly vulnerable and offering a large surface for a wide variety of constantly evolving attacks. In reality, the operational constraints in industry sectors such as energy, but also in a variety of others including defense, manufacturing, healthcare, and transport, require that an approach to Cybersecurity should include OT that belongs to the physical world and ensures the correct execution and control of all actions.

In fact, a cyber incident may affect more than one domain at the same time. For instance, in early 2009 the French navy's computer systems and internal network were reported to have been infected by a malware (Conficker virus) as a result of failure to install Microsoft updates.[1] Starting from the navy's internal network (Intramar) on 12 January, the virus reportedly spread and affected logistics and communication exchanges. Claims that the virus also affected aircraft on the ground, which were unable to download flight plans as the virus also affected databases, were denied by the French defense ministry.[2] This example demonstrates that cyber incidents could potentially create a domino effect, affecting more than one domain simultaneously.

There are already several examples of cyber-attacks that exploit the Internet-connected domestic appliances, vehicles, or smart vocal assistants, in order to, e.g., perform Denial-of-Service (DoS) or Distributed DoS (DDoS) attacks of unprecedented scales; spy on people in their office/homes and; take over (hijack) communication links, thus delivering full control of anything that is remotely controlled, such as drones, vehicles, and dams to cyber-criminals. For instance, in the health sector, potentially deadly vulnerabilities have been found in a large number of medical devices, including insulin pumps, CT scanners, implantable defibrillators, and X-ray systems. By using multiple sources to attack a victim, the mastermind behind the attack is not only able to amplify the magnitude of the attack, but can better hide their actual source IP address [3]. Although the methods and motives behind DDoS attacks have changed, their fundamental goal, namely to deny legitimate users resources or services, has not [4]. As explained in [5], the realization of these attacks is enabled by Internet Robots (Bots), software designed to perform repetitive jobs and that needs to be installed on the target system. The Bot then establishes a Command and Control (C&C) channel through which it is updated and directed. How a Bot infects a system varies with user interaction (e.g., downloading a legitimate program that has been altered to contain a malware), system misconfiguration, and system vulnerabilities and can be expressed in terms of probability [6, 7].

In the case of IoT, due to the security problems arising from embedded devices and other legacy hardware, whose flawed design (such as the use of hardcoded administrative passwords) or their poor configuration allows cyber-criminals to easily compromise them in order to form powerful botnets and launch DDoS attacks. Most importantly, there is often no efficient way to patch those devices. Many such IoT devices can be located by using new search engines, for example, SHODAN (www.shodan.io), and this offers cyber-criminals the opportunity to exploit existing vulnerabilities on a large scale. Compromised IoT devices may exhibit arbitrary behavior, and hence communication from any such device should be quarantined, or even rejected, by other systems taking over this responsibility. This will affect the performance of the service provided.

It is clear that such a technological evolution is making our society vulnerable to new forms of threats and attacks exploiting the complexity and heterogeneity of IoT networks, therefore rendering the Cybersecurity among the most important aspects of a networked world [8].

To this end, cybersecurity incidents targeting Critical Information Infrastructures (CIIs), which provide the vital functions that our society depends upon, are expected to have a significant negative economic and societal impact in the next decade and should be considered global risks.

In the United States, for example, the January edition of the *National Intelligence Strategy Report*[3] warns "Cyber threats posed an increasing risk to public health, safety and prosperity as information technologies are integrated into critical infrastructure (CI), vital national networks and consumer devices.". Addressing Congress, the US National Intelligence Director, Daniel Coats, put it even more succinctly: "The warning lights are blinking red."

A review report recently published by ENISA[4] estimates that the average annual losses due to cyber-crime among the European Union (EU) countries are 0.41% of their Gross Domestic Product (GDP); in some countries (e.g., Germany and Netherlands) the losses exceed 1.50% of the GDP leading to annual costs in the range 425 K–20 M per company. Among the CII sectors within the EU, significantly affected ones are the financial, information and communication technology (ICT), health, transport, energy, and public. DoS/DDoS, targeted, and web-based attacks, along with ransomware, have been reported to be among the most common types of cyber-attacks [9]. The availability of botnets for hire led to a noticeable increase in DDoS attacks, and it is very likely to see the IoT to further facilitate the formation of such botnet armies.

The number of zero-day [10], i.e., previously unknown and immediately exploitable, vulnerabilities discovered, has roughly doubled in the last few years; given their value, a rather mature black market has evolved that allows these vulnerabilities to be employed (until their exposure) in sophisticated targeted attacks. Hence, the deployment of proactive security and threat intelligence gathering/sharing systems could prove to be efficient in preventing such types of cyber-attacks. Web-based attacks exploit website vulnerabilities to infect users or gain access to sensitive private data, due to misconfigurations, usage of no/weak security protocols, or lack of proper patch management procedures. More than 75% of the websites have unpatched (critical) vulnerabilities, which may still be exploited several months after the vulnerability is revealed. This allows cyber-criminals to put little or even trivial effort in taking control over systems that can afterward be utilized in numerous ways [11]. This situation coupled with the capability offered by the IoE concept [12] aimed as a hierarchical tree of self-regulating and self-managing sub-systems and devices with high-density connectivity and very low system latencies, may create important stability problems. Considering possible crashes and cyber-attacks with the possibility of fast propagation of perturbances, and assuming the under-specified nature of performance conditions in the system [13], a certain level of epistemic uncertainty should be considered as a contribution to critical events [14]. This is transitioning the world away from stationarity and probability and toward a *world of possibility*. This destabilization is raising countless scientific questions around forecasting and the future evolution of critical systems, and equally challenging questions for practitioners around planning, preparedness, investment, and reliability.

The term *possibility*, in its technical signification, describes the condition of increasing unpredictability as systems become more complex and dynamic; in a colloquial sense, possibility signifies the opportunity (and perhaps, necessity) for alternative future states of the world and the range of pathways to get there. In fact, it is necessary to overcome the weakness of the current risk and efficiency-based approaches in complex systems safety and security addressing the so-called unknown unknowns [15]. Hence, to correctly understand the digital threats, it is necessary to focus the attention on events generated in the digital space (not necessarily caused by cyber-attacks), such as software glitches, hardware failures or disruptions, network failures or disruptions, loss of electrical power, accidental loss of confidential data, and, more in general, all the events similar in their consequences to cyber-attacks, in order to find and classify common causes and roots, identifying vulnerabilities with "local" effect and vulnerabilities with "systemic" effects to implement strategies for their mitigation, system recovery, and adaptation. In this respect, the concept of cyber-resilience and its operationalization can help to take these challenges.

This work is based on the current research on Cybersecurity performed in the context of Cyber-Trust EU project, where vulnerabilities and cyber-attacks in IoT are investigated to improve the response capability of a cyber-physical system (CPS) [16, 17].

8.2 CYBER-RESILIENCE

The concept of "cyber-resilience" has recently emerged in the Cybersecurity field as a result of the recognition that the traditional understanding of defense in cyberspace, built upon the notion that "a system that must defend against all possible attacks" [18], is unrealistic in what is a rapidly evolving threat landscape containing increasingly sophisticated levels of cyber-attacks. In particular, the underspecified nature of operations in complex systems not only generates potential for unforeseeable failures and cascading effects, but also it creates unexpected opportunities for intentional and unlawful acts of disturbance. Moreover, the existence of multiple and diverse subsystems with complex, nonlinear and sometimes hidden interactions combined with the new and emerging threats [19] (e.g. hybrid threats [20]) requires approaches able to cope with "unknown unknowns" as well as multiple dimensions and scales of analysis and actions.

In summary, failures in critical functions affect the level of trustworthiness on which the modern society is built on provoking unexpected side effects. Therefore, there has been an increasing interest in shifting emphasis from configuring compliance with respect to security practice and mainly focused on the reaction perspective, toward cyber-resilience. Cyber-resilience can help to answer questions as follows:

a. How to prevent large-scale vulnerabilities in IoT devices?
b. How to absorb (through adaptation) the impact of a large-scale propagation of the attacks after a vulnerability has been exploited while maintaining the network functionality at an acceptable level of performance (degraded mode)?
c. What kind of resources/assets are needed to make available to secure the system adaptive capacity?

8.2.1 RISK VS RESILIENCE

The literature most commonly defines cyber risk in terms of the likelihood of an undesirable event, and a measure of the impact of the event. The description from NIST Publication SP 800-30 (NIST) states:

> Risk is a function of the likelihood of a given threat-source's exercising a particular potential vulnerability, and the resulting impact of that adverse event on the organization. To determine the likelihood of a future adverse event, threats to an IT system must be analysed in conjunction with the potential vulnerabilities and the controls in place for the IT system.

ISO's definition of IT risk is similar: the potential that a given threat will exploit vulnerabilities of an asset or group of assets and thereby cause harm to the organization. It is measured in terms of a combination of the probability of occurrence of an event and its consequence (ISO/IEC 2008). The key components of cyber risk are relatively well understood. The likelihood of a successful cyber-attack can be empirically measured and estimated *a priori* with a degree of accuracy from known characteristics of a system or network [21]. Since cyber threats are difficult to quantify, current efforts shift from quantifying risk in specific units (like probability of failure) toward risk-based decision making using multicriteria decision analysis [22]. Moreover, Cybersecurity is usually segmented in assets and areas of concern as follows: electronic and physical security, information assurance (IA) and data security, asset management and access control, and so on. Each of the areas of concern is usually managed in its own silo, with standards, governing bodies, policies, and guidance documents geared specifically to a single area of concern. In addition, many of the specific standards, policies, and guidance that have been identified to address each area of concern are sector specific (e.g., energy, retail, banking, defense). Thus, a holistic approach that interweaves measures to address each area of concern is needed [23].

Unfortunately, "risk" and "resilience" are different concepts that are often conflated [24]. Unlike the concept of resilience, the concept of risk does not answer the questions of how well the system is able to absorb a cyber-attack, or how quickly and how completely the system is able to recover from a cyber-attack.

Thus, applying a risk-based approach to a problem that requires a resilience-based solution, or vice versa, can lead to investment in systems that do not produce the effects that stakeholders expect. In particular, risk management considers effort to prevent or defuse threats before they occur, but whenever risks are identified and actions taken to reduce risk, there still remains residual risk. As such, resilience assessment and management are, in part, an effort to address that remaining known, but unmitigated, risk as well as enhance the overall ability of the system to respond to unknown or emerging threats. In other words, it represents the low-probability/high-impact events, sometimes called "fat-tail" events, that are difficult to account for and even harder to predict.

Moreover, cyber-resilience is not only about resisting the breach, but rather, it is also about learning from the breach attempt and adapt to the changing conditions. Resilience should not be confused with another important concept as robustness, even if they are tightly correlated. Robustness denotes the degree to which a system is able to withstand unexpected internal or external threats or change without degradation in system's performance. Thus, robustness focuses specifically on performance not only under ordinary, anticipated conditions but also under unusual conditions that stress its designers' assumptions. Hence, robustness determines how much damage is incurred in response to an unexpected disturbance but does not consider how quickly the system can recover from such damage as resilience does. In particular, a system that lacks robustness will often fail beyond recovery, hence offering little resiliency. Both robustness and resiliency, therefore, must be understood together. The objectives are to identify system vulnerabilities and threats, estimating the potential impact and invest resources to set up barriers and countermeasures.

Resilience management instead, accepts the possibility of system failure and focuses on its mitigation, recovery and adaptation [25]. Moreover, as mentioned in [26] resilience analysis additionally delves into the unknown, uncertain, and unexpected at the scale of systems rather than individual components. In other words, where traditional risk assessment methods seek to harden a vulnerable component of the system based upon a snapshot in time, resilience analysis instead seeks to offer a "soft landing" (graceful degradation) for the system at hand. Resilience management is the systematic process to ensure that a significant external shock—i.e., hackers to cybersecurity—does not exhibit lasting damage to the functionality and efficiency of a given system. This philosophical difference is complex yet necessary in the face of the growing challenges and uncertainties of an increasingly global and interconnected world. Resilience analysis differs temporally from traditional risk analysis by considering recovery of the system once damage is done. Thus, in addition to considering system decline immediately after an event (i.e., risk), resilience adds consideration of longer-term horizons that include system recovery and adaptation. Traditional risk analysis can integrate recovery and adaptation (for example, by considering probability of system to recover by specific time after event or likelihood that it will be able to adapt), yet this is not necessarily the prime focus of the overall risk analytic effort. Instead, a traditional risk analysis project constructs the ideal set of policies that, given available money and resources, would offer the best path forward for risk prevention and management. Attention to longer term and lower probability threats is often neglected in favor of more intermediate and likely dangers, with only limited emphasis or focus on the need for infrastructural and organizational resilience building, in the face of uncertain and unexpected harms. In this way, traditional risk assessment may not accurately or adequately prepare for those low-probability yet high-consequence events that could dramatically impact human and environmental health or various social, ecological, and/or economic systems that have become ubiquitous within modern life.

Cybersecurity, of course, refers to protecting computer systems, networks, and other information-technology infrastructure, and the data that they all house, from disruptions, theft, modification, or damage. People have a tendency to view cybersecurity from primarily a proactive standpoint—Cybersecurity is often described and measured by how well it prevents various forms of deliberate maleficence by attackers. Major principles of Cybersecurity, therefore, traditionally focus on areas such as authenticating users, implementing least-privilege authorization, layering security countermeasures to create perimeters and zones, coding with security in mind, and the like.

Cyber-resilience, on the other hand, measures how well an entity can continue operating—and delivering its goods and services as intended and expected—regardless of cyberattacks, technical failures, and other significant cyber-disruptions of normal business processes. Major principles of cyber-resilience, therefore, tend to focus on areas such as business continuity planning, implementing secure redundancy for critical business processes, continuously examining potential attack surfaces, identifying and assessing attackers' actions within compromised computer-infrastructure, reacting to attacks, cleaning up and restoring normal operations after a breach, etc.

In short, Cybersecurity is primarily about protecting, and cyber-resilience is about surviving and thriving when that protection fails, as it inevitably will at some point in time. In this respect, Cybersecurity can be seen as the first step in cyber-resilience meaning any cyber-resilience strategy must encompass Cybersecurity.

8.2.2 Cyber-resilience Frameworks

Resilience is a multifaced and not yet standardized concept even more related to complex and adaptive socio-technical systems [27, 28], and with a number of definitions and assessment methods exist [29, 30]. However, the introduction of resilience in the cyber domain is recent.

In particular, the cyber aspect of resilience has been highlighted in the Accenture report [31] in which the NIST Cybersecurity framework [32] has been combined with resilience engineering principles [14, 33–36]. In [35], cyber-resilience is considered in the context of complex systems that comprise not only physical and information but also cognitive and social domains. Cyber-resilience ensures that system recovery occurs by considering interconnected hardware, software, and sensing components of cyber infrastructure. It thus constitutes a bridge between sustaining operations of the system while ensuring mission execution.

The notion of "resilience thinking", introduced in [37], offered a way of understanding such system complexity and a new approach to manage their resources to continually adapting through cycles of change in order to achieve sustainability. By definition, resilience is about sustaining adaptability capacities in the pursuit of system purposes. Generating, maintaining, and deploying adaptability processes relies on the allocation of a wide range of resources and at many different system levels and time scales. Hence, adaptability capacities are intrinsically related to the level of resources that a system can allocate and its ability to manage these resources in view of specific adaptive cycles.

The concept of the adaptive cycle for systems is derived by the panarchy theory [38] grounded on ecosystem dynamics. A panarchic complex system [39] includes a dynamic set of adaptive cycles that evolve, irreversibly and uniquely, as it adapts to new demands, disturbances, and crisis. An adaptive cycle is here described based on the four stages of event management cycle that a system needs to maintain to be resilient (Prepare, Absorb, Recover, Adapt), as proposed in [40]. Starting from this baseline a cyber-resilience framework has been elaborated—including those from the UK's National Cyber Security Center (NCSC)[5] and the US's National Institute of Standards and Technology (NIST)[6] —which, although worded differently, tend to revolve around five key areas—prepare, protect, absorb, recover, and adapt, such as:

- **A1—Plan/Prepare:** Lay the foundation to keep services available and assets functioning during a disruptive event (malfunction or attack). Prevention will always be better than cure, and to prevent cyber-attacks and data breaches requires a multilayered approach to cyber-resilience that includes technology, people, and processes. This will include putting in place comprehensive security policies and providing cyber-resilience training and in-work support to ensure that everyone knows their role.
- **A2—Protect:** Cybersecurity falls within the protect step of cyber-resilience. In addition to basic security software such as firewalls, more sophisticated solutions like endpoint detection and response (EDR) solution provides a far greater degree of protection. In addition to EDR tools, an Endpoint Protection Platform which delivers next-generation endpoint protection solutions and integrates with DNS protection, security awareness training, and data protection layers for even greater cyber-resilience levels
- **A3—Absorb:** Maintain the most critical asset function and service availability while repelling or isolating the disruption. One of the major end goals for

cyber-resilience is building durability into the organization when an attack occurs. At this stage of cyber-resilience, organizations often adopt a single platform for their data and content, providing a single source of the truth for all information that is easier to protect. Adding content management and cloud collaboration means data can be quickly isolated and quarantine while other systems and data remain available.
- **A4—Recover:** Restore all asset function and service availability to their pre-event functionality. Returning to normal after an attack is the ultimate goal of your cyber-resilience strategy. If a successful ransomware attack locked down all your data, the results can completely stop the business from operating. To avoid such a situation, an effective data back-up and recovery is an essential part of cyber-resilience.
- **A5—Adapt:** Using knowledge from the event, alter protocol, configuration of the system, personnel training, or other aspects to become more resilient. Adaptability is a key component of cyber-resilience. Network and security solutions that leverage up-to-the-minute threat intelligence ensures that a network can automatically adapt to the latest threats. This sort of intelligence integrated into a SIEM or other tools within your Security Operations Center also allows you to understand the current threats to your network and data, as well as make accurate predictions about likely attacks in the future.

Adaptive cycles cannot be dissociated from system performance variability, as they are simultaneously the system mechanisms to cope with variability and inevitably, an important source of variability in them. This renders variability a key system performance aspect to be managed toward resilience. As resources are inevitably and always limited, so are the capacities for adaptability, and thus the types and amplitudes of variability that a system can cope with are bounded by such limitations. The challenge for system resilience resides then in the ability to understand and continuously monitor resources and the capacities that they provide toward coping with both expected and unexpected, known and unknown sources, types, and amplitudes of performance variability.

Then a cyber system needs to be considered stochastic, dynamic, and unpredictable in nature and resilience should be in direct relation with how a system performs, and how capable it is in controlling performance and coping with changing conditions. In this sense, Hollnagel and Woods [41] consider that only the potential (adaptive capacity) for resilience can be observed and quantified and not resilience itself. In fact, according to the resilience engineering approach, resilience is more precisely defined as "the intrinsic ability of a system or organization to adjust its functioning prior to, during, or following changes, disturbances, and opportunities so that it can sustain required operations under both expected and unexpected conditions" [42]. According to this view, the potential for resilience emerges from system performance and it is assessed based on the "four resilience cornerstones" [43], namely:

- **B1—Respond** (Knowing what to do): capacity to respond to disruptions by adjusting system performance to changing conditions.

- **B2—Monitor** (Knowing what to look for): capacity to monitor both the system and the environment to reduce uncertainty.
- **B3—Anticipate** (Knowing what to expect): capacity to anticipate opportunities for changes in the system identifying sources of disruption and pressure as well as their consequences for system operation.
- **B4—Learn** (Knowing what has happened): capacity to learn from past experiences either successful or not.

Such cornerstones have been implemented in [33, 44, 45]. In particular, the capacity to monitor was based on the actual possibility to collect, reconcile, process, and mining a huge amount of multimedia heterogeneous data (Big Data) from urban IoT infrastructures such as traffic sensors, water level in the underpasses, local weather, people movement on the ground, and so forth. The data collected have been used to take informed decisions (respond) and to predict future trends (anticipate, learn). In that case, the reliability of the released Decision Support System based on a complex IoT network, relied on the possibility to assess data quality and to perform specific data cleaning techniques in real time to sustain the DSS operativity. Thus, the cyber-resilience of the systems has been operationalized in such a way to maintain the quality of the data within a proper level to continue to take reliable decisions. Significant reduction in data quality caused by sensors' malfunction or attack is managed through a graceful degradation of the decision reliability communicating the actual level of uncertainty. Data coming from Byzantine sensors are temporarily excluded from the decision computation.

In certain cases, such data can be replaced with others coming from nearby sensors that are still considered reliable.

In [46], an end to end cyber-physical attack-resilient framework for a Wide Area Monitoring, Protection, and Control (WAMPC) is reported. The framework is composed by the following steps: Risk Assessment, Prevention, Detection and Mitigation/Resilience.

- **C1—Prevention:** would be achieved through a combination of multiple approaches as quantitative risk assessment, attack-resilient measurement design, and moving-target inspired algorithms. Quantitative risk assessment involves modeling all the components of risk, namely, threats, vulnerabilities, and impacts, using approaches such as probabilistic or game-theoretic modeling. Attack-resilient measurement design involves algorithms that identify and recommend redundant measurement deployments that could be fed additionally into the WAMPAC applications, thereby increasing the difficulty of creating successful attacks. The optimal selection of redundant measurements is achieved by formulating a design problem that optimizes the placement of new sensors (e.g., PMUs) such that the accuracy, bad-data detection capability, and observability of the system improve while satisfying cost constraints. Moving-target inspired algorithms leverage a redundant measurement design and randomize the associated design parameters at the WAMPAC algorithm level while still ensuring that the functionality of the algorithm is maintained.

- **C2—Detection:** In the proposed framework, CPS model-based anomaly detection approaches that leverage sound mathematical tools from machine learning and related domains are critical for the detection of cyber-attacks beyond traditional IT intrusion detection techniques. Additionally, specification-based anomaly detection approaches serve a complementary role to CPS model-based approaches and enable the reduction of false-positive rate by capturing the normal behavior and existing security policies as part of a formal specification or language
- **C3—Mitigation:** In the proposed framework, attack mitigation/resilience would be achieved using CPS model-based mitigation, and dynamic system reconfiguration and resiliency algorithms appropriately. One of the aspects of attack mitigation is the ability of the attack-resilient algorithm to recover from faults, either partially or completely. Based on the output of the anomaly detection module, if the data are considered "anomalous," a CPS model-based mitigation method would be triggered. For example, if the SCADA measurement data used by the Automatic Generation Control (AGC) application are found to be untrustworthy, then the control signal would be calculated based on a statistical model that uses short-term load forecast information along with system generation parameters to predict its most likely value for a particular balancing authority area until a redundant, trusted source of SCADA telemetry is restored.

According to CISCO [47], Cyber-resilience is multidisciplinary and requires an organization to address multiple capabilities. The company report identifies seven major capabilities that support cyber-resilience goals such as:

- **D1—Identification:** definition of critical assets. An organization must know which devices provide its cyber competences.
- **D2—Protection:** supporting organization ability to limit or contain the impact of the cyber-attack. Formally, protection is defined as the policies, processes, and mechanisms for ensuring that a system is built and operates in a state of integrity during a cyber-attack or similar event, and that it is defended from modification by unauthorized or unauthenticated processes. During a cyber-attack, the system can be trusted to act as designed.
- **D3—Detection:** It is defined as the policies, processes, and mechanisms to measure, collect, verify, and analyze system integrity.
- **D4—Recovery:** It is defined as the policies, processes, and mechanisms for restoring a system to a state of integrity. Recovery is a fundamental distinction between Cybersecurity and cyber-resilient systems, and it supports the timely recovery to normal operations to reduce the impact of a cyber-attack. The goal of a recoverable system is to restore the normal operation of a platform, application, or service if it becomes corrupted through a cyber-attack or misconfiguration.
- **D5—Visibility:** It supports the continuous monitoring and tracking of the state of integrity and provides continuous awareness to an external entity of the system's state of integrity. Protection, detection, recovery, analytics, and forensics are all supported by a visibility mechanism.

- **D6—Analytics:** Analytics gathers and examines cyber-attacks to bring situational awareness and to identify risks
- **D7—Forensics:** It is focused on the ingestion of relevant support data, along with preserving, processing, analyzing, and presenting system-related evidence in support of recovery. Historical information is analyzed to search for anomalous events (reverse engineering).

In [48, 49], cyber resiliency enables organizations to maintain critical functions in the face of persistent, stealthy, and sophisticated attacks focused on cyber resources, as well as during the course of accidental or natural disasters. The Cyber-resilience goals are:

- **E1—Anticipate:** It includes activities for prevention and avoidance precluding the successful execution of an attack or the realization of adverse conditions. Moreover, it maintains a set of realistic courses of action that address predicted or anticipated adversity and a useful representation of mission and business dependencies and the status of resources with respect to possible adversity.
- **E2—Withstand:** It is related to the maximization of the duration and viability of essential mission or business functions during adversity and limit damage.
- **E3—Recover:** Restore as much mission or business functionality as possible after adversity.
- **E4—Evolve:** It supports transformative actions as the modification of mission or business functions and supporting processes to handle adversity and address environmental changes more effectively.

In IOSCO guidance [50], the "cyber-resilience" is considered an ability to anticipate, withstand, contain, and rapidly recover from a cyber-attack. Financial Market Infrastructures, which facilitate the clearing, settlement, and recording of monetary and other financial transactions, play a critical role in fostering financial stability.

- **F1—Identification:** It is related to identify and maintain an inventory of the information assets and understand its processes, procedures, systems, and other dependencies to strengthen its overall cyber-resilience posture. An organization should also identify its business functions and supporting processes and conduct a risk assessment in order to ensure that it thoroughly understands the importance of each function and supporting processes, and their interdependencies, in performing its functions.
- **F2—Protection:** Cyber-resilience depends on effective security controls and system and process design that protect the confidentiality, integrity, and availability of an FMI's assets and services. A protective controls should enable the monitoring and detection of anomalous activity across multiple layers of the FMI's infrastructure, which requires a baseline profile of system activity.
- **F3—Detection:** It refers to the ability of recognizing signs of a potential cyber incident or detect that an actual breach is taking place. Early detection provides valuable time to apply appropriate countermeasures against a potential breach,

and allows proactive containment of actual breaches. In fact, early containment could effectively mitigate the impact of the attack—for example, by preventing an intruder from gaining access to confidential data or exfiltration of such data. Given the sophisticated nature of cyber-attacks and the multiple entry points through which a compromise could take place, an FMI should maintain effective capabilities to continuously monitor (in real time or near real time) and detect anomalous activities and events.

- **F4—Response and recovery:** FMI's arrangements should be designed to enable it to resume critical operations rapidly, safely and with accurate data in order to mitigate the potentially systemic risks of failure to meet such obligations when participants are expecting it to meet them. Continuity planning is essential in meeting related objectives. While an organization should plan to safely resume critical operations within a predefined time of a disruption, they should also plan for scenarios in which this objective is not achieved. Organizations should analyze critical functions, transactions, and interdependencies to prioritize resumption and recovery actions, which may facilitate the processing of critical transactions, for example, while remediation efforts continue.
- **F5—Testing:** All components of a cyber-resilience framework should be rigorously tested to determine their overall effectiveness before being deployed within an organization, and regularly thereafter. This includes the extent to which the framework is implemented correctly, operating as intended and producing desired outcomes. Understanding the overall effectiveness of the cyber-resilience framework in the organization and its environment is essential in determining the residual cyber risk to the organization's operations, assets, and ecosystem.
- **F6—Situational Awareness:** Situational awareness refers to the understanding of the cyber threat environment within which an organization operates, and the implications of being in that environment for its business and the adequacy of its cyber risk mitigation measures. Strong situational awareness acquired through an effective cyber threat intelligence process can make a significant difference in the organization's ability to predict cyber events or respond promptly and effectively to them. Specifically, a keen appreciation of the threat landscape allows a better understanding of the vulnerabilities in its critical business functions, and facilitate the adoption of appropriate risk mitigation strategies. It can also enable an organization to validate its strategic direction, resource allocation, processes, procedures, and controls with respect to building its cyber-resilience. A key means of achieving situational awareness for an organization and its ecosystem is the active participation in information-sharing arrangements and collaboration with trusted stakeholders within and outside the organization.
- **F6—Learning and evolving:** Organizations should systematically identify and distill key lessons from cyber events that have occurred within and outside the organization in order to advance its resilience capabilities. Useful learning points can often be gleaned from successful cyber intrusions and near misses in terms of the methods used and vulnerabilities exploited by cyber attackers.

An organization should actively monitor technological developments and keep abreast of new cyber risk management processes that can effectively counter existing and newly developed forms of cyber-attack. An organization should consider acquiring such technology and know-how to maintain its cyber-resilience.

8.2.3 Toward a Cyber-resilience Holist View

The frameworks introduced above have wide overlaps and share similar concepts and definitions that can be accommodated in a unique view to increase the common understanding. From the analysis, it is possible to identify two main high-level concepts that sometimes are used as a synonym of resilience as Adaptive Capacity and the Coping Ability, but it is clear that they refer to two different aspects of resilience. Moreover, nevertheless these two concepts are clearly interrelated, they have not yet been integrated as a part of a unified vision of Resilience. A preliminary attempt has been provided in [45, 51] where resilience is obtained through a dynamic process between them through which successful performance is continually pursued, rather than a stable outcome obtained through a static characteristic of the system [52]. According to this view, the Adaptive Capacity represents the potentiality of the system that needs to be continuously built along the four resilience cornerstones categories in order to enable the full expression of the system Coping Ability (the adaptive cycle) in terms of survivability and adaptability (see Figure 8.1.

Moreover, in order to disambiguate and reconcile the elements composing the frameworks concept clustering (see Table 8.1) and presented a mind map able to model the domain are provided (see Figure 8.2).

Coping Ability requires specific Adaptive Capacities to be enabled and therefore, a subset of the all the possible assets available in a system has to be recognized and selected. The Adaptive Capacities identified are Anticipating, Responding, Monitoring, and Learning. The capacity of **Anticipating** is related to the presence of assets in the organization to implement risk-informed strategies and plans, to perform analysis, and exploit the results of the continuous learning activities in order to anticipate opportunities for changes. The capacity of **Responding** is related to the

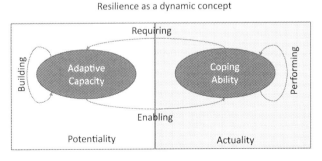

FIGURE 8.1 Resilience as an emergent property from Adaptive Capacity and Coping Ability interaction dynamics

TABLE 8.1
Framework Concepts Clustering

Adaptive Capacity				Coping Ability				
Anticipate	Respond	Monitor	Learn	Prevention	Protection	Absorption	Recovering	Adaptation
B1, F1, F5	B2, D3	B3, D5, D6 F2, F6	B4, D7, F6	A1, C1, D1, E1, F1	A2, C2, D2, D3	A3, C3, E2	A4, D4	A5, E4

presence of assets able to avoid disruption or mitigate the impact of adjusting system performance to changing conditions. The capacity of **Monitoring** is related to the presence of assets capable to collect all kind of data to allow early warnings, increase situation awareness, reducing uncertainty, and support real-time decision-making.

The **capacity of Learning** is related to the presence of assets able to collect, process, and learn from evidences generated by the system. It is based on forensic analysis as well as on a continue assessment on how the system itself behaved. This capacity is secured by using de-briefing, ex-post analysis, big data platform, and machine learning techniques, etc.

The **Asset** element models the main elements considered in the network such as services, people, infrastructures, etc. Four fundamental sub-classes of **Asset** have been identified:

- **Human:** it includes technical skills, expertise, and competencies (knowledge), as well as cognitive resources, particularly those relating to decision-making processes. These resources should be investigated within all relevant operational and managerial contexts. From an end-user perspective, both individual and collective behaviors (i.e., risk awareness and perceptions, risk aversion, among other aspects) are critical factors to be considered, as they may critically impact on the effectiveness and application of key outputs.
- **Technology:** it comprises ICT as well as built artifacts and infrastructures as implemented by utilities the energy, oil & gas and water networks, transport networks, signaling systems, engines, traffic control, and ticketing related assets.
- **Organization:** it includes hierarchical structures and formal procedures and regulations, as well as logistics elements. Information use and communication should also be taken into account, as they constitute the key resource for every decision-making process.
- **Finance:** it includes the amount of financial resources available in the system generated by the economic activities, the amount of capital invested, its value, its risk managed (assurance), and so forth.

In the reality, an **Asset** is exploited by a system function to deliver a service. In resilience, we focus the **Critical Functions** as those they are vital for the service survivability. The instance of this entity could have several levels of granularity, depending on the subject of the analysis adopted by the decision makers/analyst. For

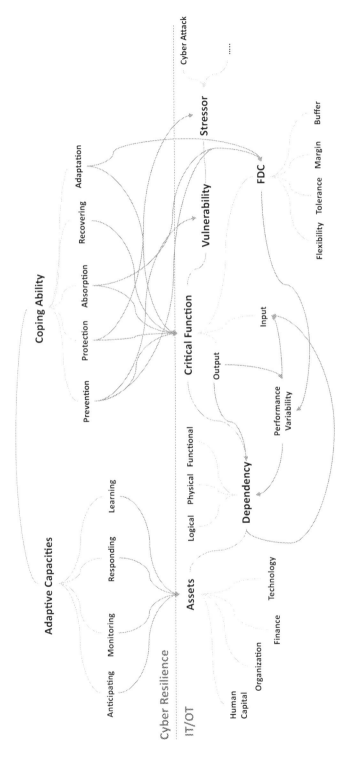

FIGURE 8.2 Mind map of Cyber-resilience Holistic View

instance, the Urban Transport System (UTS), can be considered a Critical Function adopting the city as a subject of analysis. However, within a UTS there are other critical functions that can be considered while the subject of analysis is the UTS [33, 53].

The characterization of the Critical Function has been inspired by the Resilience Engineering field and in particular from the Functional Resonance Analysis Method [41, 42] and further elaborations [53]. According to [53], the **Critical Function** has a property called **Function Damping Capacity** (FDC), considered of the key indicators to quantify resilience. FDC is defined in [44] as the capacity of the function to absorb the performance variably generated by a **stressor** that exploits **vulnerabilities** and that is propagated through the **dependency** from the upstream function **output** to the downstream function **input** [52]. The input to a function is traditionally defined as that which is used or transformed by the function to produce the output. The Input as well as the output can represent matter, energy, or information.

If the inbound input variability exceeds the FDC, the quantity not absorbed is propagated through its **Dependencies** to the other function and so on. The propagation stops as soon as the residual variability meets a function capable to absorb it and provides an output without variation. If the variability cannot be stopped, the system is going to experience the so-called *functional resonance* that leads to a disruption The FDC is a property related to the Adaptive Capacity through the functional dependency that connects some specific assets to the critical function that use them. FDC is also related to the Coping Ability, since the **Absorption** performance to damp the inbound variability is defined by the FDC boundaries. The FDC is composed of the following sup-properties: **Buffer Capacities** (BC), **Flexibility** (FX), **Margin** (FA), and **Tolerance** (TO).

The **BC** is referred to the type or amount of disruption that a system can absorb or adapt to without the appearance of a basic failure in its performance. An example of BC is the redundancy. It reduces the consequences of loss of information or services, facilitates recovery from the effects of an adverse cyber event, and limits the time during which critical services are denied or limited. However, it can degrade over time as configurations are updated or connectivity changes. Moreover, redundancy needs to be implemented in such a way that the backup is not exposed to the same threats of the primary service and can include physical as well as functional redundancy. Another aspect that contributes to the buffer capacity is the configuration of the systems. For example, modular electricity grids can have high modularity and therefore interact more among themselves than nodes in other systems. In systems with low modularity, a disturbance to one component may cascade quickly to other components and lead to the collapse of the entire or large portion of the system. In contrast, the ability of highly modular systems to "restrain" or "buffer" a shock without allowing its spread to the global network is a beneficial property [24].

The **FX** is related to the capacity of balancing between stability and flexibility, while some measure of centralized control is needed to ensure an efficient operational coordination, a certain degree of local autonomy in decision making (e.g., edge computing) is necessary to adjust tasks to ever-changing operating condition. The FX can be implemented by securing diversity. It is a well-established resilience technique; it removes single points of attack or failure and allows multiple options for reconfiguration. However, architectures and designs should take cost and manageability into consideration to avoid introducing new risks. For instance, if the use of different chips or devices reduces the risk of attack propagation through the exploitation of the same vulnerability, on the other hand, it increases the surface of attack.

The **MA** is related to how close the operating system is to its performance boundary. Margins are related to tangible and intangible resource availability (e.g., skills, time) and can be increased leveraging the assets related to the adaptive capacity. For instance, the monitor and respond capacity are secured by the implementation of specific detection and protection measures as AI-based anomaly detection algorithms and next generation of firewalls. Such measures offer the possibility of having more time to respond thanks to the attack early detection and the capacity of protection measures to resist/withstand. In this way, the system has margin of maneuver for a full re-configuration in order to mitigate the impact in case the protection measures will not be able to repel the attack. Margin is also secured by managing the trustworthiness of the system components. The limitation of the number of system components that need to be trusted to guarantee the system reliability allows an effort-effective approach for ongoing monitoring as well as respond and recovery. Moreover, continuous verification and/or validation of the integrity or correctness of data or software as well as the behavior of individual users, system components, and services can improve the early warnings effectiveness and triggers responses such as closer monitoring, more restrictive privileges, or quarantine.

The **TO** is related to how the system works near the boundary. Systems and system components, ranging from chips to software modules to running services, can be compromised for extended periods without detection. In fact, some compromises may never be detected. Systems must remain capable of meeting performance and quality requirements nonetheless. Moreover, strategies to create robust and adaptable feedback loops capable of enabling infrastructures to absorb shocks and avoid instabilities are needed.

In fact, increasing tolerance means that the system can be favorably adjusted or gracefully degrades as stress/pressure increase instead of collapsing [54]. The TO can be implemented through segmentation technique to rapidly exclude the affected components degrading instead of suspending the service functionality. Another emerging technique is represented by bioinspired approach introduced in Section 1.6, where the nodes of a system autonomously decide to allow or block communication with their neighbors on the base of a local estimation of the probability of the infection. The objective is to continue to provide functionality even if a self-reduced mode.

The **Coping ability** is divided into sub-abilities: **Prevention, Protection Absorption, Recovery, and Adaptation.** In the reality, such ability is exhibited subsequently, so that the activity of prevention is continuously performed until an event occurs. At that point, the Protection, Absorption, Recovery, and Adaptation should be executed sequentially. In particular, the **Prevention** ability it includes the identification of the critical functions and assets, the definition of policies, procedures, and mechanism to secure both, protection and resilience. **Protection** refers to the ability of detecting anomalies and applying countermeasures to avoid compromising the system. If the protection fails, the **Absorption** ability should come in to play. The system is compromised but the functionality needs to be sustained or degraded gracefully. Absorption dynamically manipulates the Functional Damping Capacity of the functions to absorb the system variability triggered by an attack in order to mitigate the impact of the event on the service performance. Once the variability generated by an event (e.g., cyber-attack) is under control, should start the recovering phase. The **Recovering (bounce back)** is related to the ability exploiting remaining system capacity to recover mission or critical functionality to the minimum accepted performance level as soon as possible in a safe way. There is a balancing aspect to be considered here: the need to restore on one side and the need of being effective in doing this. In this respect, the recovery should be safe first and then as fast as possible. Putting safety at first has two main implications: a) safe recovery means maximizing the effectiveness of the action in relation to the resources (assets) mobilized, and b) minimize the probability of a new disruption during the recovery phase that would waste the resources used and quickly bring the system into a condition of scarcity such as to be fatal for recovery operations.

The **Adaptation** (bounce forward): through the combination of restoration strategies, e.g., repairing the failed elements (bounce back) and building new elements (bounce forward) as a result of adaptation and learning process, the infrastructure can achieve a higher resilience with respect to the pre-disruption conditions. The two actions may happen consequently (recovery first then adapting) or could be mutually exclusive since can be implemented both at the same time with the same costs. In this case, a decision needs to be taken according to a high-level mid-term strategy instead of an immediate business continuity or recovery plan perspective. The quality and the effectiveness of adaptation are based on the level of Learning capacity present in the system. In fact, if old-fashioned approaches are applied (e.g., not using a data-driven approach), the adaptation will result limited exposing the system to similar stressor in the future.

8.3 EMERGING CYBER-RESILIENCE STRATEGIES APPROACHES

Resilience management should follow several approaches and strategies depending on the level of rigor required and amount of resources available considering costs, data, and time for their implementation and the related purpose [55] such as [14, 33, 44, 56–59].

In this section, we identified several emerging technologies and approaches used to operationalize cyber-resilience and associated with the related Ability.

8.3.1 Network-Based Approach (Prevention)

In the domain of IoT, the concept of resilience is very new and its understanding varies according to the perspective adopted about the nature of IoT. For instance, in [60], a resilient IoT system should deal with a number of threats occurring when a system is deployed on the Internet [61] in order to safely, and quickly recover to provide normal service. Literature on self-healing systems goes even further, not only recovering normal behavior, but also addressing the vulnerability that led to faulty behavior [62]. Thus, resilience relates to fault-tolerant systems, dependable and trustable systems, and reliable and available systems, where each category of applications adopts its specific terminology and angle of interest.

IoT can be considered as techno-social systems, combining infrastructures, devices, and people [12]. Thus, to address IoT resilience, it is necessary to study the resilience of complex systems (networks). Resilience was defined as a property of networked system in [63]. This framework adopts the definition of resilience given by the US National Academy of Sciences [64] and further discussed in [65]:

The relation between system resources, the capacities that they provide, and variability can only become meaningful when placed in the context of the interdependencies that are on the one hand used to enhance resources and capacities but also, on the other hand, require an allocation of resources to be maintained [33, 66]. Interdependencies are the means through which system functions can act in order to secure the envisaged levels of resources needed to fulfill their purposes. To this end, the principles of network science are proposed as a tool to quantify resilience by supporting system interdependencies in view of the volume and nature of available resources and the capacities that these provide. At the operational side, resilience can be quantified as an emerging property of an interdependent network system. As such, resilience assessment would identify critical functionality of the system [14, 34, 44] and evaluate the temporal profile of system recovery in response to adverse events, while resilience management would allow comparative evaluation of cross-domain management alternatives. Resilience assessment may be approached with network theoretic frameworks [63]. A network is composed of a set of nodes (N), connected by a set of links (L). The specification of N and L includes characteristics relevant to resilience (e.g., capacity, geographical location, weight, temporal response) of each node. The adverse events and failures can percolate over a network in different ways. The network control C captures the system's temporal evolution, including adaptive algorithms, and can also be defined as a way to manage temporal changes in the network. Ultimately, the system must maintain its critical functionality (CF) that reflects the state of the system as a characteristic (or a weighted sum of such characteristics) that changes with time and is of interest to the user of the model. In [62], Resilience (R) is then defined as a composite function of nodes, links, and control with respect to the critical functionality of the system and a class of disturbances as

$$R = R\_(CF,P) = f(N,L,C)$$

Due to the very complex nature of networked systems and the large number of variables defining their states, it is not possible to obtain a closed-form expression for R.

To quantify R, one of the best approaches is represented by simulations [62], each of which represents a certain network percolation [67] scenario from an infinite set of possible scenarios of the network evolution. For each simulation, it is possible to calculate the average value of the critical functionality at every time step. Resilience is then quantified as the integral of the system functionality during a disruption, normalized to its normal functionality.

8.3.2 Moving Target Defense (Protection)

In most cases, administrators are configuring their network (e.g., smart building or smart CI) in a static manner. This configuration provides great simplicity in an already complex ecosystem (IoT devices, industrial devices, etc.)[68]. Moving Target Defense (MTD) is another technique that can be used for real-time countermeasure against cyber-attacks. The main goal is to confuse and deceive the attackers by various means, such as dynamically changing the network topology while at the same time it does not affect the system's functionalities. For example, in a smart grid, it will not affect the power supply generators or the load dispatch. In this manner, the risk of the attacker tampering the data is minimized [69].

MTD techniques (or strategies) can be classified in five main categories [70, 71].

- Dynamic data: the techniques under this category can change the representation properties of data like the format and encoding.
- Dynamic software: the techniques under this category can change the code of applications dynamically.
- Dynamic platforms: the techniques under this category provide diversity by changing between platforms dynamically. To achieve that, it utilizes the characteristics of various platforms such as the storage system and the architecture of the processors
- Dynamic runtime environment: the techniques under this category can change the environment during runtime.
- Dynamic network: the techniques under this category can dynamically change the network settings like the network topology, IP addresses, etc.

To effectively apply the MTD technique, three questions must be answered, What, When, and How the "components" of the network/system, etc., will be moved [72].

- The What represents the components that will be moved when an attack occurs (e.g., IP addresses will change dynamically)
- The When represents the time that the change will occur. In most occasions, it is either on predefined time or upon the manifestation of an event (e.g., cyber-attack) and finally a combination of both.
- The How represents the method that will be employed to move a component. The three predominant methodologies that can be utilized are diversification, randomization, and redundancy based.

The application of MTD in the IoT domain is new. Due to the constraints of most IoT devices (IP cameras, home appliances, well-being and health care devices, industrial IoT devices, etc.) in terms of computational power, storage capacity, and RAM limitations the user cannot install antiviruses, IDS, or IPS and thus these devices are targeted. Due to these limitations, the most well-established technique in the IoT domain is the Dynamic network [70]. Based on current work [73–79] it has been proposed the following classification for the MTD in the IoT domain.

The dynamic network technique in the IoT ecosystem has two main categories: the identity-based randomization and the non-identity randomization. The latter is further divided into two categories: the infrastructure randomization and the traffic configuration randomization.

8.3.2.1 Identity-Based Randomization

The identity-based randomization is mainly focused at the logical and physical addresses of the IoT devices in the network. The predominant method is the address shuffling in which, the addresses of the devices are changing regularly. This confuses the attackers because they will either attack the wrong target or they will reach to a dead end. Based on the experimentation and measurements presented in [77] it is possible to shuffle the addresses in IoT ecosystems.

8.3.2.2 Non-Identity Randomization

The non-identity randomization is divided into two categories as it was depicted earlier in this chapter.

8.3.2.3 Infrastructure Randomization Technique

The infrastructure randomization technique, as the name implies it is based on the dynamic and periodic change of active infrastructure. For example, in a Wireless Sensor Network (WSN) to safeguard the base-station, it is assumed that multiple base-stations exist while only one is active each time. Periodically the active base station will change and every time the inactive stations are transmitting misleading beacons. The goal is to hide the actual location of the active base station. In the testing conducted in [78] it was proved that this method was an effective measure for safeguarding a base-station.

8.3.2.4 Traffic Configuration Randomization Technique

This technique is focused on randomizing the data transmission. It is more suitable for applications requiring scheduled data transmission [70]. A good example is the smart grid domain in which smart meters are receiving information from the power supply company while at the same time the meter is sending information to the company regarding the consumption (for the billing) and various diagnostic measurements among other information. In normal circumstances, this is happening in fixed intervals (schedule). Furthermore, smart meter is the type of IoT device that does not have the resources for utilizing proper security mechanisms (e.g., IPS). The experimentation presented in [80, 81] was focused on the protection of smart meters. The main goal was to use shuffle algorithm to randomize the data transmission of each smart meter in the network to protect against jamming attacks. The outcome was that it is very difficult to jam the smart meters when there is no specific schedule.

Based on the current research MTD can significantly enhance the cyber-resilience in IoT ecosystems as well as the overall security, providing flexibility. Nevertheless, there are several challenges and difficulties, especially in industrial environments that by default have integrated cyber-physical architecture, legacy systems, and large infrastructures.

8.3.3 GAME THEORY APPROACH (PROTECTION)

The use of the game-theoretic approach allows a better flexibility to adapt the modeling by allowing for different attacker models and behaviors in different settings and provides a pragmatic method to characterize the impacts of different types of cyberattacks. It helps to identify mitigation measures, either in terms of cyber layer security reinforcements or in terms of developing new operational planning approaches to reduce attack impacts, depending on problem formulation. A game is defined as the interactions between individual players or the cooperative behavior between them. The players will choose specific strategy which is actually a complete decision-making plan for all possible actions the player may take for any circumstance that may face. Two types of strategies exist:

- Pure strategy: in which a unique action has been specified for a situation
- Mixed strategy: which specifies a probability distribution for all possible actions in a situation (meaning in each pure strategy)

The most used solution concept in game theory is Nash equilibrium. A Nash equilibrium is a set of strategies, one for each player, in which none of the players can improve his payoffs by changing/deviating from the prescribed strategy [82, 83].

Research regarding the use of game theory for cyber-resilience and security has increased over the last decade. Games can be categorized into static and dynamic. In static, the players can play only one time while in dynamic the players can play multiple times [84].

Furthermore, games can be also classified in types (based on different game theories) as follows:

- **Cooperative/non-cooperative:** in cooperative players are creating groups due to external enforcements while in non-cooperative players are mainly competing with each other individual or they are creating groups based on their interests.
- **Symmetric/asymmetric:** In symmetric games, the payoff a player relies on the strategies used while in the asymmetric the players do not have identical strategies.
- **Zero-sum/non-zero-sum:** In zero-sum games, the gain of one player is equal to the losses of another player. In non-zero-sum games, the losses and gains do not sum to zero. This means that the gain of one player does not mean the loss of another or vice versa.
- **Simultaneous/sequential:** In simultaneous games, the players either move simultaneously or the last player that moves each time does not have knowledge regarding the actions made by previous players.
- In sequential games, the last player that moves each time has knowledge (not perfect) regarding the actions made previously.
- **Perfect information and imperfect information:** In perfect information, all players know the previous moves other players did (e.g., chess) while imperfect implies that the players are moving simultaneously [82].

- **Complete information and incomplete information:** In complete information games, the payoff functions of all players alongside with their strategies are known while in the incomplete information game, at least one player cannot monitor either the strategies or the payoffs of the other players [85].
- **Evolutionary game theory:** In the evolutionary model the number of players in the game is changing over long periods. The rationality is not so strict which means that a group of players can make employ irrational strategies while at the same time players do not have the same level of knowledge regarding the game.
- **Stochastic:** In stochastic games, the progress of the game is based on transition probabilities [86].

Several game models and approaches have been studies to solve Cybersecurity problems as depicted in [82]. Table 8.2 presents research conducted using static games while Table 8.3 presents research using dynamic games. In both tables, the security problem and the solution are been depicted.

As it is presented in the aforementioned tables, various game theory approaches can be used to solve different Cybersecurity (and cyber-physical) problems and thus, enhancing the overall resilience of an organization.

8.3.4 RISK PERCEPTION AND EPIDEMIOLOGICAL BASED APPROACH (ABSORPTION)

Similarly to epidemic spreading processes, in the mobile environment, malware could be transmitted through opportunistic networks based on devices' proximity

TABLE 8.2
Static Game Approach

Game Theoretic Approach	Security Problem	Solution
Static Prisoners' dilemma game	Selfish in Multi-hop network	Nash equilibrium
Static zero-sum game	Denial-of-service attack Hardware Trojans	Nash equilibrium
Stackelberg game	Cyber-physical security	Stackelberg equilibrium

TABLE 8.3
Dynamic Games Approach

Game Theoretic Approach	Security Problem	Solution
Zero-sum stochastic game	Cyber-physical security	Saddle-point equilibrium
Bayesian game	Denial-of-service attack Survivability	Bayesian Nash equilibrium
Dynamic game	Secure routing Cyber-physical security	Saddle-point equilibrium
Markovian game	Configuration of Intrusion Detection System Smart-grid infrastructure protection	Markov equilibrium

[87–89]. Analogous to travel restrictions for epidemic containment, actions to prevent attack propagation in a large IoT network include device lockdowns, restrictive security settings, entire sub-networks disconnection through segmentation as well as limitation of the use of equipment.

This kind of countermeasures is applied since the detection of an attack on every single device in a large IoT network to react selectively, is often computational unfeasible. Recently, a novel approach based on theoretical epidemiology field has been proposed to manage service resilience [6, 7, 90, 91]. This approach is emerging following the assumption that a cyber-attack in a Sensor or Computer networks follows similar roles of an epidemic. In fact, there is a strong relationship between the infection probability τ, the average number of contacts $\langle k \rangle$ and its variance $\langle k^2 \rangle$: the critical value τc for the onset of an epidemic is for sharp-distributed networks [90].

$$\tau\_c = k/k^2 \cong k^{(-1)}$$

In particular, an IoT is represented as a dynamical large network where devices (e.g., sensors) exchange data with each other. The current trend toward a more distribution in computing load (e.g., edge computing) requires a network resilience that can be increased by using a distributed approach, where the centralized processing of information is limited. However, relying only on peer communication can lead to security problems.

From a theoretical point of view, one can think that each node of the network processes information received from "peers" or upstream nodes, adding its own information. "Pure sensors" only add their own information, while "pure processors" do not add their own data. The received data can be "contaminated", either because a sensor is faulty, or because it is malicious (for example due to hacking). Moreover, since CPS spans four domains as physical, information, cognitive and social [92], the information can be further influenced by the contacts in social networks [93–95].

This raises the question of how to choose whether or not to trust received data, because processing wrong data means "contaminating" downstream nodes. A node can resort to communicate to a central server in order to check the validity of received information, but this will cost in terms of bandwidth and time. This approximation considers the percolation or epidemic spreading problem: what is the best strategy for stopping an epidemic process (security) while trying to maintain the network functionality (resilience) and keeping the communication costs low and performance high (efficiency). Humans develop a series of empirical rules, called heuristics that could also be useful in the computer field.

The main assumption is that the knowledge about the diffusion of the disease among neighbors (without knowing who is actually infected) effectively lowers the probability of transmission (the effective infectiousness). There are several metrics in security that can be used to derive a probability of an attack such as system vulnerabilities, defense power and so forth [91, 96]. Unfortunately, mostly of them are not suitable when faced with unknown threats. To this end,

this approach is based on the concept of "risk perception" to decide whether to keep a connection active, whether to invest resources (time) to understand if a particular acquaintance is lying, and so on [6, 7, 90, 91].

This perception of risk is" triggered" by the" global" percentage of the infected. Thus, if the probability of infection depends on the global percentage of infected, the nodes in IoT, randomly, could check the correctness of incoming information against a central server that keeps the status of the entire network. However, this procedure has a cost in terms of both time and bandwidth that during a cyber-attack cannot be acceptable. It has been shown that a memory mechanism (that implies computational capacity at the edge), it is possible to obtain a resilient behavior in an IoT network minimizing the loss of functionality during an attack. In fact, through the increment of the level of local risk perception, it is possible to stop the epidemics also in a scale-free network through a self-adaptation mechanism [97–99] . However, this effect comes at the price of large cost, so has been developed techniques for automatically detecting the infection threshold, so to apply the minimum effort needed to stop the epidemics.

8.3.5 Probabilistic and Adapting Modelling (Adapt)

Another approach to increase system resilience is the use of ad-hoc safety-net functions. Safety-net functions comprise transitioning a malfunctioning system to safe and sustainable operation thereby enabling time for human intervention. Researchers are exploring a variety of promising and formally checkable representations to defined resilience more rigorously. An example is represented by a variant of Contract-Based Design (CBD [100–103]) explored in [104] that seems to be intuitive, rigorous, and extensible in dealing with unknown-unknowns. CBD is a means for defining system requirements, constraints, behaviors, and interfaces by a pair of assertions, C = (A, G), in which A is an assumption made on the environment and G is the guarantee a system makes if the assumption is met.

Assumptions are system invariants and preconditions while guarantees are system post-conditions. Invariant contracts can be represented by deterministic Büchi automatons and by temporal logic [105]. However, most systems are at least partially non-deterministic. Moreover, invariant constructs are not well-matched with unknown and unexpected disruptions that might arise from unpredictable events. Hence, to address partial observability and need for adaptability and resilience, the research defined a mathematical construct called the "resilience contract," which extends the concept of Contract-Based Design (CBD) to address uncertainty and partial observability that contribute to non-deterministic system behavior [106, 107]. In CBD, contracts are based on "assert-guarantee" and their implementation is satisfied if it fulfills guarantees when the assumptions are true. In this respect, the statements in the contract are mathematically verifiable. However, the invariance of the assertions limits the use of CBD approach in characterizing system reliance. With a resilience contract (RC), the assert-guarantee pair is replaced by a probabilistic "belief-reward" pair. This characterization affords the requisite flexibility while assuring probabilistic verifiability of the model. The RC is a hybrid modeling construct that combines invariant and flexible assertions and is represented as a Partially Observable Markov Decision Process (POMDP). A POMDP is a special form of a Markov Decision Process that includes unobservable states and state transitions. POMDPs introduce flexibility into a traditional contract by allowing incomplete specification of legal inputs and flexible definition of post-condition corrections.

A POMDP models a decision process in which system dynamics are assumed to be a belief Markovian Decision Process (MDP), a memoryless decision process with transition rewards.

Simply stated, an RC extends a deterministic contract for stochastic systems introducing the flexibility needed for the resilience "sense-plan-act" cycle that comprises iteratively sensing the environment and system status (Sense ≡ assumption); sequencing actions that maximize the likelihood of achieving a goal (Plan) and executing those actions (Act ≡ guarantee). The environment sand system health is sensed and assessed after each action. The planning function determines whether to continue with the current plan if the actions accomplish the desired outcome, or otherwise make changes.

Moreover, in a POMDP model, some states are hidden because of uncertainties about system state or the outcomes of actions due to incomplete or imperfect information. Thus, the RC approach starts with a naïve model of system behavior comprising known states and transitions and predicted anomalous states and transitions. New states are added when an observation does not fit any of the existing states, for example, to accommodate unknown-unknowns, and the emission and transition probabilities updated.

8.4 USE CASES

The value brought by the concept of Cyber Resilience is becoming tangible in particular in critical infrastructures [108]. In this section, examples of threats and related cyber-resilience strategies in different critical infrastructures such as transport, financial, telecommunication, energy, and emerging technology as blockchain have been provided.

8.4.1 Transport Infrastructure

An example of such a critical CPS is represented by the transport infrastructures. Today's transportation networks highly depend on smart technologies [109, 110] and even more complex networks that enlarges the attack surface making it difficult to know what and how the assets are exposed; therefore, cyber-attacks have become a scary reality[7]. As highlighted in [45, 111], transportation systems are prone to suffering from various kinds and levels of disruption, which often result in lowering the level of provided service and putting at risk the security of the infrastructure and equipment, as well as of the people that use them. The ransomware attack to the Colorado Department of Transport in February 2018 (CODOT) that disrupted operations for several weeks is an example among several. In today's trend of creating "Smart" cities of which transportation systems are considered as a cornerstone [112] as well as the emerging autonomous and connected vehicle technologies, digital infrastructure is dominant and interconnections between different services and actors may become target of cyber-attacks. The core of information and communication systems is data; such data may contain information on tracking the location, status, and condition of physical assets and associated infrastructure, and thus provide the capability of control of the different assets. This implies a series of minor or major threats to the safety and security of transportation systems upon a cyber-attack such as physical asset damage and associated loss of use (e.g., traffic lights and electronic traffic signals); unavailability of IT systems and networks (e.g., interruption of ticketing services, traffic management systems); loss of operational control causing safety issues (metro control), etc.

Such risks are usually managed through cyber risk-based security strategies including detection, identification, mitigation, and protection. However, due to their complexity, transportations systems' exposure to threats is characterized by high uncertainty and unpredictability. In fact, any disturbance to these structures could lead to negative systemic consequences for society and the economy as a whole. For instance, Airports have long been a target for those wishing to cause destruction, or worse. However, the industry track record of exemplar safety and security capability falls short when it comes to the cyber arena. This doesn't only refer to highly complex, cloud-based IT system security either. Runway lighting, RTS tower operations, baggage handling equipment—this all can come to a grinding, and expensive, halt in the event of a targeted attack.

On the other hand, because the number of air travel passengers is exponentially growing each year, airports are also at the forefront of technological innovation to be able to sustain or thrive. The integration of Industrial IoT in airport facilities, such as upgraded X-ray and screening machines, and the increased use of smart devices from travelers and employees has augmented the available attack surface [113–123]. Civil Air Navigation Services Organization (CANSO) [122] developed a guide for increasing security level to Air Traffic Management (ATM), by presenting cyber threats and risks, as well as threat actors with their motives. CANSO proposed a model in order to Cybersecurity to be addressed, in combination with international standards, NIST Cybersecurity Framework, as well as a risk assessment methodology.

Although noteworthy research work has been conducted regarding Air Traffic Management (ATM) cyber risks, there is a lack of focus on cyber-resilience, especially considering that risks are now very dynamic and constantly change, as new threats and vulnerabilities evolve, along with ever-changing technology implementations and the presence of outdated systems.

In fact, according to [49], one of the primary concerns of the airline industry is the legacy software with significant planning and resources being spent on replacing them with modern alternatives that can be patched and better secured.

For the legacy systems that can't be replaced, a resilient approach requires the segmentation of these systems from the rest of the network and to put in place additional monitoring to detect and limit cyber-attacks against these outdated systems.

Moreover, organizations should review the commercial and open source products they are using and be aware of and prepare for potential attacks using legitimate vendor's software as the infection vector. On the other hand, airports have to continuously upgrade their technologies, provide optimum service in a sustainable manner, and at the same time balance among business growth, efficient services, security, and safety.

8.4.2 FINANCIAL SYSTEM

The systemic importance of the **financial system** progressively increased during the last years, and almost boosted with the advent of the Internet and mobile technologies. The electronic transfer of funds is currently the most common way to settle transactions between companies and private citizens and the role of cash is constantly declining as new payment alternatives are entering the market.

Usually, the financial infrastructures are supported by specialized Cyber infrastructures for insurance and financial aspects. They are particularly critical for the value of the information managed and thus for the risks, in terms of money and people involved. A fault on these infrastructures means to provoke a fault in several services and activities: financial (business discontinuity), government, etc.

Unfortunately, despite the huge investments in business continuity and disaster recovery and the regulatory framework imposed by national and international regulators, banks are exposed to potential catastrophic attacks affecting not only the single bank but the entire system of payments and settlements. The Carbanak attack[8] demonstrated important systemic vulnerabilities, as up to one billion US $ was stolen in about two years from 100 financial institutions worldwide, exploiting the same vulnerabilities in a restricted number of software solutions used by banks globally (as an example the criminal gang took control of the software used by one of the most important global players and replicated the attack in different countries and different banks). Banks are especially exposed to cyber risk as there are economic, political, and motivational reasons to perpetrate attacks. A specific attention should be dedicated to the use of cloud service providers and IT consortia, widely used by small banks to outsource their IT system and operations. Cloud services and IT outsourcing are reducing "local" risk but increasing "systemic" risk, with shared risk management policies that are extremely difficult to implement. The focus in solutions as Next Generation firewall that helps back segments their network, preventing breaches from spreading too far beyond the initially compromised machine, represents a reliable resilience strategy. In fact, it acts on the FDC in general and on Tolerance (TO) in particular. Moreover, the use of solutions such as the Security Orchestration Automation and Response (SOAR), which is an asset that belongs to the Responding Capacity, and whose use refers to the ability to Protect, can support routine incidents management automation increasing the Margin of maneuver due to time saving (MA).

8.4.3 TELECOMMUNICATION

Telecommunication infrastructure is vital to the transfer of information and coordination of societal effort within small communities and international organizations alike. Capacity demands have continued to increase in scale and scope as growing amounts of rich contents is being transferred between devices (M2M), apps, and users (e.g., Social Network). Network reliability represents the most critical success factor for communication service providers. However, downtown in the telecom industry is not so rare. In 2016, the phone service on Level 3 network in the USA was blocked for 1 hour because of an administrative error. In this respect, the level of competition is reducing their risk of fault since a number of redundant solutions are provided: wired network is overlapped with mobile and more than one mobile operator are present in a region, Wi-Fi has a large coverage and overlap even if they depend on different operators as well. In some extent, the energy grid is also a communication channel alternative to the main telecommunication. At the current level of Internet connectivity, both public and private network exists, and most of the public CIs are connected to both or should be connected to both. The Public Connections can maintain the connectivity in more critical conditions while they are not capillary in reaching the final users/consumers. In any way, the coverage and the redundancy of telecommunication infrastructures is not uniform and neither very clear for the other CIs connected. Each company follows their own business continuity strategies that could not be coherent with a global strategy for resilience.

The adoption of cyber-resilience approach in telecommunication consists of focusing not only on redundancy but also on diversity and dynamic adaptation. In fact, what has to be actually secured is the capacity of serving subscribers/users even when the line is out of service. Moreover, such a capacity should be sustained during the four phased composing the coping ability concept. The Ofcom in UK requires a KPI of 99,999% of resilience- this means that are allowed only 5 minutes of downtime per year. In the telecommunication network the well-established roaming mechanisms allow an agile management of emergency situation.

Adopting a resilient perspective means focusing on the network as a whole and immersed in a complex network of networks (NoNs). Thus, instead of focusing on the hardware, telco should build resilience working at FDC level and from core to edge within its own network first and then establishing operator to operator (O2O) agreement for business continuity. Moreover, some specific strategies as redundancy, alternate routes, segmentations, separacy, dual parenting, ported number services (points numbers to ISDN30 DDI during business as usual and delivered to alternate numbers during business continuity need), help organization build resilience in progressively rather than all at once.

8.4.4 ENERGY SECTOR

A similar situation happens in the **energy sector**. Although severe weather still remains the leading cause of power outages in the developed world[9], key findings on cyber threats[10] indicate some evidence of a huge increase of attacks on Supervisory Control and Data Acquisition (SCADA) systems. Also relevant is that whereas the motivation behind points-of-sale and secure web browser attacks is typically financial, the SCADA systems attacks tend to be political in nature, since they target information and operational capabilities within the power sector, factories, and refineries rather than credit card information. In this respect, a power outage forces the closure of schools, shutters businesses, and often impede emergency services, ultimately costing billions to the economy and disrupting the lives of millions of citizens. Additional to this threat, the deploying of smart grids adds millions of access points to the electric grid, which adds responsibility to the Energy Sector to develop a higher and sustained knowledge in Cybersecurity and CI resilience. Thus, the

Energy Sector, namely the electric grid, due to the fact of being one of the largest interconnected systems in Europe and that it is considered the most critical to the operation of other sectors [11], should evolve their policy and technologies to establish a secure inter-organizational protocol for collaboration in face of systemic resilience.

The adoption of a cyber-resilience perspective in the Energy sector means that operators start to consider themselves as a part of a complex system represented by the community of reference. Thus, the focus should not be only on the efficiency and redundancy of the infrastructure that is usually addressed by the business continuity. The service provides should establish a reliable communication channel with the city managers to inform them about the outage, the expected time for recovery, the area affected, the potential risks, and so forth. Such information allows city managers to promptly respond mobilizing available resources to mitigate the impact. For instance, for a patient that needs continuous caring at home through medical machine (e.g., breath support), in case of a prolonged outage, the device battery might not be sufficient, and a dedicated support is necessary. Such intervention is indeed out of the scope of the energy supplier. On the other hand, without the access to such a relevant information, a city manager cannot implement countermeasures with the needed effectiveness. The second point is related to the infrastructure implementation. For instance, in electricity distribution grids and microgrids are called resilient if they can withstand extreme events as the islanding event. The islanding events are defined as power outages lasting longer than a specified threshold of 5 hours and up to 24 hours, whereas shorter outages are categorized as reliability events. Technically the resilience in microgrid is managed using storage and re-dispatchment techniques. However, as explained in [124–126], modeling islanding as a stochastic event with uncertain starting time and duration make the microgird resilient because (1) the microgrid prepares for extreme events causing a blackout by being able to island at any time; (2) if the event occurs, the microgrid changes its operation by ramping up the distributed energy resources to supply all loads, mitigating the impact; (3) recovery is modeled with the cost of buying back depleted storage energy. Moreover, as introduced in [46], another technique implemented to mitigate the impact of an attack resides in the ability to dynamically alter the configuration of a power system to minimize the effects of the attack on an energy grid. This type of countermeasures is critical for attack-resilient protection, where the failure of protection schemes could potentially cause cascading outages in a very short time frame. Examples of such dynamic reconfiguration and resiliency are intelligent islanding of the system and generation redispatch to relieve overloads. Also, various dynamic routing and network reconfiguration strategies would be applied by quantifying the latency and service interruption caused by them versus the security benefits offered in terms of reducing the attack surface on the fly [127].

8.4.5 EMERGING TECHNOLOGIES

Emerging technologies are rife with opportunities for organizations of all shapes and sizes. Self-driving cars may be some ways into the future still, but connected devices in hospitals, factories, and homes are already sharing troves of data for better analytics and decision-making. Workloads are moving to the cloud, and digital personal assistants are becoming commonplace for both enterprises and consumers. Despite their exciting potential, security and business leaders must balance the benefits of these new technologies against the threats that subsequently emerge. In fact, since criminals are the "best" early adopters on emerging technologies,[12] it is mandatory that the introduction of a new technology in the society should be evaluated against systemic risk and resilience in advance.

It is very common that, the introduction of new technologies, designed to solve some critical issues, brings new vulnerabilities. In the case of Artificial Intelligence, an attack on the data integrity used for training will alter the expected learning result.

As remarked in [128], malicious parties can abuse this vulnerability by intentionally feeding AI with incorrect data (data poisoning). Attackers who know how a machine learning system has been trained can subtly manipulate the results, for example, by presenting a facial recognition algorithm with photos that have been manipulated with "noise." Nevertheless, the humans still see the same image, the facial recognition algorithm can be misled. This could cause applications for making medical diagnoses to arrive at incorrect conclusions on the basis of scans that have been contaminated with noise [129].

Stickers on a road can also lead the Lane Detection System of one particular Tesla model to believe that there is a diversion and cause the car to change lanes, while a human driver would simply ignore the stickers [130]. Another emerging technology that needs to be managed using a resilient approach is the Blockchain. The Blockchain technology aims to guarantee security and trustworthiness of the transactions executed in a trustless network of peers. This technology is applied in very diverse domains as IoT [5, 99, 131–134], voting system [135, 136], M2M [137], distributed trust and reputation management systems [138], persistent identifier systems [139], control rooms implementation [140], and many others. Unfortunately, new class of vulnerabilities has been introduced and then exploited. For instance, in [141], the stealing of $50 million in funds from the Decentralized Autonomous Organizations - DAO [142] has been analyzed. The attacks exploited a concurrency-based vulnerability (latency in updating the amount of an account after an operation) of a contract since the Smart Contract (SC) was designed and implemented as a simple single-threaded program. This means that that even SCs are affected by vulnerabilities. In [143], a survey of pitfalls and common bugs in SCs, which are disguised versions of common concurrency pitfalls, is provided. While in [144], several vulnerabilities in Ethereum SC design are also analyzed. As remarked in [138], these studies raise awareness and call for a general shift in BC from the enthusiastic early adoption toward a more professional and cyber risk and resilience aware approach. In fact, as stated in [129], the exploitation of the opportunities for increasing cyber-resilience created by new and existing technologies calls for specific capacity and expertise (Assets: Human Capital). Due to the chronic shortage of experts in this field, greater investment is needed to increase IT skills that are connected to the Adaptive Capacity development. Such a capacity will be employed in a specific function and will contribute to dampen the potential variability that an attack may cause to the system.

8.5 RECOMMENDATIONS

A system cannot be designed considering every possible scenario, therefore a resilient design should consider how to maintain the functionality of networks while mitigating/damping the impact of the disturbance and avoiding it from spreading across the whole system, creating systemic contagion and system-wide collapse [145]. From the frameworks considered and the scenarios presented following recommendations emerge:

- **Focus on critical assets and functionalities:** A system should know and understand which are the functionalities that need to be guarded and maintained and what assets they exploit/consume. This system understanding is necessary to prioritize actions and decisions in presence of competing objectives. For instance, the restoration of the services is one of the high-level policies of a resilient system, on the other hand, which functionality should be recovered first is a matter of a dynamic multicriteria analysis. In fact, aspects as functionality interdependencies and current assets availability should be considered to implement an effective recovery strategy as well as an investments program during preparation and planning phase.
- **Focus on Functional Damping Capacity:** the FDC represents the point in which the Coping ability meets the Adaptive Capacity, and the resilient behavior may emerge. The decision to design a decentralized or fully distributed architecture including mechanism for segmentation, modularity, local decision-making, redundancy, diversity, and so forth, may increases the survivability of the service avoiding the introduction of a single point of failure and thus a high-value target in the system. However, such resilient strategies have to be supported by providing the proper level of assets and resources and by developing the right level of ability to cope with changing conditions. For instance, the capacity of distributing and dynamically relocating functionality, system resources, and control, allowed by specific equipment and skills (assets), is the way to dampen the variability generated in the system by a critical event. An effective damping capacity supports the ability to mitigate the impact and rapidly recover from critical events. Therefore, in order to increase the system resilience, the major effort should be directed toward a continuous development of FDC that passes through a continuous development of the Adaptive Capacity and an effective implementation of adaptation cycle.
- **>> Focus on costs-benefits implication driven coping strategy implementation:** It is well documented that the suspension of network service has an immediate and quantifiable impact on the system. The loss of performance is usually directly linked to the loss of money (e.g., Service Level Agreement). However, the underspecified nature of the complex systems implies the existence of a number of hidden costs, including those derived by the human behaviors (social costs) that are usually not carefully explored. In this respect, the implementation of resilient strategies should adopt different time frame (immediate to mid/long term). For instance, the severity of communication restrictions required to achieve a real risk reduction could not be justified

unless the risk of a malware spread is particularly high. While risk management is crucial to epidemic control what is not accounted for, generally, is that the restrictions imposed by risk reduction can be detrimental to the normal IoT functioning and create new risks themselves.

8.6 CONCLUSIONS

Failures in critical functions affect the level of trustworthiness on which the modern society is built on provoking unexpected side effects. Therefore, there has been an increasing interest in shifting emphasis from configuring compliance of devices with respect to security practice and mainly focused on the reaction perspective, toward cyber-resilience.

Introducing a holistic view for Cyber-resilience (CRHV) contributes to shape the Resilience Thinking in cyber domain. The intention is to help decision-makers and security practitioners to adopt a broader perspective considering all the phases needed to manage a critical event (planning for adverse events, absorbing stress, recovering and predicting and preparing for future stressors) in order to continuously adapt to the system to potential threats in a world of limited resources and "unlimited" possibilities. In the present chapter has been revised several cyber-resilience definitions and frameworks. An integration and reconciliation into a unified vision has been proposed and used to properly classify emerging techniques with respect to their use in a cyber-resilience approach.

The key point of CRHV is represented by the consideration of the two main synonyms of resilience as Adaptive Capacity and Coping Ability like subconcepts of the cyber-resilience that represent the unique root concept. In this way has been possible to identify two building blocks of resilience that contribute to the emerging behavior in a different but synergic way.

The concept of Adaptive capacity allowed a better definition of the contribution of the assets available in the system (including ITC technologies) as well as their exploitation at the appropriate phase. The Coping ability concept allows a better definition of the phases in which different capacities have to be properly exploited. In this respect, a list of emerging technologies and approaches such as network-based approach, moving target defense, game theory approach, risk perception and epidemiological based approach, and so forth, have been introduced and mapped to their related Ability.

Moreover, a number of use cases from different sectors have been depicted to highlight the added value of adopting a cyber-resilience perspective in face of known and unknown cyber risks. Finally, three major recommendations have been provided to better guide decision-makers in their cyber-resilience program implementation.

8.7 ACKNOWLEDGMENT

This project has received funding from the European Union's Horizon 2020 research and innovation program under grant agreement No 786698. The work reflects only the authors' view, and the Agency is not responsible for any use that may be made of the information it contains.

NOTES

1. Willsher, K. (2009), 'French fighter planes grounded by computer virus', *The Telegraph*, https://www.telegraph.co.uk/news/worldnews/europe/france/4547649/French-fighter-planes-grounded-by-computer virus.html (accessed 22 Dec. 2019).
2. Merchet, J. (2009), 'Les armées attaquées par un virus informatique (actualisé)', *Libération*, 5 February 2009, http://secretdefense.blogs.liberation.fr/2009/02/05/les-armes-attaq/ (accessed 28 May 2019).
3. https://www.dni.gov/files/ODNI/documents/NationalIntelligenceStrategy2019.pdf
4. https://www.enisa.europa.eu/publications/the-cost-of-incidents-affecting-ciis/atdownload/fullReport
5. https://www.ncsc.gov.uk/blog-post/cyber-resilience-nothing-sneeze
6. https://www.nist.gov/cyberframework
7. http://www.wired.com/2015/07/hackers-remotely-kill-jeep-highway/
8. http://www.kaspersky.com/about/news/virus/2015/Carbanak-cybergang-steals-1-bn-USD-from-100-financial-institutions-worldwide
9. Economic Benefits of Increasing Electric Grid Resilience to Weather Outages, White House, 2013.
10. Security Annual Threat Report, Dell, 2015.
11. Identifying, Understanding and Analyzing Critical Infrastructure Interdependencies, Steven M. Rinaldi,James P. Peerenboom, Terrence K. Kelly, 2001.
12. Future Crimes, Marc Goodman, 2015.

REFERENCES

1. "Ericsson, mobility report: On the pulse of the networked society, interim update, sep. 2016." Available: https://www.ericsson.com/en/mobility-report/reports [Accessed: Sep. 21, 2020].
2. M. Conti, S. K. Das, C. Bisdikian, M. Kumar, L. M. Ni, A. Passarella, G. Roussos, G. Trster, G. Tsudik, and F. Zambonelli, "Looking ahead in pervasive computing: Challenges and opportunities in the era of cyberphysical convergence," *Pervasive and Mobile Computing*, vol. 8, no. 1, pp. 2–21, 2012.
3. S. Shiaeles and M. Papadaki, "*FHSD: An improved IP spoof detection method for web ddos attacks*", The Computer Journal, vol. 58, no. 4, pp. 892–903, 2015.
4. S.N. Shiaeles, V. Katos, A.S. Karakos, and B.K. Papadopoulos, "Real time DDoS detection using fuzzy estimators," *Computers & Security*, vol. 31, no. 6, pp. 782–790.
5. N. Kolokotronis, S. Shiaeles, E. Bellini, L. Charalambous, D. Kavallieros, O. Gkotsopoulou, C. Pavue, A. Bellini, and G. Sargsyan, *Cyber-Trust: The Shield for IoT Cyber-Attacks, in Resilience and Hybrid Threats*, IOS Press, pp. 76–93, 2019.
6. F. Bagnoli, E. Bellini, and E. Massaro, "A self-organized method for computing the epidemic threshold in computer networks," in S. S. Bodrunova (Ed.) *Internet Science*. vol. 11193, Cham, Switzerland: Springer International Publishing, pp. 119–130, 2018.
7. F. Bagnoli, E. Bellini, and E. Massaro, "*Risk perception and epidemics in complex computer networks*," in *2018 IEEE Workshop on Complexity in Engineering (COMPENG)*, pp. 1–5, IEEE, 2018.

8. A. Vespignani, "Complex networks: The fragility of interdependency," *Nature*, vol. 464, pp. 984–985, 2010. doi: 10.1038/464984a.
9. "Symantec, internet security threat report, vol. 21, Apr. 2016."
10. L. Yin, Y. Sun, Z. Wang, Y. Guo, F. Li, and B. Fang, "Security measurement for unknown threats based on attack preferences," *Security and Communication Networks*, vol. 2018, pp. 1–13, 2018.
11. B. Krebs, "The scrap value of a hacked PC, revisited," Oct. 2012.
12. K. A. Delic, "On resilience of IoT systems: The internet of things (ubiquity symposium)," *Ubiquity*, vol. 2016, no. February, pp. 1–7, 2016.
13. J. R. Wilson, B. Ryan, A. Schock, P. Ferreira, S. Smith, and J. Pitsopoulos, "Understanding safety and production risks in rail engineering planning and protection," *Ergonomics*, vol. 52, no. 7, pp. 774–790, 2009.
14. D. Komljenovic, M. Gaha, G. Abdul-Nour, C. Langheit, and M. Bourgeois, "Risks of extreme and rare events in asset management," *Safety Science*, 88, pp. 129–145, 2016.
15. J. Park, T. P. Seager, P. S. C. Rao, M. Convertino, and I. Linkov, "Integrating risk and resilience approaches to catastrophe management in engineering systems: perspective," *Risk Analysis*, vol. 33, no. 3, pp. 356–367, 2013.
16. S. Shiaeles, N. Kolokotronis, and E. Bellini, "Iot vulnerability data crawling and analysis," in *1st IEEE Service workshop on Cyber Security and Resilience in IoT*, 2019.
17. C. Constantinides, S. Shiaeles, B. Ghita, and N. Kolokotronis, "*A novel online incremental learning intrusion prevention system*," in *10th IFIP International Conference on New Technologies, Mobility and Security*, 2019.
18. B. Schneier, *Beyond Fear*, New York: Springer, 2006.
19. J. Ferdinand and R. Benham, "The Cyber security ecosystem: defining a taxonomy of existing, emerging and future cyber threats," 2017.
20. I. Linkov, F. Baiardi, M.V. Florin, S. Greer, J.H. Lambert, M. Pollock, J.M. Rickli, L. Roslycky, T. Seager, H. Thorisson, and B.D. Trump, "Applying resilience to hybrid threats," *IEEE Security & Privacy*, vol. 17, no. 5, pp. 78–83, Sep.–Oct. 2019.
21. N.O. Leslie, R.E. Harang, L.P. Knachel, and A. Kott, "Statistical models for the number of successful cyber intrusions," *The Journal of Defense Modeling and Simulation*, vol. 15, no. 1, pp. 49–63, 2017.
22. A. A. Ganin, D. Marchese, Z. A. Collier, P. Quach, M. Panwar, and I. Linkov, "Multicriteria decision framework for Cybersecurity risk assessment and management," *Risk Analysis*, vol. 40, no. 1, pp. 183–199, 2020.
23. D. DiMase, Z.A. Collier, K. Heffner, and I. Linkov, "Systems engineering framework for cyber physical security and resilience," *Environment Systems & Decisions*, vol. 35, pp. 291–300, 2015.
24. A. Kharrazi, Y. Yu, A. Jacob, N. Vora, and B. D. Fath, "Redundancy, diversity and modularity in network resilience: Aplications for interational trade and implications for pubic policy," *Current Research in Enviromental Sustainability*, 2020.
25. Y. Sheffi, *The Resilient Enterprise: Overcoming Vulnerability for Competitive Advantage*. Cambridge, MA: MIT Press, 2007.
26. I. Linkov, B.D. Trump, and J. Keisler, "Risk and resilience must be independently managed," *Nature*, vol. 555, p. 30, 2018.
27. M. Batty, "The size, scale, and shape of cities," *Science*, vol. 319, no. 5864, pp. 769–771, 2008.
28. D. Godschalk, "Urban hazard mitigation: Creating resilient cities," *Natural Hazards Review*, vol. 4, pp. 136–143, 2003.
29. A. Sharifi, "A critical review of selected tools for assessing community resilience," *Ecological Indicator*, vol. 69, pp. 629–647, Elsevier, Oct. 2016.

30. S. Meerow, J.P. Newell, and M. Stults, "Defining urban resilience: A review," *Landscape and Urban Planning*, vol. 147, pp. 38–49, 2016.
31. Accenture, *The Nature of Effective Defense: Shifting from Cybersecurity to Cyber-resilience*, 2018, available at: https://www.accenture.com/_acnmedia/Accenture/Conversion-Assets/DotCom/Documents/Local/en/Accenture-Shifting-from-Cybersecurity-to-Cyber-Resilience-POV.pdf [Accessed: Dec. 20, 2020]
32. National Institute of Standards and Technology, "NIST Cybersecurity framework," 2018. Available: https://www.nist.gov/cyberframework.
33. E. Bellini, P. Nesi, L. Coconea, E. Gaitanidou, P. Ferreira, A. Simoes, and A. Candelieri, "Towards resilience operationalization in urban transport system: The resolute project approach," in *Proceedings of the 26th European Safety and Reliability Conference*, ESREL 2016, 2017.
34. E. Bellini, P. Nesi, G. Pantaleo, and A. Venturi, "Functional resonance analysis method based-decision support tool for urban transport system resilience management," in *2016 IEEE International Smart Cities Conference (ISC2)*, Trento, Italy: IEEE, pp. 1–7, 2016.
35. I. Linkov and A. Kott, *Fundamental Concepts of Cyber-resilience: Introduction and Overview, Cyber-resilience of Systems and Networks*, Springer, 2018.
36. D.D. Woods, "Four concepts for resilience and the implications for the future of resilience engineering," *Reliability Engineering & System Safety*, vol. 141, pp. 5–9, 2015.
37. B. Walker, and D. Salt, *Resilience Thinking—Sustaining Ecosystems and People in a Changing World*. Washington, DC: Island Press, 2006.
38. C.S. Holling, L. Gunderson, and G. Peterson, "Sustainability and panarchies," in L.H. Gunderson and C.S. Holling (Eds.) *Panarchy: Understanding Transformations in Human and Natural Systems*, Washington, DC: Island Press, pp. 63–102, 2002.
39. B.A. Harmon, W.D. Goran, and R.S. Harmon, "Military installations and cities in the twenty-first century: Towards sustainable military installations and adaptable cities," In I. Linkov (ed) *Sustainable Cities and Military Installations*, New York: Springer, 2012.
40. I. Linkov, D.A. Eisenberg, K. Plourde, T.P. Seager, J. Allen, and A. Kott, "Resilience metrics for cyber systems," *Environment Systems and Decisions*, vol. 33, no. 4, pp. 471–476, 2013.
41. E. Hollnagel, and D. Woods, "Epilogue: Resilience engineering precepts," in E. Hollnagel, D.D. Woods, and N. Leveson (Eds.) *Resilience Engineering—Concepts and Precepts*, Aldershot: Ashgate, pp. 347–358, 2006.
42. E. Hollnagel, "The four cornerstone of resilience engineering," in C.P. Nemeth, E. Hollnagel, and S. Dekker (Eds.) *Resilience Engineering Perspective, Vol. 2: Preparation and Restoration*. Aldershot: Ashgate, 2009.
43. E. Hollnagel, *Safety-II in Practice: Developing the Resilience Potentials*, Taylor & Francis, 2017.
44. E. Bellini, L. Coconea, and P. Nesi, "A functional resonance analysis method driven resilience quantification for socio-technical systems", *IEEE Systems Journal*, vol. 14, no. 1, pp. 1234–1244, 2020. doi:10.1109/JSYST.2019.2905713
45. E. Bellini, E. Gaitanidou, E. Bekiaris, P. Ferreira, "The RESOLUTE project's European resilience management guidelines for critical infrastructure: Development, operationalisation and testing for the urban transport system," *Environment Systems and Decisions*, pp. 1–21, 2020.
46. A. Ashok, M. Govindarasu, and J. Wang, "Cyber-physical attack-resilient wide-area monitoring, protection, and control for the power grid," in *Proceedings of the IEEE*, vol. 105, no. 7, pp. 1389–1407, 2017.
47. CISCO *Cyber-resilience: Safeguarding the Digital Organization*, 2016.

48. "Deborah Bodeau Jumpstart resiliency with what you've got—MITRE," Available: https://www.mitre.org/capabilities/cybersecurity/overview/cybersecurity-blog/jumpstart-resiliency-with-what-you%E2%80%99ve-got [Accessed: Sep. 8, 2020].
49. "Cyber-resilience and Response Public Private Analytic exchange program 2018."
50. IOSCO, "Guidance on cyber-resilience for financial market infrastructures," Available: https://www.iosco.org/library/pubdocs/pdf/IOSCOPD535.pdf.
51. E. Bellini and S. Marrone, "*Towards a novel conceptualization of Cyber Resilience*," *2020 IEEE World Congress on Services (SERVICES), Beijing, China*, 2020, pp. 189–196, doi: 10.1109/SERVICES48979.2020.00048.
52. P. Ferreira, E. Bellini, "*Managing interdependencies in critical infrastructures—a cornerstone for system resilience*," in *Safety and Reliability - Safe Societies in a Changing World - Proceedings of the 28th International European Safety and Reliability Conference*, ESREL, pp. 2687–2692, 2018.
53. E. Bellini, P. Ceravolo, and P. Nesi, "Quantify resilience enhancement of UTS through exploiting connected community and internet of everything emerging technologies," *ACM Transactions on Internet Technology*, vol. 18, no. 1, pp. 1–34, 2017.
54. D.D. Woods, "Creating foresight: How resilience engineering cantransform NASA's approach to risky decision making," *Work*, vol. 4, no. 2, pp.137–144, 2003.
55. I. Linkov, C. Fox-Lent, L. Read, C. R. Allen, J. C. Arnott, E. Bellini, J. Coaffee, M.-V. Florin, K. Hatfield, I. Hyde, W. Hynes, A. Jovanovic, R. Kasperson, J. Katzenberger, P. W. Keys, J. H. Lambert, R. Moss, P. S. Murdoch, J. Palma-Oliveira, R. S. Pulwarty, D. Sands, E. A. Thomas, M. R. Tye, and D. Woods, "Tiered approach to resilience assessment: Tiered approach to resilience assessment," *Risk Analysis*, vol. 38, no. 9, pp. 1772–1780, 2018.
56. I. Haring, G. Sansavini, E. Bellini, N. Martyn, T. Kovalenko, M. Kitsak, G. Vogelbacher, K. Ross, U. Bergerhausen, K. Barker, and I. Linkov, "Towards a generic resilience management, quantification and development process: General definitions, requirements, methods, techniques and measures, and case studies," in I. Linkov and J. M. Palma-Oliveira (Eds.) *Resilience and Risk*, Dordrecht: Springer Netherlands, pp. 21–80, 2017.
57. A. A. Ganin, A. C. Mersky, A. S. Jin, M. Kitsak, J. M. Keisler, and I. Linkov, "Resilience in intelligent transportation systems (ITS)," *Transportation Research Part C: Emerging Technologies*, vol. 100, pp. 318–329, 2019.
58. V. Gisladottir, A. A. Ganin, J. M. Keisler, J. Kepner, and I. Linkov, "Resilience of cyber systems with over- and underregulation," *Risk Analysis*, vol. 37, no. 9, pp. 1644–1651, 2017.
59. E. Massaro, A. A. Ganin, N. Perra, I. Linkov, and A. Vespignani, "Resilience management during large-scale epidemic outbreaks," *Scientific Reports*, vol. 8, no. 1, 1859, 2018, doi: 10.1038/s41598-018-19706-2.
60. E. Sherratt, "Intelligent resilience in the IoT," in T. Csndes, G. Kovcs, and G. Rthy (Eds.) *SDL 2017: Model-Driven Engineering for Future Internet*, Springer International Publishing, pp. 46–60, 2017.
61. E. Sherratt, I. Ober, E. Gaudin, P. Fonseca i Casas, and F. Kristoffersen, "SDL - the IoT language," in J. Fischer, M. Scheidgen, I. Schieferdecker, and R. Reed (Eds.) *SDL 2015: Model-Driven Engineering for Smart Cities*, Springer International Publishing, pp. 27–41, 2015.
62. P. M. D. Scully, "CARDINAL-Vanilla: immune system inspired prioritization and distribution of security information for industrial networks," Ph.D. Thesis, Aberystwyth, UK: Aberystwyth University, 2016.
63. A. A. Ganin, E. Massaro, A. Gutfraind, N. Steen, J. M. Keisler, A. Kott, R. Mangoubi, and I. Linkov, "Operational resilience: Concepts, design and analysis," *Scientific Reports*, vol. 6, p. 19540, 2016.

64. National Academy of Science, *Disaster Resilience: A National Imperative*, The National Academies Press, 2012.
65. I. Linkov, T. Bridges, F. Creutzig, J. Decker, C. Fox-Lent, W. Krger, J. H. Lambert, A. Levermann, B. Montreuil, J. Nathwani, R. Nyer, O. Renn, B. Scharte, A. Scheffler, M. Schreurs, and T. Thiel-Clemen, "Changing the resilience paradigm," *Nature Climate Change*, vol. 4, no. 6, pp. 407–409, 2014.
66. P. Ferreira and E. Bellini, "*Managing interdependencies in critical infrastructures a cornerstone for system resilience*," in *Proceedings of European Safety and Reliability Conference 2018*, Trondheim, Norway: Taylor & Francis, Jun. 2018.
67. F. Bagnoli, E. Bellini, E. Massaro, and R. Rechtman, "Percolation and internet science," *Future Internet*, vol. 11, no. 2, p. 35, 2019.
68. U.S. Departement of Homeland Security, "Moving target defense," *U.S. Departement of Homeland Security*, [Online]. Available: https://www.dhs.gov/science-and-technology/csd-mtd. [Accessed: Jul. 31, 2020].
69. D. B. Rawat and K. Z. Ghafoor, Eds., "Security in smart cyber-physical systems: A case study on smart grids and smart cars," In *Smart Cities Cybersecurity and Privacy*, Elsevier, pp. 147–163, 2019.
70. N. Saputro, S. Tonyali, A. Aydeger, K. Akkaya, M. A. Rahman and S. Uluagac, "A review of moving target defense mechanisms for Internet of Things applications," in C. A. Kamhoua, L. L. Njilla, A. Kott and. S. Shetty (Eds.) *Modeling and Design of Secure Internet of Things*, Wiley, pp. 563–614, 2020.
71. H. Okhravi, T. Hobson, D. Bigelow, and W. Streilein, "Finding focus in the blur of moving-target techniques," *Finding Focus in the Blur of Moving-Target Techniques*, vol. 12, no. 2, pp. 16–26, 2014.
72. R. E. Navas, H. Sandaker, F. Cuppens, N. Cuppens, and L. Toutain, "*IANVS: A moving target defense framework for a resilient Internet of Things*," in *the 25th IEEE Symposium on Computers and Communications (ISCC)*, Rennes, 2020.
73. M. Sherburne, M. Randy, and J. Tron, "*Implementing movingtarget ipv6 defense to secure 6lowpan in the internet of things and smart grid*," in *Proceedings of the 9th Annual Cyber and Information Security Research Conference*, CISR'14, New York, 2014.
74. T. Preiss, M. Sherburne, R. Marchany, and J. Tront, "*Implementing dynamicaddress changes in Contiki OS2014*," in *International Conference on Information Society (i-Society 2014)*, 2014.
75. K. Zeitz, M. Cantrell, R. Marchany, and J. Tront, "*Designing a micro-moving targetipv6 defense for the internet of things*," in *IEEE/ACM Second InternationalConference on Internet-of-Things Design and Implementation (IoTDI)*, Pittsburgh, 2017.
76. K. Zeitz, M. Cantrell, R. Marchany, and J. Tront, "Changing the game: A micromoving target ipv6 defense for the internet of things," *IEEE WirelessCommunications Letters*, vol. 7, no. 4, pp. 578–581, 2018.
77. A. Judmayer, G. Merzdovnik, J. Ullrich, A. G. Voyiatzis and E. Weippl, "A performance assessment of network address shuffling in IoTsystems," In: R. Moreno-Díaz, F. Pichler, A. Quesada-Arencibia (eds) *Computer Aided Systems Theory—EUROCAST 2017*, Lecture Notes in Computer Science, vol 10671, 2018, Springer, Cham. doi: 10.1007/978-3-319-74718-7_24
78. T. Chin and K. Xiong, "*Mpbsd: A moving target defense approach for basestation security in wireless sensor networks*," in W. Y. A. Y. Qing Yang (Eds.) *Wireless Algorithms, Systems, and Applications*, Springer International Publishing, pp. 487–498, 2016.
79. K. Andrea, A. Gumusalan, R. Simon, and H. Harney, "*The design andimplementation of a multicast address moving target defensive system forinternet-of-things applications*," in *MILCOM 2017–2017 IEEE MilitaryCommunications Conference (MILCOM)*, Baltimore, 2017.

80. M. Q. Ali, E. Al-Shaer and Q. Duan , "*Randomizing AMI configuration forproactive defense in smart grid,*" in *2013 IEEE International Conference on SmartGrid Communications (SmartGridComm)*, Vancouver, 2013.
81. R. Algin, H. O. Tan, and K. Akkaya, "*Mitigating selectivejamming attacks in smart meter data collection using moving target defense,*" in *Proceedings of the 13th ACM Symposium on QoS and Security for Wireless andMobile Networks*, New York, 2017.
82. C. T. Do, N. H. Tran, C. Hong, C. A. Kamhoua, K. A. Kwiat, E. Blasch, S. Ren, N. Pissinou and S. S. Iyengar, "Game theory for Cybersecurity and privacy," *ACM Computing Surveys*, vol. 50, no. 2, pp. 1–37, 2017.
83. J. Y. Halpern, "Beyond nash equilibrium: Solution concepts for the 21st century," *Lecture Notes in Computer Science*, vol. 7037, pp. 1–3, 2011.
84. D. Fudenberg and J. Tirole, *Game Theory*, Cambridge: MIT Press, 1991.
85. Y. Liu, D. Feng, Y. Lian, K. Chen and Y. Zhang, "Optimal defense strategies for DDoS defender using bayesian game model," in *Information Security Practice and Experience. ISPEC 2013*, Lecture Notes in Computer Science, Lanzhou, 2013.
86. C. T. Do, N. H. Tran, C. Hong, C. A. Kamhoua, K. A. Kwiat, E. Blasch, S. Ren, N. Pissinou, and S. Sitharama Iyengar, "Game theory for Cybersecurity and privacy," *ACM Computing Surveys*, pp. 1–37, 2017.
87. S.-M. Cheng, P.-Y. Chen, C.-C. Lin, and H.-C. Hsiao, "Traffic-aware patching for Cybersecurity in mobile IoT," *IEEE Communications Magazine*, vol. 55, no. 7, pp. 29–35, 2017.
88. S. Tanachaiwiwat and A. Helmy, "*Encounter-based worms: Analysis and defense,*" in *2006 2nd IEEE Workshop on Wireless Mesh Networks*, Reston, VA, USA: IEEE, pp. 170–172, 2006.
89. P. Wang, M. C. Gonzalez, C. A. Hidalgo, and A.-L. Barabasi, "Understanding the spreading patterns of mobile phone viruses," *Science*, vol. 324, no. 5930, pp. 1071–1076, 2009.
90. E. Massaro and F. Bagnoli, "Epidemic spreading and risk perception in multiplex networks: A self-organized percolation method," *Physical Review E*, vol. 90, p. 052817, 2014.
91. F. Bagnoli, P. Li, and L. Sguanci, "Risk perception in epidemic modeling," *Physical Review E*, vol. 76, no. 6, p. 061904, 2007.
92. D.S. Alberts, "The agility advantage: A survival guide for complex enterprises and endeavors," *DOD Command and Control Research Program*, Washington, DC, 2011.
93. J. Ginsberg, M. H. Mohebbi, R. S. Patel, L. Brammer, M. S. Smolinski, and L. Brilliant, "Detecting influenza epidemics using search engine query data," *Nature*, vol. 457, no. 7232, pp. 1012–1014, 2009.
94. D. Scanfeld, V. Scanfeld, and E. L. Larson, "Dissemination of health information through social networks: Twitter and antibiotics," *American Journal of Infection Control*, vol. 38, no. 3, pp. 182–188, 2010.
95. C. Chew and G. Eysenbach, "Pandemics in the age of twitter: Content analysis of tweets during the 2009 H1n1 outbreak," *PLoS ONE*, vol. 5, no. 11, p. e14118, 2010.
96. M. Pendleton, R. Garcia-Lebron, J.-H. Cho, and S. Xu, "A survey on systems security metrics," *ACM Computing Surveys*, vol. 49, no. 4, pp. 1–35, 2016.
97. F. Bagnoli, P. Palmerini, and R. Rechtman, "Algorithmic mapping from criticality to self-organized criticality," *Physical Review E*, vol. 55, no. 4, pp. 3970–3976, 1997.
98. J. Farooq and Q. Zhu, "On the secure and reconfigurable multi-layer network design for critical information dissemination in the internet of battlefield things," *IEEE Transactions on Wireless Communications*, vol. 17, no. 4, pp. 2618–2632, 2018.
99. E. Bellini, F. Bagnoli, A.A. Ganin, and I. Linkov, "*Cyber-resilience in IoT network: Methodology and example of assessment through epidemic spreading approach,*" *2019 IEEE World Congress on Services, SERVICES*, vol. 2642-939X, pp. 72–77, 2019.

100. A. Sangiovanni-Vincentellik, W. Damm, R. Passerone, D. Taming, "Frankenstein: contract-based design for cyber-physical systems," *European Journal of Control*, vol. 18, pp. 217–238, 2012.
101. B. Meyer, "Towards more expressive contracts," *Journal of Object Oriented Program*, pp. 39–43, July 2000.
102. Y. Le Traon and B. Baudry, "Design by contract to improve software vigilance," *IEEE Transactions on Software Engineering*, vol. 32, pp. 571–586, 2006.
103. A. Cimatti and S. Tonetta. "*A property-based proof system for contract-based design*," in *Proceedings of the 38h Euromicro Conference on Software Engineering and Advanced Applications*, Sep. 5–8, 2012, Izmir, Turkey.
104. A. M. Madni, D. Erwin, and M. Sievers, "Constructing models for systems resilience: Challenges, concepts, and formal methods," *Systems*, vol. 8, no. 3, 2020. doi:10.3390/systems8010003
105. J.R. Büchi, "On a decision method in restricted second-order arithmetic," in *The 1960 Congress on Logic, Methodology and Philosophy of Science*, Stanford, CA, USA: Stanford University Press, 1962.
106. M. Sievers and A.M. Madni, "*Contract-based byzantine resilience for spacecraft swarm*," in *Proceedings of the 2016 AIAA Science and Technology Forum and Expo*, Grapevine, TX, USA, Jan. 9–13, 2017.
107. A.M. Madni, J. D'Ambrosio, M. Sievers, J. Humann, E. Ordoukhanian, and P. Sundaram, "*Model-based approach for engineering resilient system-of-systems: applications to autonomous vehicle network*," in *Proceeding of the Conference on Systems Engineering Research*, Redondo Beach, CA, USA, Mar. 23–25, 2017.
108. I. Linkov, and J.M. Palma-Oliveira, "An introduction to resilience for critical infrastructures," *NATO Science for Peace and Security Series – C: Environmental Security*, 2016
109. R. Pethuru and A. Raman, *The Internet of Things: Enabling Technologies. Platforms & Use Cases*, 3rd ed., Boca Raton, FL, USA: CRC Press, 2017. ISBN 978-1498761284.
110. E. Bellini and P. Nesi, "Exploiting smart technologies to build smart resilient cities," in P. Gardoni (Ed.) *Routledge Handbook of Sustainable and Resilient Infrastructure*, Routledge, pp. 685–705, 2018.
111. L. Save, M. Branlat, W. Hynes, E. Bellini, P. Ferreira, and J. P. Lauteritz, J.J. Gonzalez, "The Development of Resilience Management Guidelines to Protect Critical Infrastructures in Europe". In: S. Bagnara, R. Tartaglia, S. Albolino, T. Alexander, and Y. Fujita (eds) *Proceedings of the 20th Congress of the International Ergonomics Association (IEA 2018)*. IEA 2018. Advances in Intelligent Systems and Computing, vol. 819. Springer, Cham. doi:10.1007/978-3-319-96089-0_65
112. L. Coconea and E. Bellini, "Advanced traffic management systems supporting resilient smart cities," *Transportation Research Procedia*, vol. 41, pp. 556–558, 2019.
113. K. Gopalakrishnan, M. Govindarasu, D.W. Jacobson, B.M. Phares, "Cybersecurity for airports," *International Journal of Traffic and Transportation Engineering*, vol. 3, pp. 365–376, 2013.
114. J.A. Urban, "Not your granddaddy's aviation industry: the need to implement Cybersecurity standardsand best practices within the international aviation industry," *Albany Law Journal of Science & Technology*, vol. 27, pp. 62–93, 2017.
115. ENISA, "Securing smart airports," Available: https://www.enisa.europa.eu/publications/securingsmart- airports. [Accessed: Oct. 12, 2018].
116. European Commission, "Commission staff working document (SWD 318)," 2013. Available https://ec.europa.eu/home-affairs/sites/homeaffairs/files/what-we-do/policies/european-agenda-security/20180613swd-2018-331-commission-staff-working-documenten.pdf. [Accessed: Oct. 12, 2018].

117. IATA, "Fact sheet Cybersecurity," Available: https://www.iata.org/pressroom/factsfigures/factsheets/Documents/fact-sheet-cyber-security.pdf. [Accessed: Oct. 12, 2018].
118. ICAO, "Assembly resolution A39-19," Sep. 2016. Available: https://www.icao.int/Meetings/a39/Documents/Resolutions/a39resproven.pdf [Accessed: Oct. 12, 2018].
119. K. Sampigethaya, R. Poovendran, S. Shetty, T. Davis, C. Royalty, "Future E-enabled aircraft communications and security: The next 20 years and beyond," *Proceedings IEEE*, vol. 99, pp. 2040–2055, 2011.
120. M. Strohmeier, M. Schafer, M. Smith, V. Lenders, and I. Martinovic, "*Assessing the impact of aviation security on cyber power*," in *Proceedings of the 2016 8th International Conference on Cyber Conflict (CyCon)*, Tallinn, Estonia, pp. 223–241, 31 May–3 Jun. 2016.
121. National Academies of Sciences, Engineering, and Medicine, *Guidebook on Best Practices for Airport Cybersecurity*, Washington, DC, USA: The National Academies Press, 2015.
122. CANSO (Civil Air Navigation Services). "Cyber security and risk assessment guide," Available: https://www.canso.org/sites/default/files/CANSO%20Cyber%20Security%20and%20Risk%20Assessment%20Guide.pdf. [Accessed: Oct. 12, 2018].
123. G. Lykou, A. Anagnostopoulou, and D. Gritzalis, "Implementing Cybersecurity measures in airports to improve cyber-resilience," in *2018 Global Internet of Things Summit (GIoTS)*, Bilbao, pp. 1–6, 2018.
124. A. Hussain, V.-H. Bui, and H.-M. Kim, "Microgrids as a resilience resource and strategies used by microgrids for enhancing resilience," *Applied Energy*, vol. 240, pp. 56–72, Apr. 2019.
125. F. H. Jufri, V. Widiputra, and J. Jung, "State-of-the-art review on power grid resilience to extreme weather events: Definitions, frameworks, quantitative assessment methodologies, and enhancement strategies," *Applied Energy*, vol. 239, pp. 1049–1065, Apr. 2019.
126. R. Wu and G. Sansavini, "Integrating reliability and resilience to support microgrid design," *Arxiv*. Available: https://arxiv.org/abs/2004.00877.
127. H. Lin, C. Chen, J. Wang, J. Qi, D. Jin, Z. T. Kalbarczyk, and R. K. Iyer, "Self-healing attack-resilient PMU network for power system operation," *IEEE Transactions on Smart Grid*, vol. 9, no. 3, pp. 1551–1565, May 2018.
128. P. van Boheemen, G. Munnichs, L. Kool, G. Diercks, J. Hamer, and A. Vos, *Cyber-resilience with New Technology*, Rathenau Instituut, 2020.
129. S.G. Finlayson, H.W. Chung, I.S. Kohane, and A.L. Beam, "Adversarial attacks against medical deep learning systems," *arXiv:1804.05296 [cs, stat]*. 2018. Available: http://arxiv.org/abs/1804.05296.
130. E. Ackerman, "Three small stickers in intersection can cause tesla autopilot to swerve into wrong lane," 2019. Available: https://spectrum.ieee.org/cars-that-think/transportation/self-driving/three-small-stickers-on-road-can-steer-tesla-autopilot-into-oncoming-lane.
131. N. Kolokotronis, K. Limniotis, S. Shiaeles, and R. Griffiths, "Secured by blockchain: Safeguarding Internet of Things devices," *IEEE Consumer Electronics Magazine*, vol. 8, pp. 28–34, 2019.
132. M. Salek Ali, M. Vecchio, M. Pincheira, K. Dolui, F. Antonelli, and M. Husain Rehmani, "Applications of blockchains in the internet of things: A comprehensive survey," *IEEE Communication Surveys and Tutorials*, 2018.
133. A. Bellini, E. Bellini, M. Gherardelli, and F. Pirri, "*Enhancing iot data dependability through a blockchain mirror model*," Future Internet, 2019.
134. S. Brotsis, N. Kolokotronis, K. Limniotis, S. Shiaeles, D. Kavallieros, E. Bellini, and C. Pavu´e, "*Blockchain solutions for forensic evidence preservation in iot environments*," in *IEEE Conference on Network Softwarization (NetSoft)*, 2019.

135. E. Bellini, P. Ceravolo, and E. Damiani, "*Blockchain-based e-vote-asa-service*," in *IEEE 12th International Conference on Cloud Computing (CLOUD)*, pp. 484–486, 2019.
136. E. Bellini, P. Ceravolo, A. Bellini, and E. Damiani, "Designing process-centric blockchain-based architectures: A case study in e-voting as a service," In: P. Ceravolo, M. van Keulen, and M. Gómez-López (Eds.) *Data-Driven Process Discovery and Analysis. SIMPDA 2018, SIMPDA 2019.* Lecture Notes in Business Information Processing, vol. 379 Cham: Springer, 2020.
137. B. Shala, U. Trick, A. Lehmann, B. Ghita, and S. Shiaeles, "Blockchain-based trust communities for decentralized m2m application services," in F. BXhafa, F. Leu, M. Ficco, and C.-T. Yang (Eds.) *Advances on P2P, Parallel, Grid, Cloud and Internet Computing, ser. Lecture Notes on Data Engineering and Communications Technologies*, vol. 24. Cham: Springer, 2018.
138. E. Bellini, Y. Iraqui, and E. Damiani, "Blockchain-based distributed trust and reputation management systems: A survey," *IEEE Access*, 2020
139. E. Bellini, "A blockchain based trusted persistent identifier system for big data in science," *Foundations of Computing and Decision Sciences*, vol. 44, no. 4, in press, 2019.
140. E. Bellini, A. Bellini, F. Pirri, and L. Coconea, *Towards a Trusted Virtual Smart Cities Operation Center Using the Blockchain Mirror Model*, in Lecture Notes in Computer Science, vol. 11938 LNCS, pp. 283–291, Springer 2019.
141. M. Herlihy, "Blockchains from a distributed computing perspective," *Communications of the ACM*, vol. 62, no. 2, pp. 78–85, 2019.
142. U. W. Chohan, "The decentralized autonomous organization and governance issues," *Regulation of Financial Institutions eJournal: Social Science Research Network*, SSRN, 2017. http://dx.doi.org/10.2139/ssrn.3082055
143. I. Sergey and A. Hobor, "A concurrent perspective on smart contracts," In: M. Brenner et al. (eds) *Financial Cryptography and Data Security*, Lecture Notes in Computer Science, vol. 10323. Cham: Springer, pp. 478–493, 2017.
144. N. Atzei, M. Bartoletti, and T. Cimoli, "*A survey of attacks on Ethereum smart contracts SoK*," in *Proceedings of the 6th International Conference on Principles of Security and Trust*, vol. 10204, pp. 164–186, 2017.
145. G. Sansavini, *Engeenering Resilince in Critical Infrastrucutre, 2016, and IRGC, Resource Guide on Resilience*. Lausanne: EPFL International Risk Governance Center. v29-07-2016, 2016.

Index

A

absorb, 76, 149, 295, 296, 298, 307–309, 324
access control, 37, 52, 68, 71, 106, 119, 182, 233, 247, 266, 296
accidental threats, 220, 241
accuracy, 17, 18, 31, 80, 83, 84, 86, 104, 106, 128, 148, 164, 171, 227, 229, 233, 249, 257, 295, 300
adaptive capacity, 295, 299, 304, 305, 307, 308, 322–324
agriculture and farming, 5, 143, 144, 150, 152, 154, 155, 157, 163, 164
ambulance tracking system, 173
AMQP, 60, 61, 93
antivirus/antimalware, 115, 258, 262
asset management and tracking, 168, 183–186
asset tracking, 102, 168, 183–191, 194, 197
authentication, 5, 52, 68, 75, 118, 119, 130, 169, 171, 174, 183, 187, 196, 213, 214, 223, 229, 232, 242, 248, 256, 261, 283
authentication mechanisms, 213, 214
authorization, 52, 68, 206, 213, 214, 223, 231–233, 242, 273, 274, 297
authorization policy, 231
availability, 3, 4, 23, 62, 64, 69, 91, 100, 103, 105, 120, 139, 146, 159, 164, 177, 183, 206, 207, 223, 230, 233, 278, 293, 298, 299, 302, 308, 318, 323

B

backup policy, 277–280
banking, 36, 142, 154, 162, 296
biosensors, 226
Bluetooth, 6, 30, 51, 56, 68–70, 169, 172, 174, 177–179, 218
botnets, 37, 91, 116, 129, 138, 145, 196, 245, 263, 266, 270, 272, 281, 293
buffer capacity, 307

C

capillary network, 105, 106, 112, 113
chemical industries, 102
cloud computing, 36, 68, 155, 169, 178, 179, 196, 208–210
CoAP, 56–58, 92, 112, 113
communication protocols, 89, 208, 218
confidentiality, 7, 12, 17, 31, 66, 92, 100, 103, 105, 106, 146, 148, 150, 164, 183, 193, 206, 207, 221, 223, 233, 273, 281, 302
confidentiality threats, 206

connected vehicles, 5, 28, 29, 38, 105, 106
consent, 10, 13, 15, 16, 18–21, 24, 25, 27, 29, 30, 32, 206, 277
coping ability, 304–308, 320, 323, 324
Council of Europe, 7–9
countermeasures, 90, 187, 195, 196, 233, 241, 271, 277, 282, 283, 285, 296, 297, 302, 308, 315, 321
Covid-19 pandemic, 29, 38, 39, 154
critical function, 101, 120, 277, 292, 295, 302, 303, 305, 307–310, 324
critical infrastructure, 99, 101, 102, 106–108, 113, 115, 116, 118–120, 125, 126, 128, 130, 133, 246, 293, 317
critical infrastructures incidents, 106
cyber-resilience, 294–299, 301–304, 309, 312, 313, 317, 318, 320–322, 324
cybersecurity, 30, 36, 37, 98, 100, 108, 113, 133, 145, 146, 163, 210, 222, 230, 233, 260, 261, 267, 268, 270, 272, 279, 292–298, 301, 318

D

damping, 323
data breach, 7, 13, 17, 23, 100, 102, 130, 162, 174, 212, 242, 246, 260, 261, 263, 270, 276, 277, 298, 299, 314, 318, 320
data confidentiality, 92, 100, 233
data protection impact assessment, 4, 22, 29, 31, 34, 35, 38
Data protection Officer, 18, 35, 38
data protection principles, 16, 17, 35, 37
data protection, 3, 4, 6–12, 14–32, 34–39, 151, 169, 187, 298
  privacy by design, 29, 33, 35, 36, 38, 222
  privacy by default, 29
DDoS attacks, 37, 91, 141, 206, 242, 245, 247, 250, 252, 255, 257, 259, 260, 263, 266, 269–272, 276
DDS, 58, 59, 93, 163, 181, 184, 226, 297, 320
detection, 25, 80, 84, 87, 92, 93, 115, 122, 132, 138, 139, 147, 152, 158, 164, 175, 176, 179, 180, 187, 196, 213, 215, 228, 231, 233, 243, 245, 247, 249, 250, 252–255, 257, 258, 260, 266–268, 272, 282, 298, 300–302, 308, 315, 318, 322
device hijacking, 162, 247, 281
discrimination, 15, 26, 27, 34, 40
disruption, 23, 66, 76, 90–92, 99, 104, 107, 116, 123, 203, 233, 260, 294, 297–299, 303, 305, 307, 309, 310, 316, 317

335

domain, 5, 6, 62, 63, 67, 82, 88–90, 99, 102, 112, 113, 116, 129, 130, 133, 142–144, 149–154, 157, 159, 161–164, 168, 184, 194, 197, 202, 234, 256, 298, 301, 315, 322
drones, 29, 110, 120, 122, 125, 128, 133, 142, 144, 147, 152, 156, 292

E

edge computing, 49, 155, 175–177, 208, 241, 307, 315
e-privacy directive, 38
e-privacy regulation, 38

F

fleet management, 5, 102, 103, 110, 168, 169, 178–183, 197
flexibility, 72, 100, 181, 183, 190, 307, 312, 313, 316, 317

G

game, 107, 300, 313, 314, 324
General Data Protection Regulation, 3, 4, 11, 25, 26, 38
governmental authorities, 142

H

hazards, 90, 146, 148, 151, 159
health center, 168, 190, 224
healthcare provider, 225, 226, 228
hijacking, 37, 108, 130, 162, 247, 281
human-centric approach, 38
human error, 23, 261, 262, 280

I

identity theft, 129, 244, 262, 280, 281
IIoT architecture, 157
IIoT domains, 149, 164
IIoT incidents, 163
individual awareness, 241, 247
industrial IoT, 6, 161, 193, 311, 318
industrial safety, 5, 143, 144, 146, 147, 149, 150, 152, 154, 157–159, 164
information leakage, 220, 221, 252, 257, 262
infrared communication, 76, 77
injections, 107, 245, 264
insecure network services, 242, 266, 269, 272, 276
inside job, 162
Insteon, 79, 80, 93
insufficient training, 163
integrity, 17, 28, 31, 52, 92, 100, 103, 105, 106, 130, 146, 150, 164, 183, 193, 197, 206, 207, 221–223, 233, 241, 245, 261, 273, 281, 301, 302, 308

integrity threats, 206
internet of everything, 2, 4, 291
Internet of Things, 2, 4–6, 28, 36, 48, 101, 138, 173, 183, 192, 203, 241, 242
intrusion, 7, 25, 92, 93, 103, 106, 187, 196, 210–213, 215, 241, 247, 249, 250, 252–254, 256, 263, 264, 266, 268, 276, 280, 301, 303, 314
IoT applications, 2, 6, 51, 66, 69, 80, 87, 101, 103, 169, 192, 195, 202, 209

L

learning, 5, 20, 31, 49, 89, 90, 92, 93, 130, 146, 147, 179, 180, 189, 296, 301, 303–305, 309, 322
legislation, 10, 16, 17, 30, 38, 146, 273
Li-Fi, 71, 72
load redistribution, 132
logistics, 5, 6, 66, 74, 102, 103, 111, 142, 168, 169, 171, 194, 195, 197, 292, 305
logistics and transportation, 142
logistics management, 168, 169, 171, 197
  fleet management, 5, 102, 103, 110, 168, 169, 178–183, 197
  tracking, 2, 5, 13, 20, 30, 100, 102, 103, 111, 142, 157, 159, 160, 168, 169, 171, 173–179, 183–191, 194, 195, 197, 219, 227, 254, 257, 301, 317
LoRa, 73, 74, 317

M

malware, 37, 90, 91, 100, 104, 107, 108, 110, 114, 115, 138, 139, 141, 145, 153, 166, 188, 197, 206, 222, 231, 232, 243–245, 249–252, 255–260, 262, 265, 266, 268, 269, 271, 274–277, 279, 281, 314, 324
man-in-the-middle, 69, 91, 130, 161, 196, 244, 251
manufacturing, 5, 6, 63, 64, 81, 87, 101, 102, 117, 140, 142, 143, 151, 168, 169, 177, 181, 183, 192–195, 197, 279, 292
medical technologies, 203, 225, 234
middleware, 5, 19, 52, 58, 60, 180, 181, 190, 196, 207–209, 228
mitigation, 23, 34, 93, 102, 103, 130–132, 203, 233, 267, 269, 270, 272, 279, 281, 294, 296, 300, 301, 303, 313, 318
mixed dataset, 26
mobile X-ray machine, 168, 191
MQTT, 53–55, 58, 60, 92, 93, 179, 180
MQTT-SN, 55, 56, 93

N

network attacks, 115, 157

Index

O

object tracking protocol, 173
outdated systems, 162, 318

P

pastoral farming, 144
personal data, 4, 7, 9–26, 28–31, 33–36, 39, 105, 153, 162, 212, 219, 234, 242, 276
plan, 34, 86, 107, 108, 114, 126, 127, 156, 171, 185, 187, 196, 203, 215, 219, 222, 226, 227, 233, 256, 260, 265, 271, 279, 280, 292, 294, 297, 298, 303, 304, 309, 313, 317, 318, 323, 324
precision farming, 100, 144, 145
preventative maintenance, 5, 143, 144, 149, 150, 152, 154, 159, 163, 164
prevention, 9, 31, 93, 101, 139, 142, 148, 149, 157, 159, 188, 192, 196, 210, 214, 222–224, 227, 231, 249, 252, 255, 268, 297, 298, 300, 302, 305, 308, 309
profiling, 19, 21, 25, 26, 114
protection, 3, 4, 6–39, 52, 99, 101, 104–106, 115, 118, 121, 132, 138, 150, 151, 157, 158, 169, 187, 196, 214–216, 232, 233, 242, 248, 249, 251, 252, 257–260, 272, 276, 278, 279, 282, 297–302, 305, 308, 309, 311–313, 315, 318, 319, 321

R

recovery, 84, 101, 169, 233, 278–280, 294, 296–299, 301–303, 307–310, 319, 321, 323
reliability, 3, 61, 69, 70, 74, 78, 80, 83, 100, 103–106, 146, 148, 150, 163, 164, 183, 283, 294, 300, 308, 319, 321
resilience, 3, 103–106, 149, 150, 164, 294–305, 307–310, 312–324
The right to data protection, 7–9
   right to private life, 7, 9
rights of the data subject, 17
risk, 7, 14, 17, 19, 20, 22, 23, 27, 28, 30, 32–34, 36–38, 67, 92, 98–100, 103, 104, 106, 142, 144–149, 151, 152, 160–163, 211, 230, 232, 233, 241, 250, 252, 255, 257, 261–263, 269–271, 273–276, 278, 280–282, 293–297, 300, 302–305, 308, 311, 314–324
robots, 147, 151, 152, 154, 293
robustness, 296

S

safety, 3, 5, 9, 20, 23, 28, 29, 48, 62, 67, 72, 76, 88, 100, 103, 104, 106, 118, 142–152, 154, 157–161, 163, 164, 168, 173, 178, 179, 192, 194, 204, 212, 216, 218, 219, 229, 232, 234, 293, 294, 309, 316, 318

satellite, 75, 76, 111, 121, 157
SCADA, 106, 108, 113, 116, 117, 129, 133, 181, 301, 320
security attack, 69, 196
security by design, 36, 37, 233
security policy, 248, 252, 255, 273–277
security risks, 32, 38, 100, 103, 149, 250, 261–263, 274, 280–282
security threats, 195, 202, 233, 247, 263, 267, 271, 273, 279, 281, 282
sensors, 5, 6, 23, 25, 49, 50, 54, 63–65, 67, 70, 73–75, 80–84, 86–89, 93, 100, 103, 104, 106, 110, 111, 121, 122, 132, 140, 144, 147, 149–151, 154–157, 159, 160, 168, 174, 178–181, 183, 186, 189, 191–193, 195, 203, 204, 207, 209, 210, 213–217, 226–230, 232, 233, 252, 253, 257, 280, 282, 292, 300, 315
smart cities, 2, 5, 6, 62, 63, 65, 67, 75, 87, 93, 105, 160, 202–204, 206, 208–211, 234, 317
smart economy, 205
smart glasses, 6, 28, 38
smart grid, 50, 65, 104, 111, 115, 129–132, 141, 168, 311, 312, 314, 320
smart healthcare, 168, 203, 209, 210, 234
smart home, 4–6, 27, 48, 62, 63, 75, 78, 79, 87, 93, 168, 169, 172, 196, 202–204, 209, 211–223, 228, 234
   automation, 6, 212
   infrastructure, 213, 221, 223
   security, 212–214
smart infrastructure, 27, 204–206
smart manufacturing, 5, 193, 194
smart meters, 5, 62, 65, 74, 104, 105, 111, 131, 312
social engineering, 113–116, 120, 133, 244–246, 261, 262, 277
steel inventory tracking, 191
strategy, 11, 38, 210, 215, 258, 261, 278, 279, 293, 297, 299, 309, 313, 315, 319, 320, 323
sustainability, 204, 210, 212, 213, 215, 216, 219, 298

T

threat landscape, 206, 219, 241, 295, 303
threats, 6, 23, 27, 29, 37, 78, 79, 90, 93, 98–103, 110, 113, 115, 120, 122, 133, 138, 139, 141–143, 150, 151, 159, 161–163, 168, 169, 195, 202, 206, 207, 209, 218–222, 231, 233, 234, 241–244, 246–253, 257, 259–261, 263, 264, 267–269, 271–276, 279–283, 293–297, 299, 300, 302, 307, 309, 315, 317, 318, 320, 322, 324
time modification, 132

tolerance, 307, 308, 319
tracking management, 169, 171
transparency, 3, 17, 20–23, 33, 34, 38, 152, 192
trust, 3, 27, 30, 33, 35–37, 58, 114, 152, 169, 234, 244, 247, 249, 261, 264, 280, 283, 294, 295, 301, 303, 308, 310, 315, 322

U

ubiquitous computing, 49, 88–90, 93, 208
unintentional damage, 220
United Nations, 2, 7, 8, 20

V

vulnerability, 6, 21–23, 92, 99, 132, 162, 169, 187, 188, 230, 232, 242, 246, 260, 263–266, 270, 271, 281, 282, 294, 295, 308, 309, 322

W

wearables, 4–6, 20, 22, 29, 88, 143, 150, 151, 227
weightless, 74, 75
well-being, 9, 20, 89, 99, 152, 202, 203, 205, 225, 227, 234, 278, 311
Wi-Fi, 51, 70, 71, 74, 75, 78, 111, 112, 169, 174, 177, 178, 181, 192, 218, 264, 320
Wi-Max, 75, 111

X

XMPP, 59, 60, 93

Z

zero-day, 246, 258, 260, 263, 266, 280, 294
ZigBee, 51, 56, 69, 70, 78, 111, 112, 182, 218

Printed in the United States
by Baker & Taylor Publisher Services